Diagnosis, Epidemiology and Transmission Dynamics of *Cryptosporidium* spp. and *Giardia duodenalis*

Diagnosis, Epidemiology and Transmission Dynamics of *Cryptosporidium* spp. and *Giardia duodenalis*

Editors

David Carmena
David González-Barrio
Pamela Carolina Köster

MDPI • Basel • Beijing • Wuhan • Barcelona • Belgrade • Manchester • Tokyo • Cluj • Tianjin

Editors
David Carmena
Spanish National Centre for
Microbiology
Spain

David González-Barrio
Spanish National Centre for
Microbiology
Spain

Pamela Carolina Köster
Spanish National Centre for
Microbiology
Spain

Editorial Office
MDPI
St. Alban-Anlage 66
4052 Basel, Switzerland

This is a reprint of articles from the Special Issue published online in the open access journal *Pathogens* (ISSN 2076-0817) (available at: https://www.mdpi.com/journal/pathogens/special_issues/ Diagnosis_Epidemiology_Transmission_Dynamics_Cryptosporidium_spp_Giardia_duodenalis).

For citation purposes, cite each article independently as indicated on the article page online and as indicated below:

LastName, A.A.; LastName, B.B.; LastName, C.C. Article Title. *Journal Name* **Year**, *Volume Number*, Page Range.

ISBN 978-3-0365-3078-9 (Hbk)
ISBN 978-3-0365-3079-6 (PDF)

Contents

About the Editors

David Carmena (Phd.) Dr. Carmena's research focuses on zoonotic diseases caused by microeukaryotes (Cryptosporidium, Giardia, Blastocystis, Enterocytozoon) and helminth (Echinococcus) species, with a particular interest in those aspects related to the diagnosis, molecular epidemiology, transmission dynamics, and evaluation of the socioeconomic impact caused by these pathogens, both from the human and animal health perspective.

David González-Barrio (Phd.) Dr. González-Barrio's research is mainly focused on shared emerging diseases between domestic animals, wildlife, and humans, with a special interest in diagnostics, epidemiology, and molecular epidemiology in their main wildlife reservoirs.

Pamela Carolina Köster (BS) Ms. Köster's research is mainly focused on the detection and molecular characterization of diarrhoea-causing enteric protists in wildlife, with a special interest in endangered non-human primate species. Her research includes the identification and characterization of zoonotic transmission events and how anthropic activities influence the health status of free-living animal communities.

Editorial

Editorial for the Special Issue: Diagnosis, Epidemiology and Transmission Dynamics of *Cryptosporidium* spp. and *Giardia duodenalis*

Pamela C. Köster, David González-Barrio * and David Carmena

Parasitology Reference and Research Laboratory, Spanish National Centre for Microbiology, Majadahonda, 28220 Madrid, Spain; pamelakster@yahoo.com (P.C.K.); dacarmena@isciii.es (D.C.)
* Correspondence: david.gonzalezb@isciii.es; Tel.: +34-91-822-364

Citation: Köster, P.C.; González-Barrio, D.; Carmena, D. Editorial for the Special Issue: Diagnosis, Epidemiology and Transmission Dynamics of *Cryptosporidium* spp. and *Giardia duodenalis*. *Pathogens* **2022**, *11*, 141. https://doi.org/10.3390/pathogens11020141

Received: 19 January 2022
Accepted: 21 January 2022
Published: 24 January 2022

Publisher's Note: MDPI stays neutral with regard to jurisdictional claims in published maps and institutional affiliations.

Cryptosporidium spp. and *Giardia duodenalis* are major contributors to the global burden of diarrhoeal disease, primarily affecting young children in poor-resource settings. Cryptosporidiosis ranks second to rotavirus infection as a cause of life-threatening diarrhoea in children younger than two years of age, in sub-Saharan Africa and south Asia [1]. In contrast, giardiasis is rarely a cause of childhood mortality, but the disease has been consistently associated with growth faltering and cognitive impairment in malnourished children [2]. Intriguingly, large epidemiological studies, based on case-control data, have evidenced that *G. duodenalis* was more frequent among controls (individuals without diarrhoea) than in cases (individuals with diarrhoea) [1,3,4]. This finding has been interpreted by some authors as evidence in favour of a potential "protective" effect of the parasite against diarrhoea, and explains why *G. duodenalis* infections are systematically absent in global burden estimations of diarrhoeal diseases [5].

At least 44 *Cryptosporidium* species and more than 120 genotypes, as well as nine *Giardia* species are currently recognised [6]. Only three *Cryptosporidium* species (mainly anthroponotic *C. hominis* and zoonotic *C. parvum* and *C. meleagridis*) are responsible for most (circa 95%) human cases of cryptosporidiosis reported globally. *Giardia duodenalis* is the only *Giardia* species infective to humans, comprising eight (A–H) distinct genetic groups (the so-called assemblages) of which zoonotic assemblages A and B are commonly reported to infect humans [6]. Because of the large number of morphologically identical (but genetically different) species/genotypes, within both groups of protozoa, molecular-based tools, including Sanger sequencing, are needed for detection, differentiation, and subtyping purposes [7]. Molecular data are also essential to characterize the transmission dynamics of *Cryptosporidium* spp. and *G. duodenalis*, including for the identification of sources of infection and spread pathways, the tracking of virulent genetic variants or the assessment of zoonotic potential. The most common genetic markers used in subtyping analyses include the *Cryptosporidium* 60-kDa glycoprotein (*gp60*) gene, and the *G. duodenalis* β-giardin (*bg*), glutamate dehydrogenase (*gdh*), and triosephosphate isomerase (*tpi*) genes [7].

This special issue includes 14 papers that made important contributions in expanding our current knowledge, on aspects relevant to the diagnosis and epidemiology of *Cryptosporidium* spp. and *G. duodenalis*. These include the usefulness of PCR-based methods, for the first-line routine diagnosis of *Cryptosporidium* spp. in clinical settings [8], or for enlarging the available arsenal of subtyping tools, to assess *Cryptosporidium* intra-species genetic diversity in canine-adapted *C. felis* [9] and ovine-adapted *C. xiaoi* [10]. We were lucky enough to receive relevant contributions dealing with the human molecular epidemiology of *Cryptosporidium* spp. and *G. duodenalis* in a low-income country, such as Mozambique [11–13], in medium-income countries, such as Brazil and China [14,15], and in a high-income country, such as Sweden [16]. Finally, this special issue also included molecular epidemiological studies directed to assess the occurrence, genetic diversity and zoonotic potential of *Cryptosporidium* spp. and *G. duodenalis* in pet dogs and cats [17], livestock [18], and farmed rabbits and rats [19–21], mostly in China.

In their study, Costa et al. [8] evaluated the diagnostic performance of four "in-house" and four commercial PCR methods for the detection of *Cryptosporidium* spp. against a panel of selected stool samples, belonging to the collection of a French reference laboratory. The authors identified significant differences in the performance of the compared assays, highlighting the need for proper validation and standardization before routine clinical use. An asset of this study was the inclusion of less frequent or rare *Cryptosporidium* species and genotypes (*C. cuniculus*, *C. felis*, *C. meleagridis*, *C. ubiquitum*, *C.* sp. chipmunk genotype), in addition to *C. hominis* and *C. parvum*. All four commercial assays were able to identify all *Cryptosporidium* species and genotypes evaluated, but this was not always the case for the "in-house" protocols.

Li et al. [9] confirmed and expanded the usefulness of the 60-kDa glycoprotein (*gp60*) gene subtyping tool for assessing the genetic diversity within *C. felis*, initially developed by Rojas-Lopez et al. [22]. To do so, the authors identified the subtypes of 20 *C. felis* isolates, obtained from stray, sheltered, and pet cats, in Guandong province and Shanghai city in China. A high intra-species genetic diversity was observed, resulting in the identification of 13 novel and two known subtypes of the parasite. The main contribution of this survey was the demonstration that many of these genetic variants were shared between cats and humans, strongly suggesting that there could be cross-species transmission of *C. felis*. Furthermore, Fan et al. [10] developed a new *gp60* subtyping tool for *C. xiaoi*, a *Cryptosporidium* species adapted to infect sheep and goats, for which this tool was previously lacking. The authors tested 355 *C. xiaoi*-positive samples from Chinese sheep and goats and found an extremely large intra-species genetic diversity, resulting in the identification of 12 (XXIIIa to XXIII) subtype families. This study complements and expands the available *gp60* subtyping toolbox, including protocols adapted to *Cryptosporidium* species, such as *C. hominis* and *C. parvum* [23], *C. fayeri* [24], *C. meleagridis* [25], *C. ubiquitum* [26], *C. viatorum* [27], *C. felis* [22], *C. ryanae* [28], *C. canis* [29], and *C. bovis* [30].

This special issue brings together new epidemiological data that contribute to improving our understanding on the current situation of human cryptosporidiosis in Mozambique, one of the least developed countries in sub-Saharan Africa. In a seminal molecular-based study conducted in Zambézia province, Muadica et al. [11] investigated the presence and genetic diversity of *Cryptosporidium* spp., and other intestinal micro-eukaryotes, in a large population (*n* = 1093) of asymptomatic and symptomatic schoolchildren. The authors found a low prevalence of *Cryptosporidium* infections (1.6%) and managed to genotype 13 isolates of the parasite. Three species (*C. hominis*, *C. parvum*, and *C. felis*) were identified at equal rates (31% each), with *C. viatorum* being detected in 7% of cases. These preliminary results were further confirmed and expanded in a subsequent retrospective study, conducted by Messa et al. [12], taking advantage of a large panel of *Cryptosporidium*-confirmed DNA samples (*n* = 190), obtained during the Global Enteric Multicenter Study (GEMS) at the Manhiça district in the Maputo province [1]. The GEMS project was specifically designed as a case-control study to determine the etiology and population-based burden of paediatric diarrheal disease in sub-Saharan Africa and South Asia [31]. In their study, Messa et al. [12] identified three *Cryptosporidium* species including *C. hominis* (73%), *C. parvum* (23%), and *C. meleagridis* (4%). Both *C. hominis* and *C. parvum* were more prevalent among children with diarrhoea than in children without it (48% vs. 33%). A large intra-species genetic variability was observed within *C. hominis* (*gp60* subtype families Ia, Ib, Id, Ie, and If) and *C. parvum* (*gp60* subtype families IIb, IIc, IIe, and IIi) but not within *C. meleagridis* (*gp60* subtype family IIIb). Molecular genotyping data, provided by Cossa-Moiane et al. [13], in young children (*n* = 319) presenting with diarrhoea at hospital settings in the Maputo province, pointed in the same direction. In this study, a microscopy-based *Cryptosporidium* prevalence of 11% was obtained. In addition, typing results were available from a subset (*n* = 29) of these *Cryptosporidium*-positive samples, confirming the predominance of *C. hominis* (93%) over *C. parvum* (3%). A mixed infection of *C. hominis* and *C. parvum* was also detected (3%). Taken together, these three studies strongly suggest that human cryptosporidiosis in Mozambique is mainly of anthropic nature, although domestic dogs, cattle, and avian species can act as source of human infection in certain areas. This situation reflects the coexistence of different transmission pathways of cryptosporidiosis in the country.

Understanding the public health significance of emerging diarrhoea-causing microeukaryotes, including *Cryptosporidium* spp. and *G. duodenalis*, is increasingly attracting research interest in rapidly developing countries, such as Brazil and China [32,33]. In the former country, Köster et al. [14] investigated the occurrence and genetic diversity of *G. duodenalis* in a community survey of indigenous people ($n = 574$) from the Brazilian Amazon. During the four consecutive sampling campaigns of the study, *G. duodenalis* prevalence rates varied from 13–22% and primarily affected individuals younger than 15 years of age. Near 75% of the infections were attributed to the assemblage B of the parasite. Remarkably, no association between the *G. duodenalis* genotype and the occurrence of diarrhoea could be demonstrated. This finding is in agreement with the results obtained in a recent study conducted in Mozambican children younger than five years of age [34]. Considered together, these findings indicate that the parasite genotype does not suffice to explain, per se, the progression from infection to disease. Little information on the molecular variability of *Cryptosporidium* spp. and *G. duodenalis* is still available from Chinese human populations. Zhang et al. [15] attempted to overcome this gap in knowledge by analysing stool samples ($n = 507$) from randomly selected individuals, with and without gastrointestinal manifestations, seeking medical attention at hospital settings in the Yunnan Province. Interestingly, no *Cryptosporidium* infections were identified in the surveyed clinical population, whereas a low *G. duodenalis* occurrence rate (2%) was found. Sequence analyses of the *G. duodenalis*-positive isolates confirmed that assemblage A was far more prevalent than assemblage B (90% vs. 10%, respectively). The large, geographical-dependent difference in assemblage frequencies, documented in the studies mentioned above, may be indicative of different sources of infection and transmission routes.

Cryptosporidium spp. and *G. duodenalis* are a public health concern, not only in low- and medium-income countries, but also in developed nations [35]. In their study, Lebbad et al. [16] retrospectively genotyped *Cryptosporidium*-positive stool samples ($n = 398$) collected in 12 of the 21 regional laboratories that carry out routine parasitological diagnosis in Sweden. The cohort of stool samples analysed included patients that acquired the infection in the country and abroad. The authors identified 12 distinct *Cryptosporidium* species/genotypes, with *C. parvum* (75%) and *C. hominis* (12%) accounting for the majority of the cases identified. A very large intra-species genetic diversity was detected, allowing the identification of 29 *gp60* subtype families including four novel ones (*C. parvum* IIr, IIs, IIt, and *Cryptosporidium* horse genotype VIc). The authors also reported a human infection by rodent-adapted *C. ditrichi*, a *Cryptosporidium* species very rarely found in humans [36]. Another major contribution to this survey was the demonstration that almost 8% of human cryptosporidiosis cases in Sweden had a zoonotic origin. However, it is still unclear whether some of these findings (e.g., *C. erinacei*, *C. ditrichi*, *C.* horse genotype) correspond to true or spurious infections.

From the human molecular data presented above, it is clear now that zoonotic transmission is a significant route of *Cryptosporidium* spp. and *G. duodenalis*, spreading in several settings, either in developing or developed nations. Human infections can arise through direct contact with infected animals, or through accidental ingestion of contaminated water or food with the faecal material of these animals [37–39]. This special issue included five articles dealing with the potential role of companion, livestock, and farmed animals as potential sources of human cryptosporidiosis and giardiasis. Most of these studies were conducted in China. In the first one, Wang et al. [17] investigated the presence and genetic diversity of *Cryptosporidium* spp. and *G. duodenalis* in faecal samples from pet dogs ($n = 262$) and cats ($n = 171$), collected in veterinary clinics, markets and shelters, from the Yunnan province. Reported infection rates for *G. duodenalis* and *Cryptosporidium* spp. were 14% and 5% in dogs, and 1% each in cats, respectively. Sequence analyses revealed that dogs were infected only by canine-adapted *Cryptosporidium* (*C. canis*) and *G. duodenalis* (assemblages C and D) species/genotypes. Similarly, cats were infected by feline-adapted *C. felis* and *G. duodenalis* assemblage F. These data indicate that pet dogs and cats play a marginal role as sources of human cryptosporidiosis and giardiasis. It should be noted that most molecular-based surveys conducted so far, failed to demonstrate zoonotic transmission events between pet dogs/cats and humans [40,41]. In another survey, Cao et al. [18] assessed the occurrence and molecular diversity of *Cryptosporidium* spp. in faecal samples (*n* = 476) from Bactrian camels in the Xinjiang Uygur Autonomous Region. In this Chinese area, the Bactrian camel is one of the few large animal species suitable for livestock production, providing clothing, milk, meat, and transport for many people. The authors found a PCR-based prevalence of 8% and identified six different *Cryptosporidium* species circulating in the surveyed camel population. Of them, *C. andersoni* (67%) and *C. parvum* (17%) accounted for eight out of ten infections. Other, less frequent species, included *C. occultus*, *C. ubiquitum*, *C. hominis*, and *C. bovis*. The study represents the first report of *C. hominis* (*gp60* subtype family Ik) in Bactrian camels, expanding the known range of suitable hosts for this pathogen in China that already included cattle [42], donkeys [43], horses [43], and non-human primates [44], beside humans. Some authors have interpreted all this information as evidence in support of considering *C. hominis* as a zoonotic species [45]. Regarding intensive animal farming, Cui et al. [19] investigated the presence, molecular variability and zoonotic potential of *G. duodenalis* in coypus (*n* =308), reared in fur farms in six Chinese provinces/autonomous regions. The parasite was detected in all farms investigated at variable infection rates, ranging from 1–29%. Subtyping data was available for 38 isolates, of which 95% were assigned to assemblage B and the remaining 5% to assemblage A. These data indicate that giardiasis could be an occupational risk for those individuals working in close contact with the farmed coypus or their excreta. In a similar study conducted in farmed bamboo rats (*n* = 724), in Guangdong province, Li et al. [20] identified *Cryptosporidium* infections in 12% of the investigated animals. Sequence analyses of the isolates positive to the parasite allowed for the identification of five distinct *Cryptosporidium* species/genotypes, with *C. bamboo rat genotype I* (56%) and *C. parvum* (35%) being the most prevalent genetic variants found. The remaining positive samples belonged to *C. bamboo rat genotype III* (6%), *C. occultus* (2), and *C. muris* (1%). All *C. parvum* isolates were assigned to the rare *gp60* subtype families IIo and IIp (not previously identified in human cryptosporidiosis cases). Taken together, these data indicate that farmed bamboo rats were infected by rodent-adapted *Cryptosporidium* species with little, or no zoonotic potential. Finally, Naguib et al. [21] provided novel data on the presence and genetic variability of *Cryptosporidium* spp. in farmed rabbits in Egypt, a country where the molecular epidemiology of this protozoan parasite is poorly understood. The authors collected and analysed faecal samples (*n* = 235) from nine rabbit farms, located in three Egyptian provinces. *Cryptosporidium* infections were confirmed in eight out of the nine farms sampled at variable rates, ranging from 4% to 24%. All *Cryptosporidium*-positive isolates generated (*n* = 28) were identified as *C. cuniculus* and belonged to the *gp60* subtype family Vb. Although *C. cuniculus* is known to be particularly adapted to infect domestic and wild leporids, it has also been identified in clinical human cases [46] and in waterborne outbreaks of cryptosporidiosis [47] and is regarded as an emerging pathogen to humans.

Author Contributions: P.C.K., D.G.-B. and D.C. wrote the Editorial. All authors have read and agreed to the published version of the manuscript.

Funding: This research received no external funding.

Acknowledgments: We are grateful to all the authors for their exciting contributions to this Special Issue. We also thank the reviewers for their helpful recommendations. We would also like to thank the *Pathogens* editorial office staff for their assistance and support and for having given us the opportunity to propose and organize this Special Issue.

Conflicts of Interest: The authors declare no conflict of interest.

References

1. Kotloff, K.L.; Nataro, J.P.; Blackwelder, W.C.; Nasrin, D.; Farag, T.H.; Panchalingam, S.; Wu, Y.; Sow, S.O.; Sur, D.; Breiman, R.F.; et al. Burden and Aetiology of Diarrhoeal Disease in Infants and Young Children in Developing Countries (the Global Enteric Multicenter Study, GEMS): A Prospective, Case-control Study. *Lancet* **2013**, *382*, 209–222. [CrossRef]
2. Berkman, D.S.; Lescano, A.G.; Gilman, R.H.; Lopez, S.L.; Black, M.M. Effects of Stunting, Diarrhoeal Disease, and Parasitic Infection During Infancy on Cognition in Late Childhood: A Follow-up Study. *Lancet* **2002**, *359*, 564–571. [CrossRef]
3. Breurec, S.; Vanel, N.; Bata, P.; Chartier, L.; Farra, A.; Favennec, L.; Franck, T.; Giles-Vernick, T.; Gody, J.C.; Luong Nguyen, L.B.; et al. Etiology and Epidemiology of Diarrhea in Hospitalized Children from Low Income Country: A Matched Case-Control Study in Central African Republic. *PLoS Negl. Trop. Dis.* **2016**, *10*, e0004283. [CrossRef] [PubMed]
4. Becker, S.L.; Chatigre, J.K.; Gohou, J.P.; Coulibaly, J.T.; Leuppi, R.; Polman, K.; Chappuis, F.; Mertens, P.; Herrmann, M.; N'Goran, E.K.; et al. Combined Stool-based Multiplex PCR and Microscopy for Enhanced Pathogen Detection in Patients with Persistent Diarrhoea and Asymptomatic Controls from Côte d'Ivoire. *Clin. Microbiol. Infect.* **2015**, *21*, 591.e1–591.e10. [CrossRef] [PubMed]
5. GBD 2013 Mortality and Causes of Death Collaborators. Global, Regional, and National Age-sex Specific All-cause and Cause-specific Mortality for 240 Causes of Death, 1990-2013: A Systematic Analysis for the Global Burden of Disease Study 2013. *Lancet* **2015**, *385*, 117–171. [CrossRef]
6. Ryan, U.M.; Feng, Y.; Fayer, R.; Xiao, L. Taxonomy and Molecular Epidemiology of *Cryptosporidium* and *Giardia*—A 50 Year Perspective (1971–2021). *Int. J. Parasitol.* **2021**, *51*, 1099–1119. [CrossRef]
7. Xiao, L.; Feng, Y. Molecular Epidemiologic Tools for Waterborne Pathogens *Cryptosporidium* spp. and *Giardia duodenalis*. *Food Waterborne Parasitol.* **2017**, *8-9*, 14–32. [CrossRef]
8. Costa, D.; Soulieux, L.; Razakandrainibe, R.; Basmaciyan, L.; Gargala, G.; Valot, S.; Dalle, F.; Favennec, L. Comparative Performance of Eight PCR Methods to Detect *Cryptosporidium* Species. *Pathogens* **2021**, *10*, 647. [CrossRef]
9. Li, J.; Yang, F.; Liang, R.; Guo, S.; Guo, Y.; Li, N.; Feng, Y.; Xiao, L. Subtype Characterization and Zoonotic Potential of *Cryptosporidium felis* in Cats in Guangdong and Shanghai, China. *Pathogens* **2021**, *10*, 89. [CrossRef]
10. Fan, Y.; Huang, X.; Guo, S.; Yang, F.; Yang, X.; Guo, Y.; Feng, Y.; Xiao, L.; Li, N. Subtyping *Cryptosporidium xiaoi*, a Common Pathogen in Sheep and Goats. *Pathogens* **2021**, *10*, 800. [CrossRef]
11. Muadica, A.S.; Köster, P.C.; Dashti, A.; Bailo, B.; Hernández-de-Mingo, M.; Balasegaram, S.; Carmena, D. Molecular Diversity of *Giardia duodenalis*, *Cryptosporidium* spp., and *Blastocystis* sp. in Symptomatic and Asymptomatic Schoolchildren in Zambézia Province (Mozambique). *Pathogens* **2021**, *10*, 255. [CrossRef] [PubMed]
12. Messa, A., Jr.; Köster, P.C.; Garrine, M.; Nhampossa, T.; Massora, S.; Cossa, A.; Bassat, Q.; Kotloff, K.; Levine, M.M.; Alonso, P.L.; et al. Molecular Characterisation of *Cryptosporidium* spp. in Mozambican Children Younger than 5 Years Enrolled in a Matched Case-Control Study on the Aetiology of Diarrhoeal Disease. *Pathogens* **2021**, *10*, 452. [CrossRef] [PubMed]
13. Cossa-Moiane, I.; Cossa, H.; Bauhofer, A.F.L.; Chilaúle, J.; Guimarães, E.L.; Bero, D.M.; Cassocera, M.; Bambo, M.; Anapakala, E.; Chissaque, A.; et al. High Frequency of *Cryptosporidium hominis* Infecting Infants Points to A Potential Anthroponotic Transmission in Maputo, Mozambique. *Pathogens* **2021**, *10*, 293. [CrossRef] [PubMed]
14. Köster, P.C.; Malheiros, A.F.; Shaw, J.J.; Balasegaram, S.; Prendergast, A.; Lucaccioni, H.; Moreira, L.M.; Lemos, L.M.S.; Dashti, A.; Bailo, B.; et al. Multilocus Genotyping of *Giardia duodenalis* in Mostly Asymptomatic Indigenous People from the Tapirapé Tribe, Brazilian Amazon. *Pathogens* **2021**, *10*, 206. [CrossRef] [PubMed]
15. Zhang, S.-X.; Carmena, D.; Ballesteros, C.; Yang, C.-L.; Chen, J.-X.; Chu, Y.-H.; Yu, Y.-F.; Wu, X.-P.; Tian, L.-G.; Serrano, E. Symptomatic and Asymptomatic Protist Infections in Hospital Inpatients in Southwestern China. *Pathogens* **2021**, *10*, 684. [CrossRef] [PubMed]
16. Lebbad, M.; Winiecka-Krusnell, J.; Stensvold, C.R.; Beser, J. High Diversity of *Cryptosporidium* Species and Subtypes Identified in Cryptosporidiosis Acquired in Sweden and Abroad. *Pathogens* **2021**, *10*, 523. [CrossRef] [PubMed]
17. Wang, Y. G.; Zou, Y.; Yu, Z.-Z.; Chen, D.; Gui, B.-Z.; Yang, J.-F.; Zhu, X.-Q.; Liu, G.-H.; Zou, F.-C. Molecular Investigation of Zoonotic Intestinal Protozoa in Pet Dogs and Cats in Yunnan Province, Southwestern China. *Pathogens* **2021**, *10*, 1107. [CrossRef]
18. Cao, Y.; Cui, Z.; Zhou, Q.; Jing, B.; Xu, C.; Wang, T.; Qi, M.; Zhang, L. Genetic Diversity of *Cryptosporidium* in Bactrian Camels (*Camelus bactrianus*) in Xinjiang, Northwestern China. *Pathogens* **2020**, *9*, 946. [CrossRef]
19. Cui, Z.; Wang, D.; Wang, W.; Zhang, Y.; Jing, B.; Xu, C.; Chen, Y.; Qi, M.; Zhang, L. Occurrence and Multi-Locus Analysis of *Giardia duodenalis* in Coypus (*Myocastor coypus*) in China. *Pathogens* **2021**, *10*, 179. [CrossRef]
20. Li, F.; Zhao, W.; Zhang, C.; Guo, Y.; Li, N.; Xiao, L.; Feng, Y. *Cryptosporidium* Species and *C. parvum* Subtypes in Farmed Bamboo Rats. *Pathogens* **2020**, *9*, 1018. [CrossRef]
21. Naguib, D.; Roellig, D.M.; Arafat, N.; Xiao, L. Genetic Characterization of *Cryptosporidium cuniculus* from Rabbits in Egypt. *Pathogens* **2021**, *10*, 775. [CrossRef] [PubMed]
22. Rojas-Lopez, L.; Elwin, K.; Chalmers, R.M.; Enemark, H.L.; Beser, J.; Troell, K. Development of a *gp60*-subtyping Method for *Cryptosporidium felis*. *Parasites Vectors* **2020**, *13*, 39. [CrossRef] [PubMed]
23. Strong, W.B.; Gut, J.; Nelson, R.G. Cloning and Sequence Analysis of a Highly Polymorphic *Cryptosporidium parvum* Gene Encoding a 60-kilodalton Glycoprotein and Characterization of its 15- and 45-kilodalton Zoite Surface Antigen Products. *Infect. Immun.* **2000**, *68*, 4117–4134. [CrossRef] [PubMed]
24. Power, M.L.; Cheung-Kwok-Sang, C.; Slade, M.; Williamson, S. *Cryptosporidium fayeri*: Diversity Within the *GP60* Locus of Isolates from Different Marsupial Hosts. *Exp. Parasitol.* **2009**, *121*, 219–223. [CrossRef]

25. Stensvold, C.R.; Beser, J.; Axén, C.; Lebbad, M. High Applicability of a Novel Method for *gp60*-based Subtyping of *Cryptosporidium meleagridis*. *J. Clin. Microbiol.* **2014**, *52*, 2311–2319. [CrossRef]

26. Li, N.; Xiao, L.; Alderisio, K.; Elwin, K.; Cebelinski, E.; Chalmers, R.; Santin, M.; Fayer, R.; Kvac, M.; Ryan, U.; et al. Subtyping *Cryptosporidium ubiquitum*, a Zoonotic Pathogen Emerging in Humans. *Emerg. Infect. Dis.* **2014**, *20*, 217–224. [CrossRef]

27. Stensvold, C.R.; Elwin, K.; Winiecka-Krusnell, J.; Chalmers, R.M.; Xiao, L.; Lebbad, M. Development and Application of a *gp60*-Based Typing Assay for *Cryptosporidium viatorum*. *J. Clin. Microbiol.* **2015**, *53*, 1891–1897. [CrossRef]

28. Yang, X.; Huang, N.; Jiang, W.; Wang, X.; Li, N.; Guo, Y.; Kváč, M.; Feng, Y.; Xiao, L. Subtyping *Cryptosporidium ryanae*: A Common Pathogen in Bovine Animals. *Microorganisms* **2020**, *8*, 1107. [CrossRef]

29. Jiang, W.; Roellig, D.M.; Guo, Y.; Li, N.; Feng, Y.; Xiao, L. Development of a Subtyping Tool for Zoonotic Pathogen *Cryptosporidium canis*. *J. Clin. Microbiol.* **2021**, *59*, e02474-20. [CrossRef]

30. Wang, W.; Wan, M.; Yang, F.; Li, N.; Xiao, L.; Feng, Y.; Guo, Y. Development and Application of a *gp60*-Based Subtyping Tool for *Cryptosporidium bovis*. *Microorganisms* **2021**, *9*, 2067. [CrossRef]

31. Levine, M.M.; Kotloff, K.L.; Nataro, J.P.; Muhsen, K. The Global Enteric Multicenter Study (GEMS): Impetus, Rationale, and Genesis. *Clin. Infect. Dis.* **2012**, *55*, S215–S224. [CrossRef] [PubMed]

32. Coelho, C.H.; Durigan, M.; Leal, D.A.G.; Schneider, A.B.; Franco, R.M.B.; Singer, S.M. Giardiasis as a Neglected Disease in Brazil: Systematic Review of 20 Years of Publications. *PLoS Negl. Trop. Dis.* **2017**, *11*, e0006005. [CrossRef] [PubMed]

33. Guo, Y.; Ryan, U.; Feng, Y.; Xiao, L. Emergence of Zoonotic *Cryptosporidium parvum* in China. *Trends Parasitol.* 2021, in press. [CrossRef] [PubMed]

34. Messa, A., Jr.; Köster, P.C.; Garrine, M.; Gilchrist, C.; Bartelt, L.A.; Nhampossa, T.; Massora, S.; Kotloff, K.; Levine, M.M.; Alonso, P.L.; et al. Molecular Diversity of *Giardia duodenalis* in Children Under 5 Years From the Manhiça District, Southern Mozambique Enrolled in a Matched Case-Control Study on the Aetiology of Diarrhoea. *PLoS Negl. Trop. Dis.* **2021**, *15*, e0008987. [CrossRef]

35. Fletcher, S.M.; Stark, D.; Harkness, J.; Ellis, J. Enteric Protozoa in the Developed World: A Public Health Perspective. *Clin. Microbiol. Rev.* **2012**, *25*, 420–449. [CrossRef]

36. Beser, J.; Bujila, I.; Wittesjö, B.; Lebbad, M. From Mice to Men: Three Cases of Human Infection with *Cryptosporidium ditrichi*. *Infect. Genet. Evol.* **2020**, *78*, 104120. [CrossRef]

37. Cai, W.; Ryan, U.; Xiao, L.; Feng, Y. Zoonotic Giardiasis: An Update. *Parasitol. Res.* **2021**, *120*, 4199–4218. [CrossRef]

38. Ryan, U.; Fayer, R.; Xiao, L. *Cryptosporidium* Species in Humans and Animals: Current Understanding and Research Needs. *Parasitology* **2014**, *141*, 1667–1685. [CrossRef]

39. Zahedi, A.; Ryan, U. *Cryptosporidium*—An Update with an Emphasis on Foodborne and Waterborne Transmission. *Res. Vet. Sci.* **2020**, *132*, 500–512. [CrossRef]

40. de Lucio, A.; Bailo, B.; Aguilera, M.; Cardona, G.A.; Fernández-Crespo, J.C.; Carmena, D. No Molecular Epidemiological Evidence Supporting Household Transmission of Zoonotic *Giardia duodenalis* and *Cryptosporidium* spp. From Pet Dogs and Cats in the Province of Álava, Northern Spain. *Acta Trop.* **2017**, *170*, 48–56. [CrossRef]

41. Rehbein, S.; Klotz, C.; Ignatius, R.; Müller, E.; Aebischer, A.; Kohn, B. *Giardia duodenalis* in Small Animals and Their Owners in Germany: A Pilot Study. *Zoonoses Public Health* **2019**, *66*, 117–124. [CrossRef] [PubMed]

42. Chen, F.; Huang, K. Prevalence and Molecular Characterization of *Cryptosporidium* spp. in Dairy Cattle from Farms in China. *J. Vet. Sci.* **2012**, *13*, 15–22. [CrossRef] [PubMed]

43. Jian, F.; Liu, A.; Wang, R.; Zhang, S.; Qi, M.; Zhao, W.; Shi, Y.; Wang, J.; Wei, J.; Zhang, L.; et al. Common Occurrence of *Cryptosporidium hominis* in Horses and Donkeys. *Infect. Genet. Evol.* **2016**, *43*, 261–266. [CrossRef] [PubMed]

44. Karim, M.R.; Zhang, S.; Jian, F.; Li, J.; Zhou, C.; Zhang, L.; Sun, M.; Yang, G.; Zou, F.; Dong, H.; et al. Multilocus Typing of *Cryptosporidium* spp. and *Giardia duodenalis* from Non-human Primates in China. *Int. J. Parasitol.* **2014**, *44*, 1039–1047. [CrossRef]

45. Widmer, G.; Köster, P.C.; Carmena, D. *Cryptosporidium hominis* Infections in Non-human Animal Species: Revisiting the Concept of Host Specificity. *Int. J. Parasitol.* **2020**, *50*, 253–262. [CrossRef]

46. Martínez-Ruiz, R.; de Lucio, A.; Fuentes, I.; Carmena, D. Autochthonous *Cryptosporidium cuniculus* Infection in Spain: First Report in a Symptomatic Paediatric Patient from Madrid. *Enferm. Infecc. Y Microbiol. Clínica* **2016**, *34*, 532–534. [CrossRef]

47. Puleston, R.L.; Mallaghan, C.M.; Modha, D.E.; Hunter, P.R.; Nguyen-Van-Tam, J.S.; Regan, C.M.; Nichols, G.L.; Chalmers, R.M. The First Recorded Outbreak of Cryptosporidiosis Due to *Cryptosporidium cuniculus* (Formerly Rabbit Genotype), Following a Water Quality Incident. *J. Water Health* **2014**, *12*, 41–50. [CrossRef] [PubMed]

pathogens

Article

Comparative Performance of Eight PCR Methods to Detect *Cryptosporidium* Species

Damien Costa [1,2,3,*], Louise Soulieux [1], Romy Razakandrainibe [2,3], Louise Basmaciyan [4,5], Gilles Gargala [1,2,3], Stéphane Valot [4,5], Frédéric Dalle [4,5] and Loic Favennec [1,2,3]

[1] Department of Parasitology/Mycology, University Hospital of Rouen, 76000 Rouen, France; louise.soulieux@ecomail.fr (L.S.); gilles.gargala@chu-rouen.fr (G.G.); loic.favennec@chu-rouen.fr (L.F.)
[2] EA ESCAPE 7510, University of Medicine Pharmacy Rouen, 76000 Rouen, France; romy.razakandrainibe@univ-rouen.fr
[3] CNR LE Cryptosporidiosis, Santé Publique France, 76000 Rouen, France
[4] CNR LE Cryptosporidiosis Collaborating Laboratory, Santé Publique France, 21000 Dijon, France; Louise.Basmaciyan@u-bourgogne.fr (L.B.); stephane.valot@chu-dijon.fr (S.V.); frederic.dalle@u-bourgogne.fr (F.D.)
[5] Department of Parasitology/Mycology, University Hospital of Dijon, 21000 Dijon, France
* Correspondence: damien.costa@chu-rouen.fr

Abstract: Diagnostic approaches based on PCR methods are increasingly used in the field of parasitology, particularly to detect *Cryptosporidium*. Consequently, many different PCR methods are available, both "in-house" and commercial methods. The aim of this study was to compare the performance of eight PCR methods, four "in-house" and four commercial methods, to detect *Cryptosporidium* species. On the same DNA extracts, performance was evaluated regarding the limit of detection for both *C. parvum* and *C. hominis* specificity and the ability to detect rare species implicated in human infection. Results showed variations in terms of performance. The best performance was observed with the FTD® Stool parasites method, which detected *C. parvum* and *C. hominis* with a limit of detection of 1 and 10 oocysts/gram of stool respectively; all rare species tested were detected (*C. cuniculus*, *C. meleagridis*, *C. felis*, *C. chipmunk*, and *C. ubiquitum*), and no cross-reaction was observed. In addition, no cross-reactivity was observed with other enteric pathogens. However, commercial methods were unable to differentiate *Cryptosporidium* species, and generally, we recommend testing each DNA extract in at least triplicate to optimize the limit of detection.

Keywords: *Cryptosporidium*; PCR; detection; diagnosis; sensitivity; specificity

Citation: Costa, D.; Soulieux, L.; Razakandrainibe, R.; Basmaciyan, L.; Gargala, G.; Valot, S.; Dalle, F.; Favennec, L. Comparative Performance of Eight PCR Methods to Detect *Cryptosporidium* Species. *Pathogens* **2021**, *10*, 647. https://doi.org/10.3390/pathogens10060647

Academic Editor: Hans-Peter Fuehrer

Received: 29 April 2021
Accepted: 21 May 2021
Published: 23 May 2021

1. Introduction

Human cases of cryptosporidiosis were first reported in the 1970s in children and immunosuppressed adults [1]. In 2015, the Global Enteric Multicenter Study (GEMS) described *Cryptosporidium* spp. as the second leading cause (5–15%) of moderate to severe diarrhea among infants in countries of sub-Saharan Africa and South Asia, after rotavirus [2]. At the same time, *Cryptosporidium* spp. were found to be responsible for more than 8 million cases of foodborne illnesses in 2010, and they were ranked fifth out of 24 potentially foodborne parasites in terms of importance [3,4]. In 2017 in France, the National Reference Center-Expert Laboratory (CNR-LE) for cryptosporidiosis was set up, allowing the collection and interpretation of epidemiological data thanks to the participation of members of the network. Published data from the French CNR-LE for cryptosporidiosis show that: i) even with around 250 notified cases each year, cryptosporidiosis is still largely underestimated in France, ii) cryptosporidiosis is predominant in immunocompetent individuals and especially in young children and young adults, and iii) cryptosporidiosis is over-represented in the summer [5,6]. The routine diagnosis of cryptosporidiosis still relies on light microscopy examination for many laboratories [7–10]. However, light microscopy examination lacks sensitivity, is time-consuming, and requires skilled technicians, making it an inefficient method for laboratories which are able to switch to PCR analysis [8,10–12]. Currently, several PCR methods are available to screen *Cryptosporidium* spp. DNA, both "in-house" and commercial methods, sometimes incorporated into multiplex panels [13–17]. Consequently, more and more laboratories are opting for such methods based on the practicalities of economic management. However, disparities exist in the performance of these methods. DNA extraction is essential to obtain good performance in PCR analysis and especially regarding parasitological investigations on stool samples. Some studies reported different performances of DNA extraction methods, and regarding the extraction of *Cryptosporidium* oocysts, a mechanical treatment of stool samples seems essential [18–23]. In addition to the extraction method, the removal of inhibitory substances and the gene locus targeted by related primers plays a major role in the performance of the method. One of the tasks of the CNR-LE for cryptosporidiosis is to assess the performance of available diagnostic tools. A previous work already compared performances of various extraction methods on *C. parvum* oocysts from stool samples [22]. In continuity of this work and based on the most effective extraction method, we propose a comparison of the limit of detection of eight real-time PCR methods (commercial or not) on the DNA of *Cryptosporidium* species. The main aim was to provide data to select the best methods for DNA amplification in terms of sensitivity and ability to detect human pathogenic *Cryptosporidium* species (even rare ones) in routine diagnosis.

2. Results

The results obtained from the four "in-house" PCR methods are summarized in Table 1. Except for the most concentrated *C. parvum* extract (10^5 oocysts/gram), significant differences in threshold cycle (Cq) values were observed when applicable (on ANOVA test). All four "in-house" methods detected *C. parvum* DNA and *C. hominis* DNA with a limit of detection of 10^3 oocysts/gram and 10^4 oocysts/gram, respectively. The most sensitive "in-house" PCR method for both *C. parvum* and *C. hominis* was the method developed by the CNR-LE Cryptosporidiosis Collaborating Laboratory (University Hospital of Dijon) and described by Valeix et al. 2020 [22]. Cq values obtained with the method described by Mary et al. 2013 [15] were lower than other tested methods but analysis performed in triplicate was insufficient to detect 10 oocysts/gram for *C. parvum*, contrary to the method described by Valeix et al. PCR efficiencies were satisfactory only on *C. parvum* DNA amplification for the methods described by Fontaine et al. 2002 and Valeix et al. 2020 [13,22]. R^2 values were satisfactory (>0.99) only to detect *C. parvum* and for the methods described by Fontaine et al. 2002, Hadfield et al. 2011 and Mary et al. 2013 [13–15].

The results obtained from the four multiplex commercial methods are presented in Table 2. All four commercial methods detected *C. hominis* DNA and *C. parvum* DNA with a limit of detection of 10^3 oocysts/gram. The best performance was obtained with the FTD Stool parasites method, with a detection of DNA corresponding to 1 oocyst/gram for *C. parvum* and 10 oocysts/gram for *C. hominis*. The Allplex GI Parasite Assay kit was the second-best method, with a detection of DNA corresponding to 100 oocysts/gram for *C. hominis* and 10 oocysts/gram for *C. parvum* but requiring a triplicate to reach the limit of detection (only 1/3 triplicates was positive to detect *C. parvum* at 10 oocysts/gram). PCR efficiencies and R^2 values varied greatly depending on the studied method but overall were unsatisfactory (PCR efficiency < 90% or > 110% and R^2 value < 0.99).

Table 1. Limit of detection of tested "in-house" methods on *C. parvum* and *C. hominis*. PCR efficiencies were calculated based on obtained results from corresponding ranges of dilutions.

	Oocysts/Gram	Mean Cq Value (+/− Standard Deviation)				
		Fontaine et al. 2002	Valeix et al. 2020	Hadfield et al. 2011	Mary et al. 2013	*p*-Value
C. parvum	10^5	28.47 (+/− 0.54)	27.60 (+/− 0.27)	28.14 (+/− 0.42)	24.80 (+/− 1.41)	0.15
	10^4	31.78 (+/− 0.17)	29.86 (+/− 0.49)	31.02 (+/− 0.32)	26.97 (+/− 1.32)	0.002
	10^3	34.71 (+/− 1.37)	35.96 (+/− 0.31)	35.72 (+/− 0.27)	30.19 (+/− 0.27)	<0.001
	10^2	/	36.24 (+/− 0.25)	39.15	32.38 (+/− 1.8)	/
	10	/	36.64 (+/− 0.28)	/	/	/
	1	/	/	/	/	/
	Corresponding *C. parvum* PCR efficiency (%)	109	111	84.9	143	/
	R^2 value	0.998	0.878	0.994	0.994	/
C. hominis	10^5	28.14 (0.20)	28.04 (+/− 0.25)	28.76 (+/− 0.01)	27.15 (+/− 0.04)	<0.001
	10^4	30.9 (+/−0.25)	29.66 (+/− 0.47)	31.33 (+/− 0.09)	29.69 (+/− 0.06)	0.001
	10^3	38.27	36.14 (+/− 0.48)	/	36.66 (+/− 0.61)	/
	10^2	/	/	/	/	/
	10	/	/	/	/	/
	1	/	/	/	/	/
	Corresponding *C. hominis* PCR efficiency (%)	57.5	76.5	/	62.3	/
	R^2 value	0.939	0.893	/	0.932	/

Table 2. Limit of detection of tested commercial methods on *C. parvum* and *C. hominis*. PCR efficiencies were calculated based on obtained results from corresponding ranges of dilutions.

	Oocysts/Gram	Mean Cq Value (+/− Standard Deviation)				
		RIDA® GENE Parasitic Stool Panel II	FTD® Stool Parasites	Amplidiag® Stool Parasites	Allplex® GI Parasite Assay	*p*-Value
C. parvum	10^5	27.61 (+/− 0.15)	19.84 (+/− 0.25)	28.02 (+/− 0.11)	28.32 (+/− 0.12)	<0.001
	10^4	30.62 (+/− 0.25)	22.88 (+/− 0.22)	32.24 (+/− 0.15)	31.97 (+/− 0.21)	<0.001
	10^3	37.7	26.59 (+/− 0.24)	35.17 (+/− 0.95)	34.72 (+/− 0.42)	/
	10^2	/	30.50 (+/− 0.57)	44.2	37.71 (+/− 0.66)	/
	10	/	34.47 (+/− 1.77)	/	37.68	/
	1	/	34.61 (+/− 1.82)	/	/	/
	Corresponding *C. parvum* PCR efficiency (%)	57.8	104	56.4	156	/
	R^2 value	0.949	0.970	0.939	0.932	/

Table 2. *Cont.*

		Mean Cq Value (+/− Standard Deviation)				
	Oocysts/Gram	RIDA® GENE Parasitic Stool Panel II	FTD® Stool Parasites	Amplidiag® Stool Parasites	Allplex® GI Parasite Assay	*p*-Value
C. hominis	10^5	27.73 (+/− 0.05)	21.41 (+/− 0.09)	29.01 (+/− 0.16)	27.39 (+/− 0.45)	<0.001
	10^4	29.63 (+/− 0.10)	22.95 (+/− 0.09)	32.06 (+/− 0.38)	29.60 (+/− 0.09)	<0.001
	10^3	38.53 (+/− 2.74)	26.84 +(/− 0.31)	36.49 (+/− 1.10)	33.35 +(/− 0.06)	0.003
	10^2	/	29.22 (+/− 0.04)	44.28	36.78 (+/− 0.62)	/
	10	/	31.12 (+/− 0.28)	/	/	/
	1	/	/	/	/	/
	Corresponding *C. hominis* PCR efficiency (%)	53.1	145	58.1	105	/
	R^2 value	0.877	0.982	0.956	0.985	/

The ability to detect rare species implicated in human pathologies for each tested method is summarized in Table 3. All tested methods were able to detect the species *C. cuniculus*, *C. meleagridis*, *C. felis*, *C. chipmunk*, and *C. ubiquitum*, except the methods described by Mary et al. 2013 and Fontaine et al. 2002 [13,15]. Specificity tests performed in triplicate per condition, as described in the Methods section, revealed cross-reactivity only for the method described by Hadfield et al. 2011 [14] with *Encephalitozoon intestinalis* DNA.

Table 3. Detection of rare species of *Cryptosporidium* implicated in human cases by tested methods.

	C. cuniculus	*C. meleagridis*	*C. felis*	*C. chipmunk*	*C. ubiquitum*
Fontaine et al. 2002	Yes	Yes	No	No	No
Valeix et al. 2020	Yes	Yes	Yes	Yes	Yes
Hadfield et al. 2011	Yes	Yes	Yes	Yes	Yes
Mary et al. 2013	No	No	No	No	No
RIDA® GENE Parasitic Stool Panel II	Yes	Yes	Yes	Yes	Yes
FTD® Stool parasites	Yes	Yes	Yes	Yes	Yes
Amplidiag® Stool Parasites	Yes	Yes	Yes	Yes	Yes
Allplex® GI Parasite Assay	Yes	Yes	Yes	Yes	Yes

3. Discussion

This study was designed to address questions regularly raised within the framework of scientific exchanges of the CNR-LE for cryptosporidiosis. It compared the performance of eight PCR methods to detect *Cryptosporidium* species (even rare) implicated in human infection, and their limit of detection. At first, the subject appeared to be well-investigated within the scientific community. However, in most cases, PCR performances to detect *Cryptosporidium* DNA were evaluated in cohorts from microscopically positive stool samples (probably relatively highly concentrated in oocysts), or not specifically through multiplex panels and from various extraction methods, or sometimes from DNA extracts stored for a long time [24–33]. In this study, thanks to a standardized extraction procedure (selected among the best methods regarding specific *Cryptosporidium* DNA extraction from stool samples [22]), observed PCR performances were exclusively due to the DNA amplification step. The limit of each studied PCR method was determined by assessing titrations of *Cryptosporidium* oocysts in stool samples as well as testing rare species implicated in human infection. The main interest of the study was to provide data on efficient methods for the routine diagnosis of cryptosporidiosis as a complement to extraction methods already assessed [22,23].

The results obtained generally showed similar performances between commercial and "in-house" methods in terms of limit of detection, with variations between each tested kit. Regarding *C. parvum* and *C. hominis* respectively, limits of detection generally reached at least 100 and 1000 oocysts/gram regardless of the method. Nevertheless, the limit of detection appeared optimal with the FTD® method considering both *C. parvum* and *C. hominis*. Variations in limits of detection may first be explained by the genes targeted by PCR methods. Three of the four "in-house" methods target the 18S rRNA gene whose expression is estimated at 5 copies/genome (20 copies/oocyst) [15]. The "in-house" method described by Fontaine et al. 2002 targets a gene whose expression is estimated at 1 copy/genome, and indeed, its observed performance in terms of limit of detection was generally poorer than that of the three methods targeting the 18S rRNA gene. Regarding commercial methods, targeted genes were only available for FTD® (DNA J-like protein, number of copies per genome not known) and Amplidiag® methods (COWP gene; 1 copy/genome). Of note, the observed performance of the Amplidiag® method was close to that of Fontaine et al.'s "in-house" method targeting a gene also expressed in 1 copy/genome. Comparing Tables 1 and 2, the limit of detection of the Amplidiag® method appeared slightly better than that of Fontaine et al.'s "in-house" method (for both *C. parvum* and *C. hominis*) but this was only due to DNA detection in one replicate at the very end of the PCR program. It could be explained by the heterogeneous distribution of DNA in elution volume when parasite concentrations are low. To limit this bias, and to obtain optimized performance, we recommend running each DNA extract in several replicates (at least in triplicate) or until exhaustion if possible.

Regarding the results obtained to detect rare species of *Cryptosporidium* implicated in human infections, most tested methods were able to detect rare species except the "in-house" methods of Fontaine et al. 2002 and Mary et al. 2013 [13,15]. However, a limitation of this study was the use of only triplicate of each tested *Cryptosporidium* subtype due to the amount of available positive stools. The use of more numerous rare strains could potentially improve the observed results. For the method described by Fontaine et al. 2002, they highlighted the use of a specific primer-probe set supposed to be specific for a *C. parvum* genomic DNA sequence. No cross-reactivity with other *Cryptosporidium* species was expected; however, they initially reported cross-reaction with the *C. meleagridis* genotype, which was confirmed in our study [13]. For the method of Mary et al. 2013, no rare species was detected in this study. In the original article, tests on *C. felis*, *C. bovis*, *C. cuniculus*, *C. canis*, and *C. chipmunk* were evoked in the discussion. However, in reality, corresponding results were not shown [15]. Consequently, primers and probes described in the article of Mary et al. 2013 are probably very specific to *C. parvum* and *C. hominis*. Regarding specificity in this study, performances obtained were highly satisfactory for each tested condition in concordance with the literature [13–15,27,34]. Cross-reactivity with *Candida albicans* DNA was tested since Mary et al. 2013 reported potential cross-reactivity with the *C. albicans* 18S rRNA gene (based on an in silico approach) and primers and probes of the PCR method described by Hadfield et al. 2011 [14,15]. For the method described by Hadfield et al. 2011, no cross-reactivity was observed with *C. albicans* but cross-reactivity was observed with *E. intestinalis*.

Finally, out of a total of 784 PCRs performed, varying results were obtained from the same DNA samples. Commercial methods (especially FTD® and Allplex®) appeared to be valuable options for large screening to detect *Cryptosporidium* species. We recommend testing each DNA extract at least in triplicate to optimize the detection of small amounts of DNA. However, if commercial methods are able to detect rare species, results are expressed exclusively as positive or negative for *Cryptosporidium* spp. DNA detection. Consequently, to discriminate species, we recommend the use of "in-house" methods, and especially the method described by Valeix et al. 2020 [22], due to the results obtained in terms of limit of detection and the ability to detect rare species. In addition, the method described by

Valeix et al. 2020 appeared to be strongly replicable, since performances in terms of limit of detection were similar to those described here, even using different stool samples [22].

4. Materials and Methods

4.1. Strains

Cryptosporidium spp. tested strains were obtained from the French cryptosporidiosis CNR-LE stools collection. *C. parvum* IIaA15G2R1 (n = 3), *C. hominis* IbA10G2 (n = 3) gp60 subtypes, *C. cuniculus* (n = 3), *C. meleagridis* (n = 3), *C. felis* (n = 3), *C. chipmunk* (n = 3), and *C. ubiquitum* (n = 3) were tested, all previously isolated from human clinical samples. In total, per studied method, six ten-fold range points (10^5-1 oocysts/gram) were studied for both *C. parvum* and *C. hominis*. Ranges of dilutions were done from highly concentrated natural stool samples that we diluted subsequently. Ranges of dilutions were performed in liquid stool matrix exempt of *Cryptosporidium* species. Each sample was vortexed for 20 s before performing dilutions. Oocyst numeration was done microscopically using Kova cells and confirmed by immunofluorescence as described in Section 4.2.

Regarding the studied *Cryptosporidium* rare species, stools were selected from the CNR collection with oocyst concentrations that varied between 10^3 and 10^4 oocysts/gram to be easily detectable.

Other positive stool specimens were obtained from the CNR-LE collection to evaluate specificity: *Giardia intestinalis* (n = 3), *Blastocystis hominis* (n = 3), *Enterobius vermicularis* (n = 3), *Chilomastix mesnilii* (n = 3), *Entamoeba histolytica* (n = 3), *Entamoeba dispar* (n = 3), *Encephalitozoon intestinalis* (n = 3), and *Enterocytozoon bieneusi* (n = 3) positive stool samples were tested. Exact stool concentrations of these other pathogens were not calculated. Positivity was objectified by microscopy exclusively assuming relatively high concentrations. Additional tests were performed on *Candida albicans* (n = 3) and on negative stool samples (n = 20). A total of 784 PCRs were performed (98 per tested method).

4.2. Detection Limit Assays

Serial ten-fold oocyst dilutions (10^5-1 oocysts/gram of stool) were performed using a negative liquid stool human matrix. Each corresponding dilution was confirmed by counting oocysts microscopically both on Kova slides (Labellians, Nemours, France) and using Crypto-Cel FITC (Cellabs, Sydney, Australia) staining according to the manufacturer's instructions. Limits of detection were estimated considering the positivity of at least one of the three tested replicates per condition. *Cryptosporidium* negativity of the matrix was based on both microscopy and PCR investigations from the method described by Valeix et al. 2020.

4.3. DNA Extraction

Based on a previous published work [22], we chose an extraction protocol offering highly satisfying performances in *C. parvum* DNA extraction from stool matrix. Accordingly, the observed PCR performances were exclusively due to amplification methods since extraction was standardized. Consequently, DNA extraction was performed using a QIAamp PowerFecal DNA kit (Qiagen, Courtaboeuf, France) according to the manufacturer's instructions. Briefly, it is a manual extraction kit combining thermal, mechanical, and chemical lysis. The starting volume for DNA extraction was 250 µL of sample. Obtained DNA extracts (100 µL) were stored at −20 °C until use. In addition, to control DNA extraction, we used Diacontrol DNA® for each sample according to the manufacturer's instructions. Ten microliters of viral DNA control was inoculated in each sample before extraction. Control DNA was subsequently detected using ready-to-use ProbePrimer mix (DICD-CY-L100) with the following PCR protocol: 50 °C for 2 min; 95 °C for 10 min; 95 °C for 15 s and 60 °C for 60 s, repeated 45 times.

4.4. PCR Testing

Eight real-time PCR methods were tested on the same DNA extracts: 4 "in-house" PCRs already assessed [13–15,22] and 4 multiplex commercial PCRs: RIDA® GENE Parasitic Stool Panel II (R-Biopharm, Darmstadt, Germany), FTD® Stool parasites (Fast Track Diagnostics, Esch-sur-Alzette, Luxembourg), Amplidiag® Stool Parasites (Mobidiag, Paris, France), and Allplex® GI Parasite Assay (Seegene, Düsseldorf, Germany). The studied methods were selected based on methods used in France according to data collected by the CNR-LE for cryptosporidiosis. PCR was performed in triplicate for each tested condition on a CFX96 PCR detection system (Bio-rad, Marnes-la-Coquette, France) according to published data for "in-house" PCR and according to the manufacturer's instructions for commercial PCR (synthesized in Table 4). A total of 784 PCRs were performed (98 per tested method). In detail, each studied condition was extracted 3 times ($N = 3$) and run in simplicate per extract for PCR amplification. Consequently, regarding assays from range of dilution + rare species + specificity investigations respectively: $36 + 15 + 47 = 98$ DNA extracts were tested for each studied method ($N = 8$). Assays were divided into two runs per studied method (two distinct PCR plates). A total of 16 PCR runs were done.

Results were considered positive when curves were exponential in logarithmic scale until the last cycle expected by each PCR program (Table 4).

PCR efficiencies ($10^{-1/slope}\_1$) were estimated according to Bustin et al. 2009: plotting the logarithm of the initial template concentration on the x-axis and Cq on the y-axis [35]. R^2 values were obtained using graphical representation on Excel software.

Table 4. Description of tested methods.

Designation	Primers (5′-3′)	Probe (5′-3′)	Target	Amplicon Size (bp)	Thermocycling Conditions	Total Duration
Fontaine et al. 2002 [13] method	F:CGCTTCTCTAGCCTTTCATGA R: CTTCACGTGTGTTTGCCAAT	CCAATCACAGAATCAT CAGAATCGACTGGTATC	Specific *C. parvum* sequence	138	50 °C—2 min 95 °C—10 min 40 cycles: 95 °C—15 s/60 °C—1 min	62 min
Valeix et al. 2020 [22] method	F: GTTAAACTGCRAATGGCT R: CGTCATTGCCACGGTA	CCGTCTAAAGCT GATAGGTCAGAAACTTGAATG and GTCACATTAATTGT GATCCGTAAAG	18S rRNA	258	95 °C—10 min 50 cycles: 95 °C—15 s/50 °C—15 s (touchdown from 60 °C)/72 °C—15 s	48 min
Hadfield et al. 2011 [14] method	F:GAGGTAGTGACAAGAAATAACAATACAGG R:CTGCTTTAAGCACTCTAATTTTCTCAAAG	TACGAGCTTTTTAA CTGCAACAA	18S SSU rRNA	300	95 °C—10 min 55 cycles: 95 °C—15 s/60 °C—60 s	78 min
Mary et al. 2013 [15] method	F: CATGGATAACCGTGGTAAT R: TACCCTACCGTCTAAAGCTG	CTAGAGCTAATACAT GCGAAAAAA	18S rRNA	178	94 °C—10 min 45 cycles: 94 °C—10 s/ 54 °C—30 s/72 °C—10 s	48 min
RIDA®GENE Parasitic Stool Panel II	Not disclosed	Not disclosed	Not disclosed	Not disclosed	95 °C—1 min 45 cycles: 95 °C—15 s/ 60 °C—30 s	35 min
FTD®Stool parasites	Not disclosed	Not disclosed	DNA J-like protein gene	Not disclosed	50 °C—15 min 94 °C—1 min 40 cycles: 94 °C—8 s/ 60 °C—1 min	62 min
Amplidiag®Stool Parasites	Not disclosed	Not disclosed	COWP gene	Not disclosed	95 °C—10 min 45 cycles: 95 °C—15 s/ 65 °C—1 min	66 min
Allplex®GI Parasite Assay	Not disclosed	Not disclosed	Not disclosed	Not disclosed	50 °C—20 min 95 °C—15 min 45 cycles: 95 °C—10 s/60 °C—1 min/ 72 °C—30 s	110 min

F: Forward. R: Reverse. SSU: Small subunit. COWP: *Cryptosporidium* oocyst wall protein.

5. Conclusions

Recent epidemiology confirms that cryptosporidiosis is common worldwide in both immunocompetent and immunocompromised individuals. Diagnostic approaches are still mainly based on microscopy; however, PCR-based methods are increasingly used for routine diagnosis. The performance of PCR methods is variable and needs to be evaluated. In this study, based on PCR analysis of the same DNA extracts, we compared the performance of eight commonly used methods according to limit of detection (for both *C. hominis* and *C. parvum*), specificity, and rare species identification. All eight methods were able to detect *C. parvum* and *C. hominis* with a limit of detection of 1000 oocysts/gram of stool, but only one method (FTD®) was able to detect one and ten oocysts/gram for *C. parvum* and *C. hominis*, respectively. Specificity was satisfactory for each tested method. Six of the eight methods were able to detect rare species implicated in human infection.

Author Contributions: Conceptualization: D.C., R.R., L.F. and F.D.; methodology: D.C. and L.S.; investigation: D.C. and L.S.; supervision: D.C., L.F. and F.D.; writing—original draft preparation: D.C.; writing—review: G.G., S.V., L.B., L.F., R.R. and F.D. All authors have read and agreed to the published version of the manuscript.

Funding: The authors gratefully thank Santé Publique France for their funding of CNR-LE cryptosporidiosis activities.

Institutional Review Board Statement: Not applicable.

Informed Consent Statement: Not applicable.

Data Availability Statement: Not applicable.

Acknowledgments: The authors are grateful to Nikki Sabourin-Gibbs, Rouen University Hospital, for her help in editing the manuscript.

Conflicts of Interest: The authors declare no conflict of interest.

References

1. Hunter, P.R.; Nichols, G. Epidemiology and Clinical Features of Cryptosporidium Infection in Immunocompromised Patients. *Clin. Microbiol. Rev.* **2002**, *15*, 145–154. [CrossRef]
2. GBD Diarrhoeal Diseases Collaborators. Estimates of Global, Regional, and National Morbidity, Mortality, and Aetiologies of Diarrhoeal Diseases: A Systematic Analysis for the Global Burden of Disease Study 2015. *Lancet Infect. Dis.* **2017**, *17*, 909–948. [CrossRef]
3. FAO/WHO. Multicriteria-Based Ranking for Risk Management of Food-Borne Parasites. Microbiological Risk Assessment Series 23; 2014; Available online: https://apps.who.int/iris/handle/10665/112672 (accessed on 1 May 2021).
4. Ryan, U.; Hijjawi, N.; Xiao, L. Foodborne Cryptosporidiosis. *Int. J. Parasitol.* **2018**, *48*, 1–12. [CrossRef] [PubMed]
5. Costa, D.; Razakandrainibe, R.; Valot, S.; Vannier, M.; Sautour, M.; Basmaciyan, L.; Gargala, G.; Viller, V.; Lemeteil, D.; Ballet, J.-J.; et al. Epidemiology of Cryptosporidiosis in France from 2017 to 2019. *Microorganisms* **2020**, *8*, 1358. [CrossRef] [PubMed]
6. Costa, D.; Razakandrainibe, R.; Sautour, M.; Valot, S.; Basmaciyan, L.; Gargala, G.; Lemeteil, D.; Favennec, L.; Dalle, F.; French National Network on Surveillance of Human Cryptosporidiosis. Human Cryptosporidiosis in Immunodeficient Patients in France (2015–2017). *Exp. Parasitol.* **2018**, *192*, 108–112. [CrossRef]
7. Robinson, G.; Chalmers, R.M. Cryptosporidium Diagnostic Assays: Microscopy. *Methods Mol. Biol.* **2020**, *2052*, 1–10. [CrossRef]
8. Adeyemo, F.E.; Singh, G.; Reddy, P.; Stenström, T.A. Methods for the Detection of Cryptosporidium and Giardia: From Microscopy to Nucleic Acid Based Tools in Clinical and Environmental Regimes. *Acta Trop.* **2018**, *184*, 15–28. [CrossRef]
9. Khanna, V.; Tilak, K.; Ghosh, A.; Mukhopadhyay, C. Modified Negative Staining of Heine for Fast and Inexpensive Screening of *Cryptosporidium*, *Cyclospora*, and *Cystoisospora* Spp. *Int. Sch. Res. Not.* **2014**, *2014*, 165424. [CrossRef]
10. Jerez Puebla, L.E.; Núñez-Fernández, F.A.; Fraga Nodarse, J.; Atencio Millán, I.; Cruz Rodríguez, I.; Martínez Silva, I.; Ayllón Valdés, L.; Robertson, L.J. Diagnosis of Intestinal Protozoan Infections in Patients in Cuba by Microscopy and Molecular Methods: Advantages and Disadvantages. *J. Microbiol. Methods* **2020**, *179*, 106102. [CrossRef]
11. Ahmed, S.A.; Karanis, P. Comparison of Current Methods Used to Detect Cryptosporidium Oocysts in Stools. *Int. J. Hyg. Environ. Health* **2018**, *221*, 743–763. [CrossRef]
12. Laude, A.; Valot, S.; Desoubeaux, G.; Argy, N.; Nourrisson, C.; Pomares, C.; Machouart, M.; Le Govic, Y.; Dalle, F.; Botterel, F.; et al. Is Real-Time PCR-Based Diagnosis Similar in Performance to Routine Parasitological Examination for the Identification of Giardia Intestinalis, Cryptosporidium Parvum/Cryptosporidium Hominis and Entamoeba Histolytica from Stool Samples? Evaluation of a New Commercial Multiplex PCR Assay and Literature Review. *Clin. Microbiol. Infect.* **2016**, *22*, 190.e1–190.e8. [CrossRef]

13. Fontaine, M.; Guillot, E. Development of a TaqMan Quantitative PCR Assay Specific for Cryptosporidium Parvum. *FEMS Microbiol. Lett.* **2002**, *214*, 13–17. [CrossRef] [PubMed]

14. Hadfield, S.J.; Robinson, G.; Elwin, K.; Chalmers, R.M. Detection and Differentiation of Cryptosporidium Spp. in Human Clinical Samples by Use of Real-Time PCR. *J. Clin. Microbiol.* **2011**, *49*, 918–924. [CrossRef] [PubMed]

15. Mary, C.; Chapey, E.; Dutoit, E.; Guyot, K.; Hasseine, L.; Jeddi, F.; Menotti, J.; Paraud, C.; Pomares, C.; Rabodonirina, M.; et al. Multicentric Evaluation of a New Real-Time PCR Assay for Quantification of Cryptosporidium Spp. and Identification of Cryptosporidium Parvum and Cryptosporidium Hominis. *J. Clin. Microbiol.* **2013**, *51*, 2556–2563. [CrossRef] [PubMed]

16. Fischer, P.; Taraschewski, H.; Ringelmann, R.; Eing, B. Detection of Cryptosporidium Parvum in Human Feces by PCR. *Tokai J. Exp. Clin. Med.* **1998**, *23*, 309–311.

17. Gallas-Lindemann, C.; Sotiriadou, I.; Plutzer, J.; Noack, M.J.; Mahmoudi, M.R.; Karanis, P. Giardia and Cryptosporidium Spp. Dissemination during Wastewater Treatment and Comparative Detection via Immunofluorescence Assay (IFA), Nested Polymerase Chain Reaction (Nested PCR) and Loop Mediated Isothermal Amplification (LAMP). *Acta Trop.* **2016**, *158*, 43–51. [CrossRef] [PubMed]

18. Yu, Z.; Morrison, M. Improved Extraction of PCR-Quality Community DNA from Digesta and Fecal Samples. *Biotechniques* **2004**, *36*, 808–812. [CrossRef] [PubMed]

19. Müller, A.; Stellermann, K.; Hartmann, P.; Schrappe, M.; Fätkenheuer, G.; Salzberger, B.; Diehl, V.; Franzen, C. A Powerful DNA Extraction Method and PCR for Detection of Microsporidia in Clinical Stool Specimens. *Clin. Diagn. Lab. Immunol.* **1999**, *6*, 243–246. [CrossRef]

20. Lee, M.F.; Lindo, J.F.; Auer, H.; Walochnik, J. Successful Extraction and PCR Amplification of Giardia DNA from Formalin-Fixed Stool Samples. *Exp. Parasitol.* **2019**, *198*, 26–30. [CrossRef] [PubMed]

21. Panina, Y.; Germond, A.; David, B.G.; Watanabe, T.M. Pairwise Efficiency: A New Mathematical Approach to QPCR Data Analysis Increases the Precision of the Calibration Curve Assay. *BMC Bioinform.* **2019**, *20*. [CrossRef]

22. Valeix, N.; Costa, D.; Basmaciyan, L.; Valot, S.; Vincent, A.; Razakandrainibe, R.; Robert-Gangneux, F.; Nourrisson, C.; Pereira, B.; Fréalle, E.; et al. Multicenter Comparative Study of Six Cryptosporidium Parvum DNA Extraction Protocols Including Mechanical Pretreatment from Stool Samples. *Microorganisms* **2020**, *8*, 1450. [CrossRef]

23. Claudel, L.; Valeix, N.; Basmaciyan, L.; Pereira, B.; Costa, D.; Vincent, A.; Valot, S.; Favennec, L.; Dalle, F. Comparative Study of Eleven Mechanical Pretreatment Protocols for Cryptosporidium Parvum DNA Extraction from Stool Samples. *Microorganisms* **2021**, *9*, 297. [CrossRef]

24. Köller, T.; Hahn, A.; Altangerel, E.; Verweij, J.J.; Landt, O.; Kann, S.; Dekker, D.; May, J.; Loderstädt, U.; Podbielski, A.; et al. Comparison of Commercial and In-House Real-Time PCR Platforms for 15 Parasites and Microsporidia in Human Stool Samples without a Gold Standard. *Acta Trop.* **2020**, *207*, 105516. [CrossRef]

25. Autier, B.; Belaz, S.; Razakandrainibe, R.; Gangneux, J.-P.; Robert-Gangneux, F. Comparison of Three Commercial Multiplex PCR Assays for the Diagnosis of Intestinal Protozoa. *Parasite* **2018**, *25*, 48. [CrossRef]

26. Hartmeyer, G.N.; Hoegh, S.V.; Skov, M.N.; Dessau, R.B.; Kemp, M. Selecting PCR for the Diagnosis of Intestinal Parasitosis: Choice of Targets, Evaluation of In-House Assays, and Comparison with Commercial Kits. *J. Parasitol. Res.* **2017**, *2017*, 6205257. [CrossRef]

27. Paulos, S.; Saugar, J.M.; de Lucio, A.; Fuentes, I.; Mateo, M.; Carmena, D. Comparative Performance Evaluation of Four Commercial Multiplex Real-Time PCR Assays for the Detection of the Diarrhoea-Causing Protozoa Cryptosporidium Hominis/Parvum, Giardia Duodenalis and Entamoeba Histolytica. *PLoS ONE* **2019**, *14*, e0215068. [CrossRef]

28. Mero, S.; Kirveskari, J.; Antikainen, J.; Ursing, J.; Rombo, L.; Kofoed, P.-E.; Kantele, A. Multiplex PCR Detection of Cryptosporidium Sp, Giardia Lamblia and Entamoeba Histolytica Directly from Dried Stool Samples from Guinea-Bissauan Children with Diarrhoea. *Infect. Dis.* **2017**, *49*, 655–663. [CrossRef]

29. Madison-Antenucci, S.; Relich, R.F.; Doyle, L.; Espina, N.; Fuller, D.; Karchmer, T.; Lainesse, A.; Mortensen, J.E.; Pancholi, P.; Veros, W.; et al. Multicenter Evaluation of BD Max Enteric Parasite Real-Time PCR Assay for Detection of Giardia Duodenalis, Cryptosporidium Hominis, Cryptosporidium Parvum, and Entamoeba Histolytica. *J. Clin. Microbiol.* **2016**, *54*, 2681–2688. [CrossRef] [PubMed]

30. Shin, J.-H.; Lee, S.-E.; Kim, T.S.; Ma, D.-W.; Cho, S.-H.; Chai, J.-Y.; Shin, E.-H. Development of Molecular Diagnosis Using Multiplex Real-Time PCR and T4 Phage Internal Control to Simultaneously Detect Cryptosporidium Parvum, Giardia Lamblia, and Cyclospora Cayetanensis from Human Stool Samples. *Korean J. Parasitol.* **2018**, *56*, 419–427. [CrossRef]

31. Morio, F.; Poirier, P.; Le Govic, Y.; Laude, A.; Valot, S.; Desoubeaux, G.; Argy, N.; Nourrisson, C.; Pomares, C.; Machouart, M.; et al. Assessment of the First Commercial Multiplex PCR Kit (ParaGENIE Crypto-Micro Real-Time PCR) for the Detection of Cryptosporidium Spp., Enterocytozoon Bieneusi, and Encephalitozoon Intestinalis from Fecal Samples. *Diagn. Microbiol. Infect. Dis.* **2019**, *95*, 34–37. [CrossRef]

32. Morgan, U.M.; Pallant, L.; Dwyer, B.W.; Forbes, D.A.; Rich, G.; Thompson, R.C. Comparison of PCR and Microscopy for Detection of Cryptosporidium Parvum in Human Fecal Specimens: Clinical Trial. *J. Clin. Microbiol.* **1998**, *36*, 995–998. [CrossRef] [PubMed]

33. Aghamolaie, S.; Rostami, A.; Fallahi, S.; Tahvildar Biderouni, F.; Haghighi, A.; Salehi, N. Evaluation of Modified Ziehl-Neelsen, Direct Fluorescent-Antibody and PCR Assay for Detection of Cryptosporidium Spp. in Children Faecal Specimens. *J. Parasit. Dis.* **2016**, *40*, 958–963. [CrossRef] [PubMed]

34. Brunet, J.; Lemoine, J.P.; Pesson, B.; Valot, S.; Sautour, M.; Dalle, F.; Muller, C.; Borni-Duval, C.; Caillard, S.; Moulin, B.; et al. Ruling out Nosocomial Transmission of Cryptosporidium in a Renal Transplantation Unit: Case Report. *BMC Infect. Dis.* **2016**, *16*, 363. [CrossRef] [PubMed]
35. Bustin, S.A.; Benes, V.; Garson, J.A.; Hellemans, J.; Huggett, J.; Kubista, M.; Mueller, R.; Nolan, T.; Pfaffl, M.W.; Shipley, G.L.; et al. The MIQE Guidelines: Minimum Information for Publication of Quantitative Real-Time PCR Experiments. *Clin. Chem.* **2009**, *55*, 611–622. [CrossRef] [PubMed]

Article

Subtype Characterization and Zoonotic Potential of *Cryptosporidium felis* in Cats in Guangdong and Shanghai, China

Jiayu Li [1], Fuxian Yang [1], Ruobing Liang [1], Sheng Guo [1], Yaqiong Guo [1], Na Li [1], Yaoyu Feng [1,2,*] and Lihua Xiao [1,2,*]

[1]   Center for Emerging and Zoonotic Diseases, College of Veterinary Medicine, South China Agricultural University, Wushan Road, Guangzhou 510642, China; hnnydxsyljy@126.com (J.L.); FuxianYang1210@163.com (F.Y.); tuzi0411@126.com (R.L.); guosheng201911@163.com (S.G.); guoyq@scau.edu.cn (Y.G.); nli@scau.edu.cn (N.L.)
[2]   Guangdong Laboratory for Lingnan Modern Agriculture, Wushan Road, Guangzhou 510642, China
*   Correspondence: yyfeng@scau.edu.cn (Y.F.); lxiao@scau.edu.cn (L.X.)

Abstract: *Cryptosporidium felis* is an important cause of feline and human cryptosporidiosis. However, the transmission of this pathogen between humans and cats remains controversial, partially due to a lack of genetic characterization of isolates from cats. The present study was conducted to examine the genetic diversity of *C. felis* in cats in China and to assess their potential zoonotic transmission. A newly developed subtyping tool based on a sequence analysis of the 60-kDa glycoprotein (*gp60*) gene was employed to identify the subtypes of 30 cat-derived *C. felis* isolates from Guangdong and Shanghai. Altogether, 20 *C. felis* isolates were successfully subtyped. The results of the sequence alignment showed a high genetic diversity, with 13 novel subtypes and 2 known subtypes of the XIXa subtype family being identified. The known subtypes were previously detected in humans, while some of the subtypes formed well-supported subclusters with human-derived subtypes from other countries in a phylogenetic analysis of the *gp60* sequences. The results of this study confirmed the high genetic diversity of the XIXa subtype family of *C. felis*. The common occurrence of this subtype family in both humans and cats suggests that there could be cross-species transmission of *C. felis*.

Keywords: *Cryptosporidium felis*; 60-kDa glycoprotein; subtypes; zoonotic transmission

Citation: Li, J.; Yang, F.; Liang, R.; Guo, S.; Guo, Y.; Li, N.; Feng, Y.; Xiao, L. Subtype Characterization and Zoonotic Potential of *Cryptosporidium felis* in Cats in Guangdong and Shanghai, China. *Pathogens* **2021**, *10*, 89. https://doi.org/10.3390/pathogens10020089

Academic Editor: David Carmena
Received: 25 December 2020
Accepted: 19 January 2021
Published: 20 January 2021

Publisher's Note: MDPI stays neutral with regard to jurisdictional claims in published maps and institutional affiliations.

1. Introduction

Cryptosporidium spp. are important apicomplexan parasites inhabiting the gastrointestinal tract of humans and other vertebrates, causing severe diarrhea [1]. Human cryptosporidiosis has been associated with over 20 *Cryptosporidium* species, but *C. hominis*, *C. parvum*, *C. meleagridis*, *C. felis*, and *C. canis* are the most common ones [2]. Among them, *C. felis* mainly infects cats and is therefore considered a host-adapted species [3]. Human *C. felis* infections, however, are common in developing countries [4–7], and at least one possible zoonotic transmission of *C. felis* between a household cat and the owner has been reported [8]. Nevertheless, the limited number of reports of zoonotic infections with this species has raised questions on the importance of zoonotic transmission in the epidemiology of human *C. felis* infections [9].

Subtyping tools based on sequence analysis of the 60-kDa glycoprotein (*gp60*) gene have been developed for human-pathogenic *Cryptosporidium* spp. to track infection sources [10]. Currently, the *gp60*-based subtyping tools are available for *C. hominis*, *C. parvum*, *C. meleagridis*, *C. ubiquitum*, *C. viatorum*, *Cryptosporidium* skunk genotype, and *Cryptosporidium* chipmunk genotype I [11–16]. These subtyping methods have been used in characterizing the transmission of these *Cryptosporidium* spp. in humans and animals [3].

A *gp60* subtyping tool has been developed recently for genetic characterizations of *C. felis* [17]. Thus far, nearly 200 *C. felis* isolates have been examined, which has led to the identification of approximately 100 subtypes in five subtype families (XIXa, XIXb, XIXc, XIXd, and XIXe) worldwide [17,18]. Most of the isolates, however, were from humans, and only two isolates from a human and a rhesus macaque have been characterized from China [18]. As *C. felis* has been identified in children and immunocompromised patients in China [19–22], we sought to examine the subtype identity of cat-derived *C. felis* isolates from Guangdong and Shanghai for assessment of their zoonotic potential.

2. Results

2.1. Amplification of the gp60 Gene

Among the 30 DNA preparations that were positive for *C. felis* based on nested PCR analysis of the small subunit (SSU) rRNA gene, 20 (66.7%) generated the expected products in the *gp60* PCR. Among them, 9 were from cats in pet shelters, 5 were from cats visiting animal hospitals, 4 were from cats in pet shops, and 2 were from stray cats (Table 1). These PCR products differed slightly in size (Figure 1) and were all sequenced successfully.

Table 1. Sources of *Cryptosporidium felis* samples used in the study and their *gp60* subtype identity.

Sample ID	Host	Region	Sample Source	Subtype
SCAU320	Cat	Guangzhou	Animal hospital	XIXa-90
SCAU1149	Cat	Guangzhou	Animal hospital	XIXa-89
SCAU1850	Cat	Guangzhou	Animal hospital	XIXa-87
SCAU1851	Cat	Guangzhou	Animal hospital	XIXa-87
SCAU1857	Cat	Guangzhou	Animal hospital	XIXa-88
SCAU2396	Cat	Shenzhen	Animal hospital	N/A
SCAU252	Cat	Guangzhou	Pet shop	XIXa-81
SCAU343	Cat	Guangzhou	Pet shelter	XIXa-88
SCAU356	Cat	Guangzhou	Pet shelter	N/A
SCAU392	Cat	Guangzhou	Pet shelter	XIXa-39
SCAU405	Cat	Guangzhou	Pet shelter	XIXa-89
SCAU1309	Cat	Shantou	Pet shelter	XIXa-86
SCAU4854	Cat	Guangzhou	Pet shelter	N/A
SCAU4905	Cat	Guangzhou	Pet shelter	N/A
SCAU143	Cat	Guangzhou	Stray animal	XIXa-84
SCAU731	Cat	Guangzhou	Stray animal	N/A
SCAU732	Cat	Guangzhou	Stray animal	N/A
SCAU754	Cat	Guangzhou	Stray animal	XIXa-40
ECUST19937	Cat	Shanghai	Pet shop	XIXa-81
ECUST26245	Cat	Shanghai	Pet shop	N/A
ECUST26246	Cat	Shanghai	Pet shop	XIXa-82
ECUST26248	Cat	Shanghai	Pet shop	XIXa-81
ECUST19997	Cat	Shanghai	Pet shelter	N/A
ECUST20283	Cat	Shanghai	Pet shelter	N/A
ECUST20286	Cat	Shanghai	Pet shelter	XIXa-91
ECUST20309	Cat	Shanghai	Pet shelter	N/A
ECUST26244	Cat	Shanghai	Pet shelter	XIXa-92
ECUST26249	Cat	Shanghai	Pet shelter	XIXa-93
ECUST26250	Cat	Shanghai	Pet shelter	XIXa-85
ECUST26251	Cat	Shanghai	Pet shelter	XIXa-83

N/A: PCR negative.

2.2. Nucleotide Sequence Variations in the gp60 Gene of C. felis

A comparison of the nucleotide sequences generated led to the identification of 15 sequence types. They differed from each other by both nucleotide substitutions and copy numbers of repetitive sequences. The sequence alignments showed the presence of numerous single nucleotide substitutions (SNPs) over the partial *gp60* gene among the 20 *C. felis* isolates. Three types of simple tandem repeats were detected in the *gp60* genes (Table 2). Among them, 2–5 copies of the 33-bp repeat sequence 5′-CCACCTAGTGGCGGTAGTGGCGTGTCCCCTGCT-3′ and 2–4 copies of the 39-bp repeat sequence 5′-CCACCTAGTGGCGGTAGTGGCGTGTCCCCTGCT-3′

were observed at nucleotides 460–577 and 778–911 of the sequence alignment, respectively. In both repeat types, the last copy had only half of the usual length. In addition, 5 copies of the trinucleotide repeat GTT were found in the *gp60* sequences at nucleotides 1143-1154.

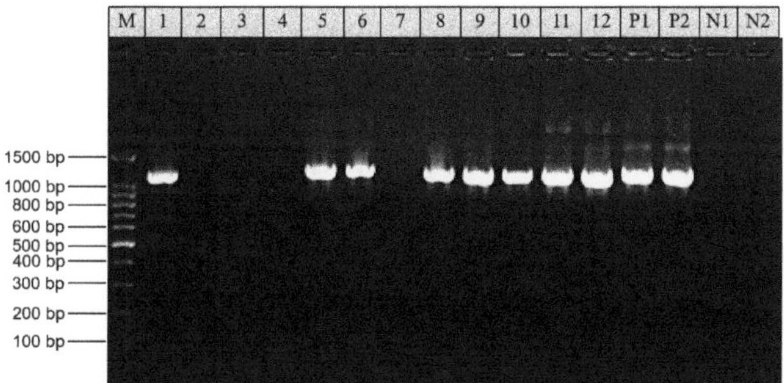

Figure 1. Analysis of the 60-kDa glycoprotein (*gp60*) gene in *Cryptosporidium felis* by nested PCR: lanes 1 and 2, SCAU320; lanes 3 and 4, SCAU2396; lanes 5 and 6, SCAU1149; lanes 7 and 8, SCAU1850; lanes 9 and 10, SCAU1851; lanes 11 and 12, SCAU1857; lane M, 100-bp molecular marker; P1 and P2, positive control (*C. felis* DNA); and N1 and N2, negative control (reagent-grade water).

Table 2. Tandem repeats in nucleotide sequences of the *gp60* gene of *Cryptosporidium felis*.

Sample ID [a]	GenBank Accession no.	Subtype	33-bp Repeat (No.) at 460–577 bp	39-bp Repeat (No.) at 778–911 bp	GGT Repeat (No.) at 1143–1154
SCAU392	MH240852	XIXa-39	R1 [b] (2)	R2 [c] (3)	4
SCAU754	MH240853	XIXa-40	R1 (3)	R2 (3)	4
SCAU252	MW351820	XIXa-81	R1 (2)	R2 (4)	4
ECUST19937	MW351820	XIXa-81	R1 (2)	R2 (4)	4
ECUST26248	MW351820	XIXa-81	R1 (2)	R2 (4)	4
ECUST26246	MW351821	XIXa-82	-	R2 (2)	4
ECUST26251	MW351822	XIXa-83	-	R2 (3)	4
SCAU143	MW351823	XIXa-84	R1 (5)	R2 (3)	4
ECUST26250	MW351824	XIXa-85	-	R2 (3)	4
SCAU1309	MW351825	XIXa-86	-	R2 (3)	4
SCAU1850	MW351826	XIXa-87	R1 (2)	R2 (2)	4
SCAU1851	MW351826	XIXa-87	R1 (2)	R2 (2)	4
SCAU343	MW351827	XIXa-88	R1 (2)	R2 (2)	4
SCAU1857	MW351827	XIXa-88	R1 (2)	R2 (2)	4
SCAU405	MW351828	XIXa-89	R1 (2)	R2 (3)	4
SCAU1149	MW351828	XIXa-89	R1 (2)	R2 (3)	4
SCAU320	MW351829	XIXa-90	R1 (2)	R2 (4)	4
ECUST20286	MW351830	XIXa-91	R1 (2)	R2 (3)	4
ECUST26244	MW351831	XIXa-92	R1 (?)	R2 (3)	4
ECUST26249	MW351832	XIXa-93	R1 (2)	R2 (3)	4

[a] Sample IDs labelled with ECUST were from cats in Shanghai, while those with SCAU were from cats in Guangdong; [b] 33-bp tandem repeat (5′-CCACCTAGTGGCGGTAGTGGCGTGTCCCCTGCT-3′) with a partial copy at the end; [c] 39-bp tandem repeat (5′-AGCACAACTGCGGCTACAGCGAGCACTGCGAGTTCGACA-3′) with a partial copy at the end and 0–2 nucleotide differences.

2.3. Cryptosporidium felis Subtypes Identified

Altogether, 15 subtypes were identified, with two subtypes in samples SCAU392 and SCAU754 having nucleotide sequences identical to the XIXa-39 and XIXa-40 subtypes (GenBank reference sequences MH240852 and MH240853) from humans in the United Kingdom, respectively (Table 1). According to the nomenclature system of the XIXa subtypes [18], the 13 new subtypes were named XIXa-81 (3), XIXa-82 (1), XIXa-83 (1), XIXa-84 (1), XIXa-85 (1), XIXa-86 (1), XIXa-87 (2), XIXa-88 (2), XIXa-89 (2), XIXa-90 (1), XIXa-91 (1), XIXa-92 (1), and XIXa-93 (1).

2.4. Phylogenetic Analysis of C. felis Subtypes

The *gp60* sequences from the 20 *C. felis* isolates were all placed in the XIXa subtype family in the phylogenetic tree (Figure 2). They formed several subclusters with strong bootstrap support with *C. felis* subtypes from humans. Among the novel subtypes identified in the study, 4 subtypes (XIXa-90, XIXa-91, XIXa-92, and XIXa-93) formed a subcluster with several subtypes identified from humans and cats in several European countries and another subtype (XIXa-82) formed a subcluster with several subtypes from humans and cats in the United Kingdom and Sweden. In addition, two cat-derived subtypes (XIXa-87 and XIXa-88) in 4 isolates from this study formed their own subcluster. As expected, the known subtypes (XIXa-39 and XIXa-40) identified in the present study and previously in humans from the United Kingdom and Indonesia formed a subcluster (Figure 2).

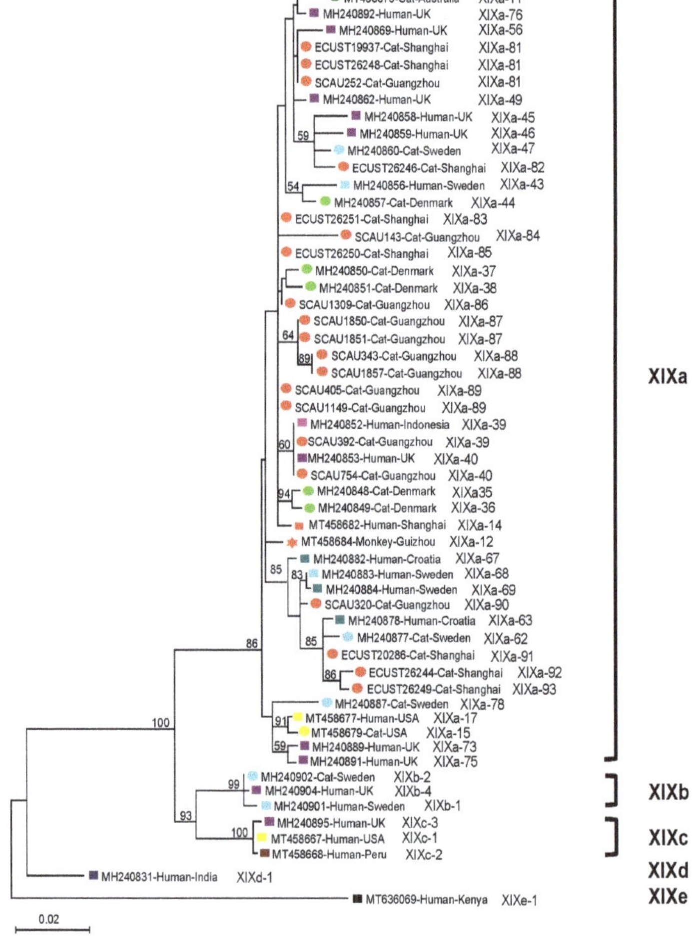

Figure 2. Phylogenetic relationship among XIXa subtypes of *Cryptosporidium felis* identified in this study and references from GenBank based on maximum likelihood analysis of the sequences of partial *gp60* gene: bootstrap values over 50 percent are shown on the branches of the phylogenetic tree. The human-, monkey-, and cat-derived isolates are indicated by rhombus, star, and round labels, respectively. Sequences from different countries are shown in different colors. The names of known subtype families of *C. felis* are shown on the right side of the phylogenetic tree. The subtype name of each isolate is labeled at the end of the sequence.

3. Discussion

The *gp60* subtyping tools have been widely used in assessing the intra-species diversity and zoonotic transmission of human-pathogenic *Cryptosporidium* spp. Several genes including *gp60*, other genetic loci with simple tandem repeats, and double-stranded viral RNA have been used to further differentiate the *Cryptosporidium* species into different subtypes [10]. Among them, the *gp60* gene is highly polymorphic in all *Cryptosporidium* spp. examined thus far and therefore can be categorized into multiple subtype families by nucleotide sequence differences [4]. Moreover, the *gp60* is an invasion-related protein and the subtype families identified have been linked to differences in host ranges and with virulence in *C. parvum* and *C. hominis* [3,23,24]. In this study, the newly developed *gp60* subtyping tool was employed to characterize the *C. felis* and to understand its zoonotic potential in China.

This represents the first subtyping study of *C. felis* in China. The previous two studies on *C. felis* subtypes mostly examined *C. felis* isolates from other countries, with five subtype families (XIXa, XIXb, XIXc, XIXd, and XIXe) being identified [17,18]. In the present study, 20 cat-derived isolates from China were subtyped and the results of the phylogenetic analysis showed that all of them belonged to the subtype family XIXa. The PCR amplification efficiency (66.7%) of this study was similar to that of a previous study (67.0%) [18]. The light infections of *C. felis* in healthy cats could be one of the reasons for the relatively low PCR amplification efficiency.

An analysis of the *gp60* gene confirmed the high genetic diversity of *C. felis* isolates. Like observations in previous studies [17,18], 15 subtypes were seen in 20 *C. felis* isolates successfully analyzed from cats. However, all *C. felis* isolates in this study belonged to the subtype family XIXa. As 13 of the 15 XIXa subtypes identified were novel, there could be geographic isolation among some of the XIXa subtypes, as previously suggested by us [18].

The results of the present study suggest that cross-species transmission of *C. felis* could be possible. In this study, among the 15 XIXa subtypes, two were identical to subtypes previously found in humans, while others clustered with human-derived subtypes from other countries. Importantly, one of the subclusters formed included two sequences from an owner and the household cat (GenBank reference sequences MH240883 and MH240884) with possible zoonotic transmission [17]. Moreover, we previously subtyped two *C. felis* isolates from one child in Shanghai (XIXa-14) and one rhesus macaque (XIXa-12) in Guizhou, China [18], while the *gp60* sequences were different with the 15 XIXa subtypes identified in the present study. They clustered together in a large clade within the XIXa subtype family in the phylogenetic tree (Figure 2). The subtype characteristics of *C. felis* identified in this suggest that the zoonotic XIXa subtype family could be the dominant subtypes in pet cats in China. As the sequences from *C. felis* in Guangzhou and Shanghai are dispersed throughout the large clade within the XIXa subtype family, they could be good representatives of *C. felis* in cats in China. Since *C. felis* infections have been reported in cats and humans in several locations in China [19,25–28], an analysis of the *gp60* sequences of human- and cat-derived *C. felis* isolates from the same location is needed to understand the epidemiological importance of zoonotic transmission of *C. felis*.

Currently, the significance of the divergent *C. felis* subtypes in cats is not clear. As mentioned above, different *C. parvum* and *C. hominis* subtypes have been linked to differences in virulence. Therefore, the hyper-transmissible and virulent *C. hominis* IbA10G2 and *C. parvum* IIaA15G2R1 subtypes have caused numerous outbreaks of human cryptosporidiosis worldwide [10] and infections with the former have induced more clinical symptoms than other *C. hominis* subtypes [23]. In *C. felis*, although previous studies and results of the present study suggest that the XIXa is the dominant subtype family in humans and cats [18], the relationship between the pathogenicity and its high transmission is not yet clear. More subtyping studies at additional genetic loci are needed to understand the differences in pathogenicity among *C. felis* subtypes.

4. Materials and Methods

4.1. Ethics Statement

This research was reviewed and approved by the Ethics Committee of the South China Agricultural University. The fecal specimens were collected with the permission of the owners of the pets. Each sample was collected and placed into a 50-mL plastic centrifuge tube containing 2.5% potassium dichromate and was transferred to the laboratory for storage at 4 °C. The DNA extraction was completed within one week. The DNA was stored at −20 °C for less than three years before being used in the PCR analysis in the present study.

4.2. C. felis Isolates

The nested PCR targeting the small subunit (SSU) rRNA gene was used to detect *Cryptosporidium* spp. [29]. DNA preparations of 30 *C. felis*-positive fecal samples from China were used in the present study. Among them, 18 samples were obtained from cats in Guangdong province, including 7 from pet shelters, 6 from animal hospitals, 4 from stray cats, and 1 from a pet shop. The remaining 12 samples were obtained from cats in Shanghai, including 8 from pet shelters and 4 from pet shops. They were from two previous studies of molecular epidemiology of cryptosporidiosis in cats [27,28].

4.3. PCR Amplification

The newly developed nested PCR targeting the conserved region of the *gp60* gene was employed to identify the subtypes of *C. felis* in this study [17]. Briefly, primers GP60-Felis-F1 (5′-TTT CCG TTA TTG TTG CAG TTG CA-3′) and GP60-Felis-R1 (5′-ATC GGA ATC CCA CCA TCG AAC-3′) were used in primary PCR, while GP60-Felis-F2 (5′-GGG CGT TCT GAA GGA TGT AA-3′) and GP60-Felis-R2 (5′-CGG TGG TCT CCT CAG TCT TC-3′) were used in secondary PCR. The sizes of primary and secondary PCR products were approximately 1200 and 900 bp, respectively. The PCR reaction and cycling program were described recently [18]. Each DNA preparation was analyzed in duplicate, with the inclusion of both positive (*C. felis* DNA) and negative (reagent-grade water) controls in each PCR run. The positive products from the secondary PCR were identified by 1.5% agarose electrophoresis.

4.4. DNA Sequence Analyses

All positive products of the expected size were sequenced bidirectionally on an ABI3730 autosequencer by the Sangon Biotech (Shanghai, China) using secondary PCR primers. The DNA sequences generated were assembled using ChromasPro 2.1.6 (http://technelysium.com.au/wp/) and edited using BioEdit 7.1.3.0 (http://www.mbio.ncsu.edu/BioEdit/bioedit.html). These and the reference sequences from GenBank were aligned with each other using the MUSCLE program implemented in MEGA 6 (https://www.megasoftware.net/). Tandem Repeats Finder 4.09 (http://tandem.bu.edu/trf/trf.html) was used to identify repetitive sequences within them. A maximum-likelihood tree was constructed using MEGA 6 based on substitution rates calculated using the general time reversible model and gamma distribution. The bootstrap method with 1000 replicates was used to assess the reliability of the phylogenetic clusters formed. Representative nucleotide sequences of the *C. felis* subtypes identified in the present study were deposited in the GenBank database under accession numbers MW351820-MW351832.

5. Conclusions

The present study reported the subtype characteristics of *C. felis* isolates from cats in China for the first time. The results of the phylogenetic analysis suggested the potential zoonotic transmission of this pathogen. More isolates from diverse areas and hosts should be analyzed to confirm this conclusion.

Author Contributions: Conceptualization, L.X. and Y.F.; data curation, J.L.; formal analysis, J.L.; funding acquisition, L.X. and Y.F.; investigation, J.L., F.Y., R.L. and S.G.; methodology, L.X. and Y.F.; project administration, N.L.; resources, Y.F.; software, L.X. and Y.F.; supervision, Y.G.; validation, L.X. and Y.F.; visualization, L.X. and Y.F.; writing—original draft, J.L.; writing—review and editing, L.X. All authors have read and agreed to the published version of the manuscript.

Funding: This work was funded by the Guangdong Major Project of Basic and Applied Basic Research (2020B0301030007), by the National Natural Science Foundation of China (U1901208 and 31820103014), by the 111 Project (D20008), and by the Innovation Team Project of Guangdong University (2019KCXTD001).

Institutional Review Board Statement: Not applicable.

Informed Consent Statement: Not applicable.

Data Availability Statement: Data is contained within the article.

Conflicts of Interest: The authors declare no conflict of interest.

References

1. Checkley, W.; White, A.C., Jr.; Jaganath, D.; Arrowood, M.J.; Chalmers, R.M.; Chen, X.M.; Fayer, R.; Griffiths, J.K.; Guerrant, R.L.; Hedstrom, L.; et al. A review of the global burden, novel diagnostics, therapeutics, and vaccine targets for *Cryptosporidium*. *Lancet Infect. Dis.* **2015**, *15*, 85–94. [CrossRef]
2. Zahedi, A.; Ryan, U. *Cryptosporidium*—An update with an emphasis on foodborne and waterborne transmission. *Res. Vet. Sci.* **2020**, *132*, 500–512. [CrossRef]
3. Feng, Y.; Ryan, U.M.; Xiao, L. Genetic diversity and population structure of *Cryptosporidium*. *Trends Parasitol.* **2018**, *34*, 997–1011. [CrossRef]
4. Xiao, L. Molecular epidemiology of cryptosporidiosis: An update. *Exp. Parasitol.* **2010**, *124*, 80–89. [CrossRef] [PubMed]
5. de Lucio, A.; Merino, F.J.; Martinez-Ruiz, R.; Bailo, B.; Aguilera, M.; Fuentes, I.; Carmena, D. Molecular genotyping and sub-genotyping of *Cryptosporidium* spp. isolates from symptomatic individuals attending two major public hospitals in Madrid, Spain. *Infect. Genet. Evol.* **2016**, *37*, 49–56. [CrossRef]
6. Cieloszyk, J.; Goni, P.; Garcia, A.; Remacha, M.A.; Sanchez, E.; Clavel, A. Two cases of zoonotic cryptosporidiosis in Spain by the unusual species *Cryptosporidium ubiquitum* and *Cryptosporidium felis*. *Enferm. Infecc. Microbiol. Clin.* **2012**, *30*, 549–551. [CrossRef]
7. Caccio, S.; Pinter, E.; Fantini, R.; Mezzaroma, I.; Pozio, E. Human infection with *Cryptosporidium felis*: Case report and literature review. *Emerg. Infect. Dis.* **2002**, *8*, 85–86. [CrossRef]
8. Beser, J.; Toresson, L.; Eitrem, R.; Troell, K.; Winiecka-Krusnell, J.; Lebbad, M. Possible zoonotic transmission of *Cryptosporidium felis* in a household. *Infect. Ecol. Epidemiol.* **2015**, *5*, 28463. [CrossRef]
9. Lucio-Forster, A.; Griffiths, J.K.; Cama, V.A.; Xiao, L.; Bowman, D.D. Minimal zoonotic risk of cryptosporidiosis from pet dogs and cats. *Trends Parasitol.* **2010**, *26*, 174–179. [CrossRef] [PubMed]
10. Xiao, L.; Feng, Y. Molecular epidemiologic tools for waterborne pathogens *Cryptosporidium* spp. and *Giardia duodenalis*. *Food Waterborne Parasitol.* **2017**, *8–9*, 14–32. [CrossRef] [PubMed]
11. Yan, W.; Alderisio, K.; Roellig, D.M.; Elwin, K.; Chalmers, R.M.; Yang, F.; Wang, Y.; Feng, Y.; Xiao, L. Subtype analysis of zoonotic pathogen *Cryptosporidium* skunk genotype. *Infect. Genet. Evol.* **2017**, *55*, 20–25. [CrossRef] [PubMed]
12. Stensvold, C.R.; Elwin, K.; Winiecka-Krusnell, J.; Chalmers, R.M.; Xiao, L.; Lebbad, M. Development and application of a gp60-based typing assay for *Cryptosporidium viatorum*. *J. Clin. Microbiol.* **2015**, *53*, 1891–1897. [CrossRef] [PubMed]
13. Guo, Y.; Cebelinski, E.; Matusevich, C.; Alderisio, K.A.; Lebbad, M.; McEvoy, J.; Roellig, D.M.; Yang, C.; Feng, Y.; Xiao, L. Subtyping novel zoonotic pathogen *Cryptosporidium* chipmunk genotype I. *J. Clin. Microbiol.* **2015**, *53*, 1648–1654. [CrossRef] [PubMed]
14. Stensvold, C.R.; Beser, J.; Axen, C.; Lebbad, M. High applicability of a novel method for gp60-based subtyping of *Cryptosporidium meleagridis*. *J. Clin. Microbiol.* **2014**, *52*, 2311–2319. [CrossRef] [PubMed]
15. Li, N.; Xiao, L.; Alderisio, K.; Elwin, K.; Cebelinski, E.; Chalmers, R.; Santin, M.; Fayer, R.; Kvac, M.; Ryan, U.; et al. Subtyping *Cryptosporidium ubiquitum*, a zoonotic pathogen emerging in humans. *Emerg. Infect. Dis.* **2014**, *20*, 217–224. [CrossRef] [PubMed]
16. Sulaiman, I.M.; Hira, P.R.; Zhou, L.; Al-Ali, F.M.; Al-Shelahi, F.A.; Shweiki, H.M.; Iqbal, J.; Khalid, N.; Xiao, L. Unique endemicity of cryptosporidiosis in children in Kuwait. *J. Clin. Microbiol.* **2005**, *43*, 2805–2809. [CrossRef]
17. Rojas-Lopez, L.; Elwin, K.; Chalmers, R.M.; Enemark, H.L.; Beser, J.; Troell, K. Development of a gp60-subtyping method for *Cryptosporidium felis*. *Parasites Vectors* **2020**, *13*, 39. [CrossRef]
18. Jiang, W.; Roellig, D.M.; Lebbad, M.; Beser, J.; Troell, K.; Guo, Y.; Li, N.; Xiao, L.; Feng, Y. Subtype distribution of zoonotic pathogen *Cryptosporidium felis* in humans and animals in several countries. *Emerg. Microbes Infect.* **2020**, *9*, 2446–2454. [CrossRef]
19. Feng, Y.; Wang, L.; Duan, L.; Gomez-Puerta, L.A.; Zhang, L.; Zhao, X.; Hu, J.; Zhang, N.; Xiao, L. Extended outbreak of cryptosporidiosis in a pediatric hospital, China. *Emerg. Infect. Dis.* **2012**, *18*, 312–314. [CrossRef]

20. Chen, S.; Ai, L.; Tian, L.; Zhang, Y.; Tong, X.; Li, H.; Chen, J. Investigation and fecal specimens detection of cryptozoite and other protozoon infection from patients with diarrhea. *Chin. J. Zoonoses* **2012**, *28*, 815–819.
21. Hung, C.C.; Tsaihong, J.C.; Lee, Y.T.; Deng, H.Y.; Hsiao, W.H.; Chang, S.Y.; Chang, S.C.; Su, K.E. Prevalence of intestinal infection due to *Cryptosporidium* species among Taiwanese patients with human immunodeficiency virus infection. *J. Formos. Med. Assoc.* **2007**, *106*, 31–35. [CrossRef]
22. Wang, R.; Zhang, X.; Zhu, H.; Zhang, L.; Feng, Y.; Jian, F.; Ning, C.; Qi, M.; Zhou, Y.; Fu, K.; et al. Genetic characterizations of *Cryptosporidium* spp. and *Giardia duodenalis* in humans in Henan, China. *Exp. Parasitol.* **2011**, *127*, 42–45. [CrossRef] [PubMed]
23. Cama, V.A.; Bern, C.; Roberts, J.; Cabrera, L.; Sterling, C.R.; Ortega, Y.; Gilman, R.H.; Xiao, L. *Cryptosporidium* species and subtypes and clinical manifestations in children, Peru. *Emerg. Infect. Dis.* **2008**, *14*, 1567–1574. [CrossRef] [PubMed]
24. Cama, V.A.; Ross, J.M.; Crawford, S.; Kawai, V.; Chavez-Valdez, R.; Vargas, D.; Vivar, A.; Ticona, E.; Navincopa, M.; Williamson, J.; et al. Differences in clinical manifestations among *Cryptosporidium* species and subtypes in HIV-infected persons. *J. Infect. Dis.* **2007**, *196*, 684–691. [CrossRef] [PubMed]
25. Liu, A.; Gong, B.; Liu, X.; Shen, Y.; Wu, Y.; Zhang, W.; Cao, J. A retrospective epidemiological analysis of human *Cryptosporidium* infection in China during the past three decades (1987–2018). *PLoS Negl. Trop. Dis.* **2020**, *14*, e0008146. [CrossRef]
26. Li, W.; Liu, X.; Gu, Y.; Liu, J.; Luo, J. Prevalence of *Cryptosporidium*, *Giardia*, *Blastocystis*, and *trichomonads* in domestic cats in East China. *J. Vet. Med. Sci.* **2019**, *81*, 890–896. [CrossRef]
27. Li, J.; Dan, X.; Zhu, K.; Li, N.; Guo, Y.; Zheng, Z.; Feng, Y.; Xiao, L. Genetic characterization of *Cryptosporidium* spp. and *Giardia duodenalis* in dogs and cats in Guangdong, China. *Parasites Vectors* **2019**, *12*, 571. [CrossRef]
28. Xu, H.; Jin, Y.; Wu, W.; Li, P.; Wang, L.; Li, N.; Feng, Y.; Xiao, L. Genotypes of *Cryptosporidium* spp., *Enterocytozoon bieneusi* and *Giardia duodenalis* in dogs and cats in Shanghai, China. *Parasites Vectors* **2016**, *9*, 121. [CrossRef]
29. Xiao, L.; Morgan, U.M.; Limor, J.; Escalante, A.; Arrowood, M.; Shulaw, W.; Thompson, R.C.; Fayer, R.; Lal, A.A. Genetic diversity within *Cryptosporidium parvum* and related *Cryptosporidium* species. *Appl. Environ. Microbiol.* **1999**, *65*, 3386–3391. [CrossRef]

Article

Subtyping *Cryptosporidium xiaoi*, a Common Pathogen in Sheep and Goats

Yingying Fan [1,2,†], Xitong Huang [1,†], Sheng Guo [1], Fang Yang [1], Xin Yang [3], Yaqiong Guo [1], Yaoyu Feng [1,2], Lihua Xiao [1,2,*] and Na Li [1,2,*]

1   Center for Emerging and Zoonotic Diseases, College of Veterinary Medicine, South China Agricultural University, Guangzhou 510642, China; fanyingying@webmail.hzau.edu.cn (Y.F.); huangxitong2021@163.com (X.H.); guosheng201911@163.com (S.G.); yf1432277601@163.com (F.Y.); guoyq@scau.edu.cn (Y.G.); yyfeng@scau.edu.cn (Y.F.)
2   Guangdong Laboratory for Lingnan Modern Agriculture, Guangzhou 510642, China
3   College of Veterinary Medicine, Northwest A&F University, Yangling, Xianyang 712100, China; xinyang@nwafu.edu.cn
*   Correspondence: lxiao1961@gmail.com (L.X.); nli@scau.edu.cn (N.L.)
†   These authors contributed to the research equally.

Abstract: Cryptosporidiosis is a significant cause of diarrhea in sheep and goats. Among the over 40 established species of *Cryptosporidium*, *Cryptosporidium xiaoi* is one of the dominant species infecting ovine and caprine animals. The lack of subtyping tools makes it impossible to examine the transmission of this pathogen. In the present study, we identified and characterized the 60-kDa glycoprotein (*gp60*) gene by sequencing the genome of *C. xiaoi*. The GP60 protein of *C. xiaoi* had a signal peptide, a furin cleavage site of RSRR, a glycosylphosphatidylinositol anchor, and over 100 O-glycosylation sites. Based on the *gp60* sequence, a subtyping tool was developed and used in characterizing *C. xiaoi* in 355 positive samples from sheep and goats in China. A high sequence heterogeneity was observed in the *gp60* gene, with 94 sequence types in 12 subtype families, namely XXIIIa to XXIIIl. Co-infections with multiple subtypes were common in these animals, suggesting that genetic recombination might be responsible for the high diversity within *C. xiaoi*. This was supported by the mosaic sequence patterns among the subtype families. In addition, a potential host adaptation was identified within this species, reflected by the exclusive occurrence of XXIIIa, XXIIIc, XXIIIg, and XXIIIj in goats. This subtyping tool should be useful in studies of the genetic diversity and transmission dynamics of *C. xiaoi*.

Keywords: *Cryptosporidium xiaoi*; 60-kDa glycoprotein; *gp60*; subtyping; genetic diversity; host adaptation

Citation: Fan, Y.; Huang, X.; Guo, S.; Yang, F.; Yang, X.; Guo, Y.; Feng, Y.; Xiao, L.; Li, N. Subtyping *Cryptosporidium xiaoi*, a Common Pathogen in Sheep and Goats. *Pathogens* 2021, 10, 800. https://doi.org/10.3390/pathogens10070800

Academic Editor: Luiz Shozo Ozaki

Received: 2 June 2021
Accepted: 22 June 2021
Published: 24 June 2021

Publisher's Note: MDPI stays neutral with regard to jurisdictional claims in published maps and institutional affiliations.

1. Introduction

Cryptosporidium spp. are important diarrheal pathogens in humans and various animals [1]. Currently, 45 *Cryptosporidium* species and over 100 genotypes have been recognized [2]. Among them, *C. parvum*, *C. ubiquitum*, and *C. xiaoi* are common species in sheep and goats. *C. parvum* and *C. ubiquitum* are zoonotic species that infect a wide range of hosts, while *C. xiaoi* appears to be adapted to ovine and caprine animals [3]. *C. xiaoi*, previously known as the *C. bovis*-like genotype, is the most common species in sheep and goats in most areas except Europe [4–8].

Sequence analysis of the 60-kDa glycoprotein (*gp60*) gene has been used extensively in subtyping *C. parvum*, *C. ubiquitum*, and other zoonotic species due to its high sequence heterogeneity and relevance to parasite biology. The unique distribution of subtype families and subtypes have significantly improved our understanding of host adaptation and transmission dynamics within these *Cryptosporidium* spp. [2,9–15]. Recently, *gp60* gene-based subtyping tools have been developed for molecular epidemiological studies of some non-human pathogenic *Cryptosporidium* spp., such as the bovine-adapted *C. ryanae* and the marsupial-adapted *C. fayeri* [16,17]. However, such a tool is not available for *C. xiaoi*, which is occasionally found in humans [18].

In this study, we sequenced the genome of *C. xiaoi*, identified its *gp60* gene, and developed a subtyping tool for genetic characterizations of isolates from sheep and goats.

2. Materials and Methods

2.1. Samples

DNA extracts from 434 *C. xiaoi*-positive samples were used in this study, including those from Small Tail Han sheep (*Ovis aries*), Hu sheep (*Ovis aries*), Tibetan sheep (*Ovis aries*), Huanghuai goats (*Capra hircus*), and Black goats (*Capra hircus*) on 11 farms in Qinghai, Henan, Anhui, and Guangdong, China (Table 1). The *C. xiaoi*-positive samples were obtained from previous and ongoing studies of molecular epidemiology of cryptosporidiosis in sheep and goats in China [19]. All the samples were identified as positive for *C. xiaoi* by PCR and sequence analysis of an ~830-bp fragment of the small subunit (*SSU*) rRNA gene [20].

2.2. Identification of the gp60 Gene of C. xiaoi

To obtain the nucleotide sequence of the *gp60* gene of *C. xiaoi*, we conducted whole-genome sequencing of one isolate (SCAU2942) from a Hu sheep in Anhui, China using the established procedures [21]. The genome was sequenced using Illumina HiSeq 2500 analysis of an Illumina TruSeq (v3) library with 250-bp paired-end reads. The sequence reads were assembled de novo using the SPAdes version 3.13 (http://cab.spbu.ru/software/spades/, accessed on 21 November 2019) with a K-mer size of 63. The *gp60* gene of *C. xiaoi* was identified by the blastn analysis of the genome assembly with the *gp60* (cgd6_1080) sequence of *C. parvum*. The coding region and amino acid sequence of the *gp60* gene were predicted using the combination of FGENESH (http://www.softberry.com/berry.phtml, accessed on 15 December 2019) and blastp search of the NCBI database.

2.3. Subtyping of C. xiaoi

Based on the sequence of the *C. xiaoi gp60* gene, nested PCR primers were designed for the subtyping analysis. The primers used in primary and secondary PCR were Xiaoi-*gp60*-F1 (5′-CCTCTCGGCACTTATTGCCCT-3′) and Xiaoi-*gp60*-R1 (5′-ATACCTGAGATCAAAT GCTGATGAA-3′), and Xiaoi-*gp60*-F2 (5′-CCTCTTAGGGGTTCATTGTCTA-3′) and Xiaoi-*gp60*-R2 (5′-TACCTTCAAAGATGACATCAC-3′), respectively. Each PCR was performed in a 50 μL-reaction containing 1×PCR master mix (Thermo Scientific, Waltham, MA, USA), 0.25 μM primary PCR primers or 0.5 μM secondary PCR primers, and 1 μL of DNA (primary PCR) or 2 μL of the primary PCR product (secondary PCR). To reduce PCR inhibitors, 400 ng/μL of nonacetylated bovine serum albumin (Sigma-Aldrich, St. Louis, MO, USA) was used in the primary PCR. The PCR amplification consisted of an initial denaturation at 94 °C for 5 min; 35 cycles of 94 °C (denaturation) for 45 s, 55 °C (annealing) for 45 s, and 72 °C (extension) for 90 s; and a final extension of 72 °C for 10 min. The secondary PCR products were visualized by 1.5% agarose gel electrophoresis.

Table 1. *Cryptosporidium xiaoi* subtype families identified in sheep and goats in China.

Host	Region	Breed	Farm ID	No. of *C. xiaoi*-Positive Samples	No. of Samples Positive at the *gp60* Locus (%)	No. of Samples with Divergent *gp60* PCR Banding Patterns (%)		Subtype Family (no.)
						One Band	Two Bands	
Sheep	Henan	Han	1	71	58 (81.7)	49 (84.5)	9 (15.5)	XXIIIb (13), XXIIId (3), XXIIIf (1), XXIIIh (13), XXIIIl (22), XXIIId + XXIIIk (1), XXIIId + XXIIIl (1)
			2	18	18 (100.0)	18 (100.0)	-	XXIIIb (1), XXIIIf (1), XXIIIk (5), XXIIIl (3)
	Anhui	Hu	3	84	64 (76.2)	57 (89.0)	7 (11.0)	XXIIIb (8), XXIIIe (2), XXIIIf (1), XXIIIh (17), XXIIIk (3), XXIIIl (17), XXIIIh + XXIIIk (2), XXIIIb + XXIIIk (1), XXIIId + XXIIIl (1)
	Qinghai *	Tibetan	4	39	8 (20.5)	8 (100.0)	-	XXIIId (1), XXIIIe (1), XXIIIh (1), XXIIIi (3)
Goats	Anhui	Huanghuai	5	77	77 (100.0)	66 (84.7)	11 (15.3)	XXIIIa (6), XXIIIb (4), XXIIIc (1), XXIIId (2), XXIIIe (5), XXIIIh (7), XXIIIi (4), XXIIIj (4), XXIIIk (5), XXIIIl (18), XXIIIb + XXIIIl (1), XXIIIh + XXIIIi (1), XXIIIh + XXIIIj (1), XXIIIh + XXIIIk (1), XXIIIi + XXIIIl (1)
			6	51	46 (90.2)	43 (93.5)	3 (6.5)	XXIIIb (3), XXIIIc (1), XXIIId (3), XXIIIe (5), XXIIIf (7), XXIIIh (4), XXIIIi (3), XXIIIj (1), XXIIIk (7), XXIIIl (4), XXIIIh +XXIIIk (1)
			7	33	33 (100.0)	31 (93.9)	2 (6.1)	XXIIIa (20), XXIIIg (5), XXIIIa + XXIIIg (1)
			8	20	16 (80.0)	16 (100.0)	-	XXIIId (1), XXIIIg (14)
	Guangdong	Black	9	11	8 (72.7)	8 (100.0)	-	XXIIIa (8)
			10	18	16 (88.9)	16 (100.0)	-	XXIIIa (16)
			11	12	11 (91.7)	11 (100.0)	-	XXIIIa (11)
Total	-	-	-	434	355 (81.8%)	323 (90.1)	32 (9.9)	XXIIIa (61), XXIIIb (29), XXIIIc (2), XXIIId (10), XXIIIe (13), XXIIIf (10), XXIIIg (19), XXIIIh (42), XXIIIi (10), XXIIIj (5), XXIIIk (20), XXIIIl (64), XXIIIa + XXIIIg (1), XXIIIb + XXIIIk (1), XXIIIb + XXIIIl (1), XXIIId + XXIIIk (1), XXIIId + XXIIIl (2), XXIIIh + XXIIIi (1), XXIIIh + XXIIIj (1), XXIIIh + XXIIIk (4), XXIIIi + XXIIIl (1)

* Samples from a previous study [19].

2.4. DNA Sequence Analysis

All secondary *gp60* PCR products were sequenced in both directions using Sanger sequencing by Sangon Biotech (Shanghai, China). For the samples yielding double PCR bands with different sizes, PCR products of each band were excised from the agarose electrophoresis gel and purified using the E.Z.N.A.® Gel Extraction Kit (Omega bio-tek, Norcross, GA, USA) before sequencing. The sequences obtained were assembled using ChromasPro 1.5 (http://technelysium.com.au/wp/chromaspro/, accessed on 20 March 2020), edited using BioEdit 7.1 (http://www. mbio.ncsu.edu/bioedit/bioedit, accessed on 20 March 2020), and aligned with reference sequences from GenBank using MUSCLE in MEGA 7.0 (https://www.megasoftware.net/, accessed on 20 March 2020). Short tandem repeats in the gene were identified using the Tandem Repeat Finder (http://www.tandem.bu.edu/trf/trf, accessed on 21 March 2020). The signal peptide and glycosylphosphatidylinositol (GPI) anchor were predicted using PSORT II (http://psort.hgc.jp/form2.html, accessed on 22 March 2020). N-glycosylated sites, O-glycosylated sites, and furin proteolytic cleavage sites were predicted using NetNGlyc 1.0 (http://www.cbs.dtu.dk/services/NetNGlyc/, accessed on 22 March 2020), YinOYang 1.2 (http://www.cbs.dtu.dk/services/YinOYang/, accessed on 22 March 2020), and ProP 1.0 (http://www.cbs.dtu.dk/services/ProP/, accessed on 22 March 2020), respectively. To assess the genetic relationship of *C. xiaoi* subtype families, a phylogenetic tree was conducted using the maximum likelihood (ML) analysis in MEGA 7.0 based on substitution rates calculated with the general time-reversible model. DnaSP 5.10 (www.ub.es/dnasp/, accessed on 25 March 2020) was used to calculate the recombination rates among subtype families of *C. xiaoi*.

2.5. Nucleotide Sequence Accession Numbers

Representative nucleotide sequences of the *C. xiaoi gp60* gene generated in this study were deposited in GenBank under accession numbers MW589389, MW815183-MW815276.

3. Results

3.1. Features of the gp60 Gene of C. xiaoi

A total of 25.85 million paired-end reads were obtained from the *C. xiaoi* isolate SCAU2942, and assembled into 334,080 contigs. The full *gp60* gene (MW589389) was identified in contig 1122 (8944 bp). The gene was 1437 bp in length and encoded 478 amino acids. Although it shared sequence similarities with the *gp60* gene of *C. parvum* (AF022929), *C. hominis* (FJ839883), *C. ubiquitum* [12], and *C. ryanae* [17] in the 5′ and 3′ regions at the amino acid level, the full sequence similarity was only 19.9 to 41.6% between *C. xiaoi* and the other four species (Figure 1). The GP60 protein of *C. xiaoi* had classic features of *Cryptosporidium* GP60 proteins, including an N-terminal signal peptide, a furin cleavage site (RSRR), two potential N-glycosylation sites, nearly 100 O-glycosylation sites in the GP40 region, and a GPI anchor at the C terminus. Nevertheless, the serine repeats (TCA/TCG/TCT) commonly seen in *C. parvum*, *C. hominis*, and related species, were absent in the 5′ region of the *gp60* gene of *C. xiaoi*.

3.2. Sequence Polymorphisms in the gp60 Gene of C. xiaoi

Among the 434 samples positive for *C. xiaoi* in this study, the *gp60* gene in 355 samples (81.8%) was successfully amplified by PCR. PCR products of 323 samples generated one expected band in gel electrophoresis. However, 32 samples yielded two PCR bands with different sizes, including 16 sheep samples and 16 goat samples (Table 1 and Figure 2). All PCR products with either one or two bands were sequenced, generating 298 *gp60* nucleotide sequences with length ranging from 800 to 1170 bp. Nucleotide sequences from 18 samples were identical to the reference sequence (SCAU2942) from the whole-genome sequencing, while the remaining sequences were highly divergent and displayed nucleotide differences of 24.0–68.3% (Table 2). Altogether, 94 sequence types were identified among the 298 *gp60* sequences obtained. In addition to the numerous nucleotide substitutions observed over the partial *gp60* gene, there was a significant length polymorphism among the 94 sequence types mostly due to the presence of repetitive sequences.

```
                  10         20         30         40         50         60         70         80
C. parvum     MRLSLIIVLL SVIVSAVFSA PAVPLRGTLK ---------- ---------- ---------- ---------- ----------
C. hominis    MRFLLAIVSL SVFISVVFSA PGVPLRGTLK ---------- ---------- ---------- ---------- ----------
C. ubiquitum  MRFLLAIVSL SVFISVVFSA PGVPLRGTLK ---------- ---------- ---------- ---------- ----------
C. xiaoi      MRLPLYVTLL GALIALVLSA PSVPLRGSLS SSQLGNSRSA PAPAPAPAST TTSGADTDTD AGSDSSRAEG EVDQTTVEGG
C. ryanae     MKPLLLASLC LAFLALVFSA P-VPLRGSLA VRQSAVSE-- -EPAGSQGSQ TSAQPGTP-G SVSQPTGAES ETEVTGQHSN

                  90        100        110        120        130        140        150        160
C. parvum     ---------- ---------- ---------- ---------- ---------- ---------- ---------- ------DVPV
C. hominis    ---------- ---------- ---------- ---------- ---------- ---------- ---------- ------DVSV
C. ubiquitum  ---------- ---------- ---------- ---------- ---------- ---------- ---------- ------EDDS
C. xiaoi      SGKGEDSGAG VNTGSGVSDA TEPSDGQNTS VGAQAPSSSA ASSSTQNGGN GHGQTGAVGG VTVTGDAAHS SGSENTHDEQ
C. ryanae     DGKDENS--- -DSQDGKENS NNVSSGQGV- --AQESSDSH VSESTS---- GHTQSSAASD STVQTEASGA GSSDSSNGDS

                 170        180        190        200        210        220        230        240
C. parvum     EGSSSSSSSS SSSSSSSSSS SSSSSTS--T VAPANKARTG -EDAEGSQDS SGTEASGSQ- GSEEEGSEDD G-------Q
C. hominis    EGSSSSSSSS SSSSTVAPAP KKERTVEGGT EGKNGESSPG SEEQDGGKED GGKEDGGKED GGKEDGGKEN GEGDTVDGVQ
C. ubiquitum  TNVSTTTAAP KKIIVRSTEE GTTPTAP--T TTPSTTAPTA APTTVSTTAP SGSGVDPTST DGDEKTDTGS G---------
C. xiaoi      H-TSHSTNTE ASG-TPAGDS EQETTGTGTG EATGEEGNEE TDNTTQAAPG TGSQGSSGTQ DGAENSGQED GSGGDGGAGG
C. ryanae     HGDSSSSTTP PSGSSSESTS DNEQSSSSDQ VGNGVGGGNG GGNGGGNGGG GGSGGGGGGG DGVDGSAQGN GGSGAGDHTS

                 250        260        270        280        290        300        310        320
C. parvum     TSAASQP--- ---------- -TTPAQSEG- ATTETIEATP K--EECGTSF VMWFGEGTPA ATLKCGAYTI VYAPIKDQTD
C. hominis    TGSGSQ---- ---------- -VTPSESAGT ATESTATTTP K--EECGTSF VMWFEKGTPV ATLKCGDYTI VYAPIKDQTD
C. ubiquitum  ---------- ---------- ---------- TTDETVTTTP DPMEKCGTSF VMWFVSGVPV TTLECGSYTM VYGPVENETN
C. xiaoi      AGAGAGG-AG AGGAGAG--- --AGGAGGND GNDGNDGS-S TAPVVCGEKF VVWFSDGVPV TTVSCGKYTG IYYPSADG-N
C. ryanae     GGSGSDGDAA TGEDGAGDDT DVDGTVPSTG PSEPQGGSGS GAGVICGEKF KVWFKSGTPV TTVNCGAYTG IYFPSKSG-G

                 330        340        350        360        370        380        390        400
C. parvum     PAPRYISGEV TSVTFEKSDN TVKIKVNGQD FSTLSANS-- ---------- ---------- SSPTENGG-- SAGQAS----
C. hominis    PAPRYISGEV TSVSFEKSES TVTIKVNGKE FSTLSANS-- ---------- ---------- SSPTKDNGES SDSHVQ----
C. ubiquitum  PAARYVSGTV TTVTYDASN- -KKLMVNNQE FATLSTDS-- ---------- ---------- SQPTTATT-- -APAAR----
C. xiaoi      PGPKYISGEV TAVVVEEGK- ---IKVNNQE LSTISVTP-- ---------- ---------- NENTGDDSVS AAAASKGTDK
C. ryanae     NGPKYISGDV TSVDVEEEK- ---IKVNGQE LSSIPVDPTK TTTIATSTVS TATTTETSTT SETSTPTSSP STTASQDAEE

                 410        420        430        440        450        460        470        480
C. parvum     ---------- --------SRS RRSLSEET-- -----SEAAA TVDLFAFTLD GGKRIEVAVP NVEDASKRDK YSLVADDKPF
C. hominis    ---------- --------SRS RRSLAEEN-- -----GETVA TVDLFAFTLD GGRRIEVAVP KDENADKRSE YSLVADDKPF
C. ubiquitum  ---------- --------LLA EDTVTEA--- --------VT MTDLYTFTLK GGKAISVGVP ANQDESKRDK YSLSADNQVF
C. xiaoi      S--------- -----AKSRS RRSLQQG--- ---TENQTTE VVDVYSFTAG G-KTFSVKLP NEKEAEKRNK YFLADDGGDV
C. ryanae     DEEDEEEAAV AAAAKAKSRS RRSLQEEGQE EGREESQTTE IADVYSFSAG G-KVFVVKLP KEESADKRNK YMLADSDGDV

                 490        500        510        520        530        540
C. parvum     YTG--ANSGT TNGVYRLNEN GDLVDKDNTV LL-KDAG-SS AFGLRYIVPS VFAIFAALFVL
C. hominis    YTG--ANSGI TNGVYKLDEN GNLVDKDNEV LL-KDAG-SS AFGFKYIVPS VFAIFAALFVL
C. ubiquitum  YTGTASNSGV TSGIFKLNEN GDLVDPSNTV VL-KDAD-SA AFGFRYIIPS VFAIFAAFFMF
C. xiaoi      IFE----GNK -KEEFHFDGE GDLLDSEGKV ILESGD-SSS AFDLKYIIPS ISAFVLSILLF
C. ryanae     IFE----GSK EQEEFKFDDK GDLLDSEGKV ILANDSGSSS AFGYRYVIPS ILAFILTVFLF
```

Figure 1. Deduced amino acid sequence of the *gp60* gene of *Cryptosporidium xiaoi* compared with sequences of *C. parvum* (AF022929), *C. hominis* (FJ839883), *C. ubiquitum* [12], and *C. ryanae* [17]. Potential *N*-glycosylation sites are indicated in bold and italic letters, and predicted O-glycosylation sites are indicated in bold and underlined letters. Amino acid sequences of the N-terminal signal peptide, furin cleavage site (RSRR), and C-terminal glycosylphosphatidylinositol anchor are highlighted in green, red, and blue, respectively. Dashes represent amino acid deletions.

Table 2. Pairwise nucleotide sequence similarity among subtype families of *Cryptosporidium xiaoi* in the *gp60* gene.

	XXIIIa	XXIIIb	XXIIIc	XXIIId	XXIIIe	XXIIIf	XXIIIg	XXIIIh	XXIIIi	XXIIIj	XXIIIk	XXIIIl
XXIIIa												
XXIIIb	76.0%											
XXIIIc	71.2%	74.0%										
XXIIId	70.1%	72.7%	68.8%									
XXIIIe	71.8%	72.7%	68.3%	71.6%								
XXIIIf	74.3%	69.0%	66.3%	70.1%	75.8%							
XXIIIg	68.3%	66.2%	66.0%	66.4%	69.8%	71.4%						
XXIIIh	54.1%	56.0%	50.4%	55.0%	56.4%	57.1%	57.6%					
XXIIIi	41.4%	38.2%	37.9%	42.6%	40.9%	41.4%	39.4%	38.5%				
XXIIIj	47.6%	45.3%	43.6%	46.1%	44.8%	45.2%	44.3%	42.4%	47.1%			
XXIIIk	37.9%	35.7%	35.5%	37.3%	37.9%	39.4%	37.6%	34.7%	49.8%	50.5%		
XXIIIl	33.9%	31.7%	32.2%	35.4%	33.6%	35.3%	33.9%	31.8%	42.3%	42.5%	58.3%	

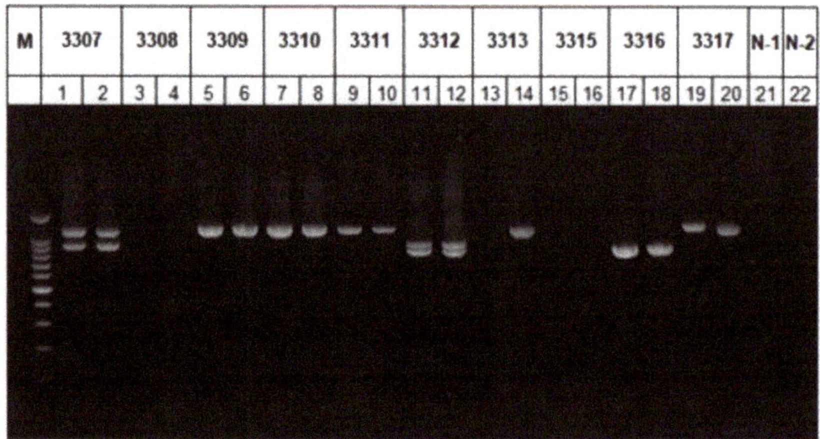

Figure 2. Nested PCR amplification of the partial *gp60* gene of *Cryptosporidium xiaoi* in sheep and goat samples. M: 100-bp DNA ladder. Lanes 1–20: Replicate PCR of 10 samples with divergent binding patterns. N-1 and N-2: No-template controls in the primary and secondary PCR, respectively.

3.3. Subtype Families and Subtypes of C.xiaoi

A total of 94 *gp60* sequences, including one sequence of each sequence type, were used in a phylogenetic analysis of the *gp60* gene. The ML tree generated comprised 12 clusters of sequences (Figure 3). They were named as subtype families XXIIIa–XXIIIl in concordance with the established nomenclature of *gp60* subtype families of *Cryptosporidium* spp. [22]. Subtype families XXIIIa–XXIIIh formed a group highly divergent from the other group of XXIIIi–XXIIIl (Figure 3). The nucleotide sequence differences among 12 subtype families ranged from 24.0 to 68.3% (Table 2). Among these subtype families, XXIIIl had the shortest nucleotide sequences and contained some unique AGC/AGT trinucleotide repeats encoding serine, leading to the occurrence of a long stretch of highly O-glycosylated amino acids. Subtypes within XXIIIl differed from each other mostly in the number of AGC/AGT trinucleotide repeats. The DnaSP analysis of the *gp60* nucleotide sequences revealed the presence of 71 potential recombination events among all 12 subtype families (Table 2). In addition, mosaic sequence patterns were observed among these subtype families (Figure 4).

At the amino acid level, extensive sequence polymorphism was found among 12 subtype families, mostly in the GP40 region (Figure 4). Despite the extensive sequence difference, all subtype families had the furin cleavage site of RSRR. There were one to four N-glycosylation sites in these subtype families, except for XXIIIk, which had none. In addition, the number of O-glycosylation sites was divergent among subtype families, with XXIIIb and XXIIIl having more O-glycosylation sites than other subtype families.

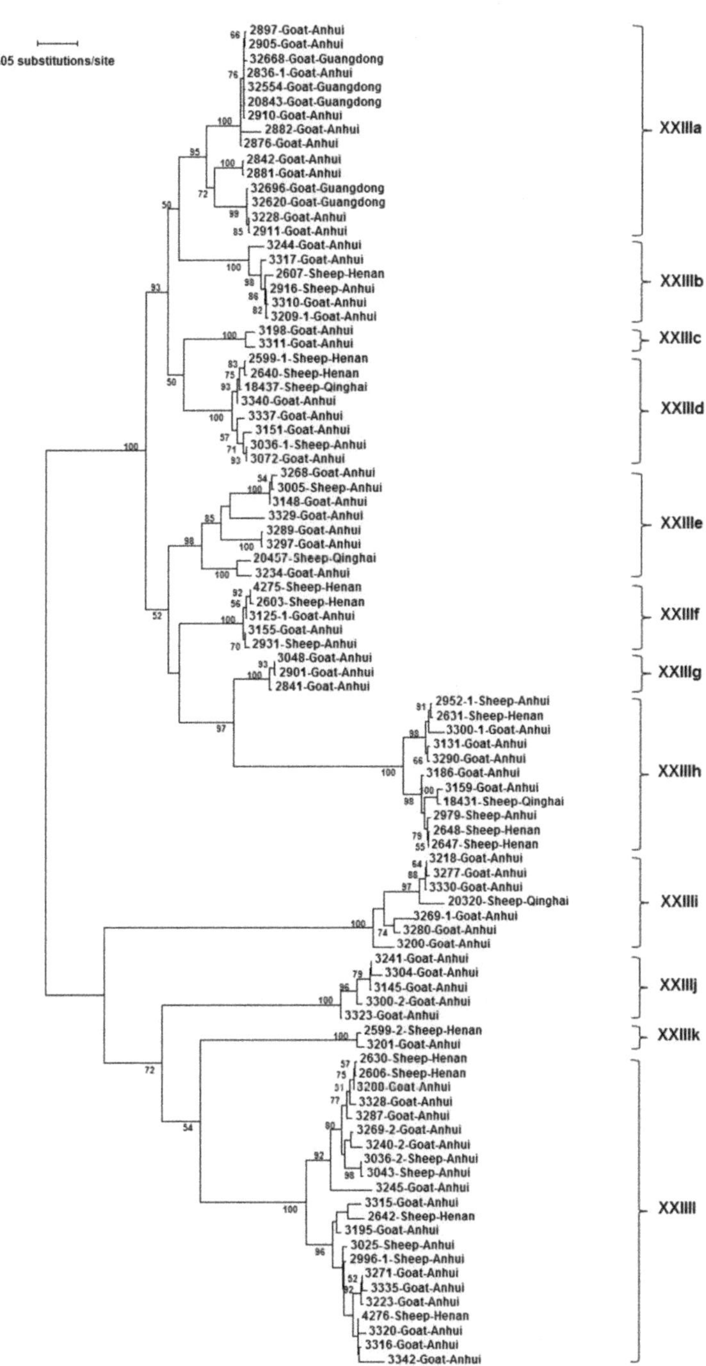

Figure 3. Phylogenetic relationship among 12 *Cryptosporidium xiaoi* subtype families (XXIIIa–XXIIIl) based on the maximum likelihood analysis of the partial *gp60* gene. General time-reversible model and Gamma distribution were used in the calculation of substitution rates. Bootstrap values lower than 50% are not displayed.

```
                     10         20         30         40         50         60         70         80         90        100
XXIIIa-20843 ---------- -----GRSSV QPRDAPAEPG VEGSGENGEN GESEESEVKE KEEKEEKEEK PEEQVTTENS TSNTVQSSDS GNTV--TQTP SASTST-SQQ
XXIIIb-2607  ---------- --GVNGSSSV QRRDS----- --TETTESAV GGSEG--SEV EGQTEEQNTT E-DSSSSSSS SSTTVQSSNS GSTD--AQTS SSSAG--TQQ
XXIIIc-3198  ---------S TTAVNGSSSV QRRDSPE--- --TPAKPSEV GGSGGSVVSE EGEQGDEQEG E-QPSTGETT VSSTTSSSSS GSTD--DETS SSSTS--APQ
XXIIId-2640  ---------- --AADGPSSP QRRDSPEPTT VPSEVEEGE  GGTEG--QEQ E-QVTGQGTT G-DSSSSSSS SSTTAQSPGS ESTD--AETP PADAETSTQQ
XXIIIe-3329  ------SSSS TTDVNGSGSV QRRDSTE--- --TTTEPSET SESAED---- EVEEQAEGQG TTEDSTS-TS TSSTVQSSGS ESTGSTQQTS STSTSA-SQQ
XXIIIf-2603  ---------- ----NGHSPV QPRDVPAESV VDGSGE---- --SEES---- KEGKEKAEE- -EGHVTTENG NSDSVQSDSG NEED--AQTS STGTVP-STQ
XXIIIg-3048  ---------- -----GRSSV QPRDSAA--- --TPTEETAA GGGVG---GE EGEQEEESKG EGEQASTVDG GSGTVQPSGA ESTESTQQTP STNAETSTHQ
XXIIIh-2648  ---------- --PAPASTTT SGADTDTDAG SDSSRAEGEV EQTTVG---- --SEGEKGEE SGAEVNTGSG VSGATEPSAG QNTSVGAQTP SSSAAS-SST
XXIIIi-3200  IVYSVKWFNR CLPVSGVRNA AKPSALS--- ---EEPKEEK PKEEG----- --SEVASGDG VVTHPTSDSE AQETPVT--- -----ETQQQ QSQEEQTEDQ
XXIIIj-3145  ---------- ---VHCLVPQ LGNSRSA--- ---PASDPSH EGDGV----- --ESTEAQEG DSQNSTTHEG QQEQPETEVG AS---TTQST VTTGSSGNGD
XXIIIk-3201  ---------S TTTVNGHSSV QRRASTE--- ---STTDSGV SGEDV----- --ESTVAQEG ASQSPTTQEG ASQSPTTQEG ASQSPTTQEG ASQSPTTQEG
XXIIIl-4276  ---------- ---------- QRRASTE--- ---STTESEV SGGGG----- --EDEGEGGG EVEEETTDDG SSSS------ ------SSSS SSSSSSSSSS

                    110        120        130        140        150        160        170        180        190        200
XXIIIa-20843 SNGETQNTGS TGSTTETPSS -TTSPPTSES TPGGEQESSS SAGT------ ---------S DTDNQEHSTT ---------- --------GG TGT----DGS
XXIIIb-2607  DSGATQNTES TSSTGETTAS STTSPSSSES TSDGQQTSST TAVTSTQQGS GGAHTSEDAS STDETTASTT QPSSSVSTSG QQTSTPGTGT SGTGDATNES
XXIIIc-3198  ES-----TES TSSTDENTAS -TTPPSNSES TSSDQETSTP STETPG---- --------TG DQDGQDHNTT ---------- --------GD TGTGDATVES
XXIIId-2640  GSDGAHTSES TGSTGGTTVS -TTQPSTSES TPGGQPTSTP SAG------- --------AS GTNNHGDSTT ---------- --------GV TGTGDANNES
XXIIIe-3329  SNSETHSTES TGSTSETASS -TTSPSSSES TTNGEQTSST STEASS---- ---------T PTGGSEQETT ---------- ---------G TGT-----ES
XXIIIf-2603  QGDGEQNTEV SDSTSETASS -TTPPSSSES TTGGEQTSAP SAGT------ ---------S GTDDHEHGAT ---------- --------GD NGT----VGS
XXIIIg-3048  GSGEAHTSEN ASSTGDTTAS STASPSSSES TSNGQHTSTP GTET------ ---------S GTHGQDHSTT ---------- --------DG TGTDNATDES
XXIIIh-2648  QNGGNGHGQT EAVGGVTVTG DAAHSSGSES THDEQDTTHS TSTEAP---- -------SST PTGGSEQGTT G--------- -----TESGS SGTGEAEGEE
XXIIIi-3200  REDSTTPGQG VAGDGHTPTG SEADVSHTEN SGAQTENGEG SDASG----- ---------- -------SVS G--------- ---------- ------ETGE
XXIIIj-3145  NSGTVDSNHS TSADSTSQPT GSGSGTTTES GSGEDDVEAG SSTS------ ---------- ---------- ---------- ---------- ---------T
XXIIIk-3201  ASQSPTTQEG ASQSPTTQEG ASQSPTTQEG ASHSPTTQEG ASHSPT---- ---------T QEGASHSPTT Q--------- ---------- ------EGAS
XXIIIl-4276  SSSSSSTSTS TSATSSS--- --SSSSSSSS S--------- SSSS------ ---------- ---------- ---------- ---------- ----------

                    210        220        230        240        250        260        270        280        290        300
XXIIIa-20843 GNN-TDHTDS SGAGTEADST EGDDTTDSNT HGTTTPGSGS QGGSTNGS-A SGNGQGNE-S GDSG-----A ENGGSAPG-- -IVCGEEFTI WFDSGAIPVT
XXIIIb-2607  EST-TDH-VS SGTSTEADST EGGDTTDSNT QEGTTPGSGS QGSSTNGS-D NGNGQG---S GSGG-----S GVGNGGS-TA GIVCGEEFTI WFDSGAVPVT
XXIIIc-3198  GST-TGN--- --TGSGAGS- AGGGSTNSGA QEGTTTETGS QGSSTDGS-A NGDSQGG--S GSGG-----A GAGNDGS-TA GIVCGEEFTI WFS-GGVPVT
XXIIId-2640  EST-TDHDAS SGTSTNTGSE EGGDTTDGNT QEGTTPGSGS PGSSTNGS-D NGNGQG---S GSSG-----S GAGNGGS-TA GIVCGEEFTI WFDSGAIPVT
XXIIIe-3329  GSG-TGETE- --------GE EGNEQTDTT- --QAASGTGS QGSSTSGS-A GGNGDG---S GGAG-----A GAGNDSSGTA PVVCGEKFVV WFSDG-VPVT
XXIIIf-2603  GST-TDHTG- SGTG----ST EGGDATDSDS QDGTTTGSGS HGSSTNGG-G NGNGQGSG-S GGAG-----A GAGNDGSGTT PVICGEKFVV WFSDG-IPVT
XXIIIg-3048  EST-TDHDVS SGTSTDTGS- TGGDNTNSGA VEGTTPETGS QDSSTNGS-A GGAGAGAGNGA GAGNGGSGTA PVVCGEKFVV WFS-DGIPVT
XXIIIh-2648  GNGETDTTQA AGTG----SQ GSSGTQDGTE NNGQENGSGG AGGPGPGAGA GGAGAG---- GAGG-----N DGGNDGSSTA PVICGEKFVV WFSDG-VPVT
XXIIIi-3200  HNGTTGDNAH DQTNNSGVHT DAESSTGSST GSSTSSDAVT NQGGDQSDVS TGAGSTTDGS GSGS------ ---HDAPGSP TVVCGEKFTI WFSGG-VPVT
XXIIIj-3145  GSPTSG---- ---------- --SESQEEPT TAPTS----- ---------- TSTGSPT--S GSAS------ ---QEGPTTA PTVCGEKFAI WFSGG-VPVT
XXIIIk-3201  HSPTTQEGAS HSPTTQGQQE QHSETEVESS SSSSSSTGTN TGTNTDTNTN TSTGLPT--S GSAP------ ---QEGPTTA AIVCGEEFTI WFSGG-VPVT
XXIIIl-4276  ---------- ---------- --------S STSTS----- ---------- -ATSSSS--S STST------ ---STGSTTA ATVCGERFTI WFSDG-VPVT

                    310        320        330        340        350        360        370        380        390        400
XXIIIa-20843 TVSCGDYTGI YYPAIGGNSE PKYISGEVTT VVVTENKIKV NGQELSTIPV SPGEATGSSG AV-------- ---------- TIPEVAASGA AGKSASRSRR
XXIIIb-2607  TVSCGDYTGI YYPSADGNSD PKYISGKVTS VDVTDNKIKV NGQELSSIPV SPGDAT-VSS PA-------- ---------- TTLEVAASED SGRSGSRSRR
XXIIIc-3198  TVDCGPYTGI YYPAIGGNSE PKYISGEVTS VDVTDDKIKV NGHDLSSIPV NPGDAV-SST PA-------- ---------- TIPEVVASKG SGESASRSRR
XXIIId-2640  TVSCGDYTGI YYPSAGGNPG PKYISGKVTS VDATDNKIKV NGQELSSMPV NPNDATGSST PA-------- ---------- TIPEVVSSGG AGKSASRSRR
XXIIIe-3329  TVSCGEYTGI YYPTADSNPG PKYISEVVSS VEVEDGKIKV NGQELSSIPV KPGDAV-SSA AV-------- ---------- AIPEVVASGG AGESASRSRR
XXIIIf-2603  TVSCGEYTGI YYPSANGNPG PKYISGGVTS VEVEDGKIKV NGEELSSIPV KPGDAASSG- ---------- ---------- --SEVAASGA AGKSASRSRR
XXIIIg-3048  TVSCGDYTGI YYPSAGGNPG PKYITGGVDS VEVKDGKIMV NDHELSSIPV NPGDPV-SST PV-------- ---------- TIPELVASED SGKSGSRSRR
XXIIIh-2648  TVSCGEYTGI YYPSADGNHG PKYISGEVTT VVVEGGKIKV NNQELSSISV TPNENTGDDS ---------- ---------- -VSAAAASKG TDKSASRSRR
XXIIIi-3200  TVDCGPYTGI YYPTTSGGGE PKYTSETVTD VKVENNVIKV NGKELSSLSV ------GSGA EG-------- ---------- ---------E TEGVASRSRR
XXIIIj-3145  TVNCGPYTGI YYPTTSGGGE PKYISGSVES VEVTGGVIKV NGKDLSSIPV KPEAAGGNGA ES-------- ---------- ----EGGGVG GEGPASRSRR
XXIIIk-3201  TVDCGPYTGI YYPAIDGNSE PKYISGKVTS VDVTDNIIKV NNQELSSIPV KPEAPGGSGA ESGAGAGAGA GAGGGGGAGV DGGEEGGEEG GEVTASRSRR
XXIIIl-4276  TVDCGPYTGI YYPTTEGSGG PKYISGPVES VVVTDGVIKV NDKDLSTIPV KPEAAGGSGA ES-------- ---------- ----GGGGDG EEGLASRSRR

                    410        420        430        440        450 454
XXIIIa-20843 SLQEGVV--T TEVADVYSFT AGGKTFTVKL PKEEE----- ---------- ----
XXIIIb-2607  SLQEEVV--T TEVADAYSFS AGGRVFTVKL PKETEAS--- ---------- ----
XXIIIc-3198  SLEQETVNQT TEVADV---- ---------- ---------- ---------- ----
XXIIId-2640  SLQQETENQT TEVADAYTFT AGGKTF---- ---------- ---------- ----
XXIIIe-3329  SLQEEVV--T TEVADVYSFS AGGKVFTVKL PKETEA---- ---------- ----
XXIIIf-2603  SLQEEVV--T TEVADVYSFT AGGKSFTVKL PKEEEAS--- ---------- ----
XXIIIg-3048  SLEEGVV--T TEVADVYSFT AGGKSFTVKL PNEKE----- ---------- ----
XXIIIh-2648  SLQQGTEDQT TEVVDVYSFT AGGKTFSVRL PNE------- ---------- ----
XXIIIi-3200  SLQEEVV--T TEVADVYSFT AGGKTFTVKL PKEADEKKRN KYFLADD--- ----
XXIIIj-3145  SLQDVET--V AQIVDAYSFT ADGKTFVVKL PKEADEEKRN KYFLADDGGD VIF-
XXIIIk-3201  SLQEDVV--T TEVADVYSFT AGGKTFTVKL PKEAEA---- ---------- ----
XXIIIl-4276  SLQEGVE--T TEVADAYTFP AGGKTFTVKL PNETEADKRN KYFLADDGGD VIFE
```

Figure 4. Deduced amino acid sequences of the partial *gp60* gene of 12 subtype families (XXIIIa–XXIIIl) in *Cryptosporidium xiaoi*. N-glycosylation sites are indicated in bold and italic letters, and O-glycosylation sites are indicated in bold and underlined letters. The furin cleavage site "RSRR" is highlighted in red. Dashes represent amino acid deletions (except those at both ends of the sequences).

3.4. Distribution of C. xiaoi Subtype Families by Host

Among the 12 subtype families of *C. xiaoi*, XXIIIa (61), XXIIIc (2), XXIIIg (19), and XXIIIj (5) were detected only in goats, while the remaining eight subtype families were found in both sheep and goats. In addition, a common occurrence of co-infections with multiple subtype families was observed in these animals. Among the three breeds of sheep, 64 samples from Han sheep were successfully subtyped, yielding XXIIIb (14), XXIIId (3), XXIIIf (2), XXIIIh (13), XXIIIk (5), XXIIIl (25), XXIIId + XXIIIk (1), and XXIIId + XXIIIl (1); 52 samples from Hu sheep were successfully subtyped, yielding XXIIIb (8), XXIIIe (2), XXIIIf (1), XXIIIh (17), XXIIIk (3), XXIIIl (17), XXIIIh + XXIIIk (2), XXIIIb + XXIIIk, (1) and XXIIId + XXIIIl (1); only a few samples from Tibetan sheep were successfully subtyped, yielding XXIIId (1), XXIIIe (1), XXIIIh (1), and XXIIIi (3). Between the two breeds of goats, only XXIIIa was identified in Black goats, while more divergent subtype families were detected in Huanghuai goats, yielding XXIIIa (26), XXIIIb (7), XXIIIc (2), XXIIId (6), XXIIIe (10), XXIIIf (7), XXIIIg (19), XXIIIh (11), XXIIIi (7), XXIIIj (5), XXIIIk (12), and XXIIIl (22). Noticeably, co-infections of various subtype families were detected in Huanghuai goats, including XXIIIa + XXIIIg (1), XXIIIb + XXIIIl (1), XXIIIh + XXIIIi (1), XXIIIh + XXIIIj (1), XXIIIh + XXIIIk (2), and XXIIIi + XXIIIl (1) (Table 1).

3.5. Distribution of C. xiaoi Subtype Families by Farm

One to 10 subtype families were found on each farm. As shown in Table 1, three farms had only one subtype family, two farms had two, two farms had four, one farm had six, one farm had seven, and two farms had 10. On Farms 9, 10, and 11 in Guangdong, all *gp60* sequences obtained belonged to the subtype family XXIIIa. In contrast, although only eight samples were subtyped on Farm 4 in Qinghai, they belonged to four subtype families (XXIIId, XXIIIe, XXIIIh, and XXIIIi). In addition, co-infections of different subtype families were observed in animals on Farms 1, 3, 5, 6, and 7, mostly with prevalent subtype families on the farm (Table 1).

4. Discussion

In the present study, we conducted whole genome sequencing of *C. xiaoi* and identified its *gp60* gene. Based on the sequence data, we established a *gp60*-based subtyping tool to assess the genetic diversity of *C. xiaoi*. The application of this subtyping tool in the analysis of *C. xiaoi*-positive samples from various breeds of sheep and goats has identified high genetic diversity within the species and possible differences in the distribution of subtypes between the two types of hosts.

The *gp60* gene sequence of *C. xiaoi* is highly divergent from that of other *Cryptosporidium* spp. Similar to the *gp60* gene of *C. ryanae* (~1548 bp), the *C. xiaoi gp60* gene (~1437 bp) is much longer than those in *C. parvum*, *C. hominis*, and *C. ubiquitum* (~873–1035 bp). Both the nucleotide and amino acid sequences of the *C. xiaoi gp60* gene showed low identity to those of other *Cryptosporidium* spp. This may explain the inability of the commonly used *gp60* primers to amplify DNA of *C. xiaoi* [23]. Similar to *C. ubiquitum*, *C. canis*, *C. felis*, and *C. ryanae*, the trinucleotide repeats of TCA/TCG/TCT encoding a polyserine tract at the 5′ end of the *gp60* gene and widely used to differentiate subtypes within subtype families, were absent in the *gp60* sequence of *C. xiaoi* [12,13,17,24]. However, a polyserine tract encoded by AGC/AGT repeats was observed in the *gp60* gene of the subtype family XXIIIl, and subtypes within XXIIIl differed mostly in the number of AGC/AGT repeats. Similar to most *Cryptosporidium* spp., the GP60 protein of *C. xiaoi* has a classic furin cleavage site "RSRR" between GP40 and GP15, which is absent in *C. ubiquitum*, *C. viatorum*, *Cryptosporidium* chipmunk genotype I and skunk genotype [9,11,12,14].

Based on the sequence analysis, the *gp60* gene of *C. xiaoi* displays an extremely high genetic diversity. The analysis of 298 sequences obtained led to the identification of 94 sequence types in 12 subtype families, including significant length polymorphism and sequence variability. The high sequence heterogeneity in this gene, nevertheless, has made PCR amplification difficult, which together with the large amplicon could be responsible for the poor amplification efficiency. In addition, some samples (13/355) produced double bands in *gp60* PCR, indicating the presence of concurrent infection with different subtypes in sheep and goats. This may facilitate the occurrence of genetic recombination among *C. xiaoi* subtypes, illustrated by the identification of mosaic sequence patterns and 71 potential recombination events in the overall sequence data. Thus, genetic recombination might be responsible for high sequence heterogeneity in the *gp60* gene of *C. xiaoi*. Genetic recombination at the *gp60* locus was observed in *C. parvum*, *C. hominis*, *C. ubiquitum*, and *C. ryanae* [2,12,17].

The *gp60* subtyping results suggest the presence of host adaptation within *C. xiaoi*. Among the 12 subtype families, XXIIIa, XXIIIc, XXIIIg, and XXIIIj were observed only in goats thus far. For the two breeds of goats, Huanghuai goats in Anhui harboured all subtype families of XXIIIa–XXIIIl. In contrast, all 35 samples from Black goats in Guangdong belonged to XXIIIa. The latter could be due to the reduced genetic diversity of *C. xiaoi* in the province. Previously, host-adapted *gp60* subtype families had been identified in other *Cryptosporidium* spp., such as *C. parvum*, *C. hominis*, *C. felis*, *C. ubiquitum*, *C. tyzzeri*, and *C. ryanae* [2,12,17,25–27].

No obvious correlation was found between the distribution of *C. xiaoi* subtype families and geographic locations in this study. Even though all *C. xiaoi* isolates from three farms in Guangdong belonged to XXIIIa, this subtype family was found in goats on two farms in Anhui. Subtyping data of *C. xiaoi* from more geographic locations and diverse animals are needed for better understanding of the distribution of *C. xiaoi* subtypes. Previously, geographical differences had been reported in the subtype distribution of *C. hominis*, *C. parvum*, *C. felis*, *C. ubiquitum*, *C. ryanae*, and *Cryptosporidium* chipmunk genotype I, indicating possible differences in the transmission of these pathogens [9,12,17,25,27,28].

5. Conclusions

In the present study, we conducted whole-genome sequencing of *C. xiaoi* and developed a subtyping tool based on the *gp60* gene. The application of this new tool in the analysis of fecal samples from sheep and goats has revealed a high genetic diversity within the species, and likely identified the occurrence of host adaptation at the subtype family level. Further studies with extensive sampling of various hosts in diverse areas are needed to improve our understanding of the transmission characteristics of *C. xiaoi*.

Author Contributions: L.X. and N.L. conceived and designed the experiments; Y.F. (Yingying Fan), X.H. and S.G. performed the experiments; F.Y. X.Y., Y.G. and Y.F. (Yaoyu Feng) provided technical assistance; Y.F. (Yingying Fan), L.X. and N.L. analyzed the data; Y.F. (Yingying Fan), L.X. and N.L. wrote the paper. All authors have read and agreed to the published version of the manuscript.

Funding: National Natural Science Foundation of China (U1901208 and 32030109), 111 Project (D20008), and Innovation Team Project of Guangdong University (2019KCXTD001).

Institutional Review Board Statement: Not applicable.

Informed Consent Statement: Not applicable.

Data Availability Statement: Data is contained within the article.

Conflicts of Interest: The authors declare no conflict of interest.

References

1. Kotloff, K.L. The burden and etiology of diarrheal illness in developing countries. *Pediatr. Clin. N. Am.* **2017**, *64*, 799–814. [CrossRef]
2. Feng, Y.; Ryan, U.M.; Xiao, L. Genetic diversity and population structure of *Cryptosporidium*. *Trends Parasitol.* **2018**, *34*, 997–1011. [CrossRef]
3. Feng, Y.; Xiao, L. Molecular epidemiology of cryptosporidiosis in China. *Front. Microbiol.* **2017**, *8*, 1701. [CrossRef] [PubMed]
4. Fayer, R.; Santin, M. *Cryptosporidium xiaoi* n. sp. (Apicomplexa: Cryptosporidiidae) in sheep (*Ovis aries*). *Vet. Parasitol.* **2009**, *164*, 192–200. [CrossRef] [PubMed]
5. Kaupke, A.; Michalski, M.M.; Rzezutka, A. Diversity of *Cryptosporidium* species occurring in sheep and goat breeds reared in Poland. *Parasitol. Res.* **2017**, *116*, 871–879. [CrossRef] [PubMed]

6. Yang, R.; Jacobson, C.; Gardner, G.; Carmichael, I.; Campbell, A.J.; Ng-Hublin, J.; Ryan, U. Longitudinal prevalence, oocyst shedding and molecular characterisation of *Cryptosporidium* species in sheep across four states in Australia. *Vet. Parasitol.* **2014**, *200*, 50–58. [CrossRef]

7. Parsons, M.B.; Travis, D.; Lonsdorf, E.V.; Lipende, I.; Roellig, D.M.; Kamenya, S.; Zhang, H.; Xiao, L.; Gillespie, T.R. Epidemiology and molecular characterization of *Cryptosporidium* spp. in humans, wild primates, and domesticated animals in the Greater Gombe Ecosystem, Tanzania. *PLoS Negl. Trop. Dis.* **2015**, *9*, e0003529. [CrossRef] [PubMed]

8. Mi, R.; Wang, X.; Huang, Y.; Zhou, P.; Liu, Y.; Chen, Y.; Chen, J.; Zhu, W.; Chen, Z. Prevalence and molecular characterization of *Cryptosporidium* in goats across four provincial level areas in China. *PLoS ONE* **2014**, *9*, e111164. [CrossRef]

9. Guo, Y.; Cebelinski, E.; Matusevich, C.; Alderisio, K.A.; Lebbad, M.; McEvoy, J.; Roellig, D.M.; Yang, C.; Feng, Y.; Xiao, L. Subtyping novel zoonotic pathogen *Cryptosporidium* chipmunk genotype I. *J. Clin. Microbiol.* **2015**, *53*, 1648–1654. [CrossRef]

10. Stensvold, C.R.; Beser, J.; Axen, C.; Lebbad, M. High applicability of a novel method for gp60-based subtyping of *Cryptosporidium meleagridis*. *J. Clin. Microbiol.* **2014**, *52*, 2311–2319. [CrossRef]

11. Yan, W.; Alderisio, K.; Roellig, D.M.; Elwin, K.; Chalmers, R.M.; Yang, F.; Wang, Y.; Feng, Y.; Xiao, L. Subtype analysis of zoonotic pathogen *Cryptosporidium* skunk genotype. *Infect. Genet. Evol.* **2017**, *55*, 20–25. [CrossRef] [PubMed]

12. Li, N.; Xiao, L.; Alderisio, K.; Elwin, K.; Cebelinski, E.; Chalmers, R.; Santin, M.; Fayer, R.; Kvac, M.; Ryan, U.; et al. Subtyping *Cryptosporidium ubiquitum*, a zoonotic pathogen emerging in humans. *Emerg. Infect. Dis.* **2014**, *20*, 217–224. [CrossRef]

13. Rojas-Lopez, L.; Elwin, K.; Chalmers, R.M.; Enemark, H.L.; Beser, J.; Troell, K. Development of a *gp60*-subtyping method for *Cryptosporidium felis*. *Parasites Vectors* **2020**, *13*, 39. [CrossRef] [PubMed]

14. Stensvold, C.R.; Elwin, K.; Winiecka-Krusnell, J.; Chalmers, R.M.; Xiao, L.; Lebbad, M. Development and application of a *gp60*-based typing assay for *Cryptosporidium viatorum*. *J. Clin. Microbiol.* **2015**, *53*, 1891–1897. [CrossRef] [PubMed]

15. Rahmouni, I.; Essid, R.; Aoun, K.; Bouratbine, A. Glycoprotein 60 diversity in *Cryptosporidium parvum* causing human and cattle cryptosporidiosis in the rural region of Northern Tunisia. *Am. J. Trop. Med. Hyg.* **2014**, *90*, 346–350. [CrossRef] [PubMed]

16. Power, M.L.; Cheung-Kwok-Sang, C.; Slade, M.; Williamson, S. *Cryptosporidium fayeri*: Diversity within the gp60 locus of isolates from different marsupial hosts. *Exp. Parasitol.* **2009**, *121*, 219–223. [CrossRef]

17. Yang, X.; Huang, N.; Jiang, W.; Wang, X.; Li, N.; Guo, Y.; Kváč, M.; Feng, Y.; Xiao, L. Subtyping *Cryptosporidium ryanae*: A common pathogen in bovine animals. *Microorganisms* **2020**, *8*, 1107. [CrossRef] [PubMed]

18. Adamu, H.; Petros, B.; Zhang, G.; Kassa, H.; Amer, S.; Ye, J.; Feng, Y.; Xiao, L. Distribution and clinical manifestations of *Cryptosporidium* species and subtypes in HIV/AIDS patients in Ethiopia. *PLoS Negl. Trop. Dis.* **2014**, *8*, e2831. [CrossRef]

19. Li, P.; Cai, J.; Cai, M.; Wu, W.; Li, C.; Lei, M.; Xu, H.; Feng, L.; Ma, J.; Feng, Y.; et al. Distribution of *Cryptosporidium* species in Tibetan sheep and yaks in Qinghai, China. *Vet. Parasitol.* **2016**, *215*, 58–62. [CrossRef]

20. Xiao, L.; Escalante, L.; Yang, C.; Sulaiman, I.; Escalante, A.A.; Montali, R.J.; Fayer, R.; Lal, A.A. Phylogenetic analysis of *Cryptosporidium* parasites based on the small-subunit rRNA gene locus. *Appl. Environ. Microbiol.* **1999**, *65*, 1578–1583. [CrossRef]

21. Guo, Y.; Li, N.; Lysen, C.; Frace, M.; Tang, K.; Sammons, S.; Roellig, D.M.; Feng, Y.; Xiao, L. Isolation and enrichment of Cryptosporidium DNA and verification of DNA purity for whole genome sequencing. *J. Clin. Microbiol.* **2015**, *53*, 641–647. [CrossRef]

22. Xiao, L.; Feng, Y. Molecular epidemiologic tools for waterborne pathogens *Cryptosporidium* spp. and *Giardia duodenalis*. *Food Waterborne Parasitol.* **2017**, *8–9*, 14–32. [CrossRef] [PubMed]

23. Feng, Y.; Lal, A.A.; Li, N.; Xiao, L. Subtypes of *Cryptosporidium* spp. in mice and other small mammals. *Exp. Parasitol.* **2011**, *127*, 238–242. [CrossRef]

24. Jiang, W.; Roellig, D.M.; Guo, Y.; Li, N.; Feng, Y.; Xiao, L. Development of a subtyping tool for zoonotic pathogen *Cryptosporidium canis*. *J. Clin. Microbiol.* **2020**, *59*, e02474-20. [CrossRef]

25. Xiao, L. Molecular epidemiology of cryptosporidiosis: An update. *Exp. Parasitol.* **2010**, *124*, 80–89. [CrossRef] [PubMed]

26. Kváč, M.; McEvoy, J.; Loudová, M.; Stenger, B.; Sak, B.; Květoňová, D.; Ditrich, O.; Rašková, V.; Moriarty, E.; Rost, M.; et al. Coevolution of *Cryptosporidium tyzzeri* and the house mouse (*Mus musculus*). *Int. J. Parasitol.* **2013**, *43*, 805–817. [CrossRef] [PubMed]

27. Jiang, W.; Roellig, D.M.; Lebbad, M.; Beser, J.; Troell, K.; Guo, Y.Q.; Li, N.; Xiao, L.H.; Feng, Y.Y. Subtype distribution of zoonotic pathogen *Cryptosporidium felis* in humans and animals in several countries. *Emerg. Microbes Infect.* **2020**, *9*, 2446–2454. [CrossRef] [PubMed]

28. Caccio, S.M.; de Waele, V.; Widmer, G. Geographical segregation of *Cryptosporidium parvum* multilocus genotypes in Europe. *Infect. Genet. Evol.* **2015**, *31*, 245–249. [CrossRef]

Article

Molecular Diversity of *Giardia duodenalis*, *Cryptosporidium* spp., and *Blastocystis* sp. in Symptomatic and Asymptomatic Schoolchildren in Zambézia Province (Mozambique)

Aly S. Muadica [1,2], Pamela C. Köster [1], Alejandro Dashti [1], Begoña Bailo [1], Marta Hernández-de-Mingo [1], Sooria Balasegaram [3] and David Carmena [1,*]

[1] Parasitology Reference and Research Laboratory, National Centre for Microbiology, Health Institute Carlos III, Majadahonda, 28220 Madrid, Spain; amuandica@unilicungo.ac.mz (A.S.M.); pamkoste@ucm.es (P.C.K.); adashti@ucm.es (A.D.); begobb@isciii.es (B.B.); martaher1@hotmail.com (M.H.-d.-M.)

[2] Departamento de Ciências e Tecnologia, Universidade Licungo, 106 Quelimane, Zambézia, Mozambique

[3] Field Epidemiology Services, National Infection Service, Public Health England, London SE1 8UG, UK; Sooria.Balasegaram@phe.gov.uk

* Correspondence: dacarmena@isciii.es; Tel.: +34-91-822-3641

Abstract: Infections by the protist enteroparasites *Giardia duodenalis*, *Cryptosporidium* spp., and, to a much lesser extent, *Blastocystis* sp. are common causes of childhood diarrhoea in low-income countries. This molecular epidemiological study assesses the frequency and molecular diversity of these pathogens in faecal samples from asymptomatic schoolchildren (*n* = 807) and symptomatic children seeking medical attention (*n* = 286) in Zambézia province, Mozambique. Detection and molecular characterisation of pathogens was conducted by polymerase chain reaction (PCR)-based methods coupled with Sanger sequencing. *Giardia duodenalis* was the most prevalent enteric parasite found [41.7%, 95% confidence interval (CI): 38.8–44.7%], followed by *Blastocystis* sp. (14.1%, 95% CI: 12.1–16.3%), and *Cryptosporidium* spp. (1.6%, 95% CI: 0.9–2.5%). Sequence analyses revealed the presence of assemblages A (7.0%, 3/43) and B (88.4%, 38/43) within *G. duodenalis*-positive children. Four *Cryptosporidium* species were detected, including *C. hominis* (30.8%; 4/13), *C. parvum* (30.8%, 4/13), *C. felis* (30.8%, 4/13), and *C. viatorum* (7.6%, 1/13). Four *Blastocystis* subtypes were also identified including ST1 (22.7%; 35/154), ST2 (22.7%; 35/154), ST3 (45.5%; 70/154), and ST4 (9.1%; 14/154). Most of the genotyped samples were from asymptomatic children. This is the first report of *C. viatorum* and *Blastocystis* ST4 in Mozambique. Molecular data indicate that anthropic and zoonotic transmission (the latter at an unknown rate) are important spread pathways of diarrhoea-causing pathogens in Mozambique.

Keywords: *Giardia*; *Cryptosporidium*; *Blastocystis*; enteric parasites; children; diarrhoea; PCR; molecular epidemiology; genotyping; Mozambique

Citation: Muadica, A.S.; Köster, P.C.; Dashti, A.; Bailo, B.; Hernández-de-Mingo, M.; Balasegaram, S.; Carmena, D. Molecular Diversity of *Giardia duodenalis*, *Cryptosporidium* spp., and *Blastocystis* sp. in Symptomatic and Asymptomatic Schoolchildren in Zambézia Province (Mozambique). *Pathogens* **2021**, *10*, 255. https://doi.org/10.3390/pathogens10030255

Academic Editor: Siddhartha Das

Received: 4 February 2021
Accepted: 22 February 2021
Published: 24 February 2021

1. Introduction

Diarrhoea has long been recognized as a major cause of child morbidity and mortality globally [1]. The main diarrhoea-causing agents include viral (adenovirus, rotavirus), bacterial (*Campylobacter*, enterotoxigenic *Escherichia coli*, *Shigella*) and parasitic (*Giardia duodenalis*, *Cryptosporidium* spp.) pathogens. Recent data by the Global Burden of Diseases, Injuries, and Risk Factors Study (GBD) revealed that diarrhoea was the fifth leading cause of death among children younger than 5 years (446,000 deaths) only in 2016 [2]. In this group of age, diarrhoea causes more deaths than acquired immunodeficiency syndrome (AIDS), malaria, and measles combined [3]. Additionally, near 90% of diarrhoea-associated deaths are attributable to unsafe water, inadequate sanitation, and insufficient hygiene [4]. There is, therefore, a clear link between diarrhoea and poverty, particularly in deprived areas of low-income countries [5].

The protozoa *G. duodenalis* and *Cryptosporidium* spp. and, to a lesser extent, the Stramenopile *Blastocystis* sp. are among the most important diarrheal pathogens of parasitic nature affecting humans [6]. Indeed, there are more than 200 million cases of symptomatic giardiosis worldwide each year only in developing countries [7]. The Global Enteric Multicenter Study (GEMS) has demonstrated that *Cryptosporidium* is second only to rotavirus in causing morbidity and mortality in young children living in South East Asian and sub-Saharan African countries [8]. *Blastocystis* sp. is the most prevalent intestinal microbial eukaryote colonising/infecting the human gut. Although its clinical significance remains controversial, *Blastocystis* carriage has been linked with intestinal (diarrhoea, irritable bowel syndrome) and extra-intestinal (urticarial) disorders [9]. In addition, childhood cryptosporidiosis and giardiosis frequently lead to stunted growth and cognitive impairment, particularly in poor-resource areas [10,11].

As with other enteric protists, *G. duodenalis*, *Cryptosporidium* spp., and *Blastocystis* sp. are transmitted via the faecal-oral route through diverse pathways including person-to-person or animal-to-person direct contact and consumption of contaminated water or food. Indeed, waterborne and foodborne illnesses constitute a major area of concern [12,13]. Molecular epidemiological studies involving human, animal, and environmental samples are essential to determine the genetic diversity of these pathogens in a given host or geographical area, and to ascertain the exact contribution of the above-mentioned transmission routes to human infections.

Giardia duodenalis (the only *Giardia* species able to infect humans) is currently regarded as a multi-species complex comprising eight (A to H) distinct assemblages, of which zoonotic assemblages A and B (particularly the latter) are the most prevalent in humans [14]. The genus *Cryptosporidium* comprises at least 40 valid species and a similar number of genotypes of unclear taxonomic status, of which more than 20 have been reported to cause human infections. Only two species, *Cryptosporidium hominis* and *C. parvum*, account for up to 90% of the human cases of cryptosporidiosis documented globally [15]. Similarly, 22 distinct subtypes (ST) have been identified within *Blastocystis* sp. to date (ST1-17, ST21, ST23-26), of which ST1–9 and ST12 have been reported in humans [16].

The epidemiology of giardiosis, cryptosporidiosis, and blastocystosis in African countries is poorly understood [17–20]. Prevalence estimations have been frequently based on low-sensitive (e.g., microscopy) methods, were restricted to certain populations and geographical areas or are in need of update. When compared, polymerase chain reaction (PCR) methods outperformed microscopy in terms of sensitivity and range of parasite species detected [21]. Although recent large-scale epidemiological surveys such as GEMS or the Etiology, Risk Factors, and Interactions of Enteric Infections and Malnutrition and the Consequences for Child Health (MAL-ED) Study have considerably improved our knowledge of certain species (particularly *Cryptosporidium* spp.), they did not consider others (e.g., *Blastocystis*) and did not conduct genotyping analyses on positive samples [8,22]. In Mozambique, GEMS-derived samples from young children positive for *G. duodenalis* and *Cryptosporidium* spp. in the Manhiça District (Maputo province) have been recently genotyped and sub-genotyped [23,24]. Earlier studies have molecularly characterised much smaller numbers of *Giardia*- and *Cryptosporidium*-positive isolates in diarrhoeal individuals and patients with human immunodeficiency virus (HIV) and/or tuberculosis in Maputo and Gaza provinces, respectively [25,26]. However, the current epidemiological situation of human blastocystosis in the country is completely unknown. This study aims to investigate the frequency and molecular diversity of *G. duodenalis*, *Cryptosporidium* spp., and *Blastocystis* sp. in asymptomatic and symptomatic children in the Zambézia province in the central coastal region of Mozambique, an area where no studies on the presence of intestinal protist parasites have been previously conducted in human or animal populations.

2. Results

2.1. Occurrence of Enteric Parasites

A total of 1093 children aged 3–14 years participated in the present study, of which 807 children were enrolled from 18 schools and resided in 66 neighbourhoods in 10 districts of the Zambézia province. The other 286 children were enrolled from six primary healthcare centres and a hospital clinic and resided in 37 different neighbourhoods in six districts. Overall, *G. duodenalis* was the most prevalent enteric parasite found [41.7%, 95% confidence interval (CI): 38.8–44.7%], followed by *Blastocystis* sp. (14.1%, 95% CI: 12.1–16.3%), and *Cryptosporidium* spp. (1.6%, 95% CI: 0.9–2.5%). The prevalence rates of these pathogens in each participating school and healthcare centre are summarized in Table 1. Estimates did not consider the clustered nature of the data, as this task was thoroughly conducted elsewhere [27].

Table 1. Molecular-based prevalence rates of *Giardia duodenalis*, *Cryptosporidium* spp., and *Blastocystis* sp. in the surveyed paediatric population by school or medical centre of origin in the Zambézia province, Mozambique.

Centre	Children (*n*)	*Giardia duodenalis* Positive (*n*)	Positive (%)	*Cryptosporidium* spp. Positive (*n*)	Positive (%)	*Blastocystis* sp. Positive (*n*)	Positive (%)
School							
1	44	23	52.3	1	2.3	1	2.3
2	50	13	26.0	0	0.0	23	46.0
3	22	9	40.9	0	0.0	0	0.0
4	22	12	54.5	0	0.0	0	0.0
5	24	11	45.8	0	0.0	0	0.0
6	31	13	41.9	0	0.0	1	3.2
7	88	36	40.9	4	4.5	9	10.2
8	60	8	13.3	2	3.3	0	0.0
9	47	23	48.9	0	0.0	0	0.0
10	49	20	40.8	0	0.0	1	2.0
11	47	13	27.7	0	0.0	4	8.5
12	50	12	24.0	0	0.0	4	8.0
13	50	25	50.0	1	2.0	19	38.0
14	30	9	30.0	0	0.0	0	0.0
15	30	5	16.7	0	0.0	0	0.0
16	40	16	40.0	0	0.0	0	0.0
17	75	46	61.3	1	1.3	47	62.7
18	48	42	87.5	2	4.2	34	70.8
Sub-total	807	336	41.6	11	1.4	143	17.7
Clinic							
1	50	32	64.0	3	6.0	2	4.0
2	25	11	44.0	0	0.0	0	0.0
3	15	4	26.7	0	0.0	2	13.3
4	52	28	53.8	0	0.0	4	7.7
5	42	10	23.8	2	4.8	2	4.8
6	29	4	13.8	1	3.4	1	3.4
7	73	31	42.5	0	0.0	0	0.0
Sub-total	286	120	42.0	6	2.1	11	3.8
Total	1093	456	41.7	17	1.6	154	14.1

2.2. Prevalence and Molecular Characterization of G. duodenalis

A total of 456 DNA isolates tested positive for *G. duodenalis* by real-time polymerase chain reaction (qPCR, 336 and 120 in asymptomatic and symptomatic children, respectively). Generated cycle threshold (Ct) values had median values of 31.6 (range: 18.0–42.1) in asymptomatic children, and of 31.2 (range: 19.7–41.4) in symptomatic children. Overall, 45.8% (209/456) had qPCR values $\geq$32, and 39.7% (181/456) had qPCR Ct values <30. Based on previous molecular studies conducted by our research team using this very same method in human populations from other African countries including Mozambique [23,28,29], only DNA isolates with qPCR Ct values <32 (*n* = 247) were assessed for genotyping and sub-genotyping purposes in order to optimise available resources.

Assemblage/sub-assemblage assignment was conducted by direct comparison of the sequencing results obtained at the three loci (*gdh*, *bg*, and *tpi*) investigated. Sequences presenting double peak positions that could not be assigned unequivocally to a given assemblage/sub-assemblage were reported as ambiguous sequences. Out of the 247 DNA isolates investigated, 15.0% (37/247), 10.1% (25/247), and 6.1% (15/247) yielded amplicons at the *gdh*, *bg*, and *tpi* loci, respectively (Table 2). Overall, 17.4% (43/247) were amplified at least at a single locus, whereas multi-locus genotyping data at the three loci were available for 4% (10/247) of them. Most (88.4%, 38/43) of the isolates successfully amplified at any of the three markers assessed had qPCR Ct values <30. Sequence analyses revealed the presence of assemblages A (7.0%, 3/43) and B (88.4%, 38/43). Two additional sequences (4.6%, 2/43) corresponded to mixed A+B infections. No infections caused by host-restricted canine (C, D), feline (F), or ruminant (E) assemblages were detected. All genotyped isolates except one were obtained in asymptomatic children.

Table 2. Multilocus genotyping results of the *G. duodenalis*-positive children (*n* = 43) successfully genotyped at any of the three loci investigated in the Zambézia province, Mozambique.

Sample ID	Ct Value in qPCR	*gdh*	*bg*	*tpi*	Assigned Genotype
4	24.49	BIV	Negative	Negative	BIV
17	25.8	BIV	Negative	Negative	BIV
24 [1]	31.6	BIV	Negative	Negative	BIV
57	25.4	BIV	Negative	Negative	BIV
62	20.9	AII	Negative	AII	AII
67	23.9	BIII/BIV	Negative	Negative	BIII/BIV
69	21.9	BIV	B	BIII	BIII/BIV
70	24.3	BIII/BIV	B	Negative	BIII/BIV
74	22.7	BIII	B	Negative	BIII
79	26.1	AII	Negative	Negative	AII
88	20.9	BIII/BIV	Negative	BIII	BIII/BIV
97	26.1	BIV	Negative	Negative	BIV
100	25.9	BIII/BIV	B	Negative	BIII/BIV
103	27.5	BIII/BIV	Negative	Negative	BIII/BIV
104	21.7	BIII/BIV	B	BIII	BIII/BIV
105	24.0	BIII/BIV	B	Negative	BIII/BIV
118	26.8	BIV	B	Negative	BIV
122	20.0	BIII/BIV	B	BIII/BIV	BIII/BIV
124	30.8	Negative	B	Negative	B
128	31.8	Negative	B	Negative	B
140	30.5	Negative	B	Negative	B
141	19.9	BIII	B	BIII/BIV	BIII/BIV
143	25.3	BIII	Negative	Negative	BIII
164	20.9	AII	AII	AII	AII
165	24.1	BIII	B	Negative	BIII
170	28.7	BIII/BIV	Negative	Negative	BIII/BIV
172	25.6	AII	B	Negative	AII+B
173	29.5	BIV	Negative	Negative	BIV
175	23.7	BIII/BIV	B	Negative	BIII/BIV
176	28.5	BIII/BIV	B	Negative	BIII/BIV
178	27.1	BIV	Negative	Negative	BIV
179	25.9	BIII/BIV	Negative	Negative	BIII/BIV
180	22.7	BIII/BIV	B	AII	AII+BIII/BIV
183	26.5	Negative	B	Negative	B
186	26.6	BIII/BIV	Negative	BIII	BIII/BIV
187	25.2	BIII	Negative	BIII	BIII
190	19.8	BIII/BIV	B	BIII	BIII/BIV
191	30.7	Negative	B	Negative	B
194	20.0	BIII/BIV	B	BIII	BIII/BIV
195	23.8	BIII/BIV	B	BIII	BIII/BIV
196	26.4	Negative	Negative	BIII	BIII
203	27.1	BIII/BIV	B	Negative	BIII/BIV
206	22.9	BIII/BIV	B	BIII	BIII/BIV

[1] Symptomatic child, *bg*: ß-giardin, *gdh*: glutamate dehydrogenase, qPCR: real-time PCR, *tpi*: triose phosphate isomerase.

All three A sequences were assigned to the sub-assemblage AII of the parasite. Out of the 40 B sequences, 12.5% (5/40) were identified as sub-assemblage BIII, 20.0% (8/40) as sub-assemblage BIV, 52.5% (21/40) as ambiguous BIII/BIV sequences, and the remaining 15.0% (6/40) were only genotyped at the assemblage level.

The diversity, frequency, and main features of the *G. duodenalis* sequences generated at the *gdh* locus are shown in Table S1. Briefly, all four AII sequences were identical to reference sequence L40510. In contrast, a much higher level of genetic diversity was observed within the 34 sequences assigned to assemblage B at this locus. Indeed, the five sequences identified as BIII differed by 1–9 single nucleotide polymorphisms (SNPs) from reference sequence AF069059, most of them associated to heterozygous C/T peaks at positions 99, 147, 150, 309, and 336. The nine sequences unambiguously assigned to BIV differed by 2–8 SNPs from reference sequence L40508. Six of them presented nucleotide substitutions (mainly C↔T transitions) at positions 183, 387, 396, and 423, but not ambiguous positions in the form of double peaks. SNPs present in the three remaining BIV sequences combined mutations and heterozygous positions at different proportions. Remarkably, virtually all ambiguous BIII/BIV sequences different among them and by 5–15 SNPs from reference sequence L40508. Most of these SNPs involved heterozygous C/T (and, to a lesser extent, A/G) peaks. In contrast, SNPs associated to transition C↔T or A↔G mutations were rare or non-existent. Some of these ambiguous BIII/BIV sequences presented clear double peaks at defined positions specific for BIII (e.g., 99, 147, 150, 309, and 336) and BIV (e.g., 183, 387, 396, and 423) sequences, suggesting the occurrence (at an unknown rate) of true BIII+BIV intra-assemblage mixed infections. Several heterozygous positions within BIII and BIII/BIV (particularly the latter) sequences were potentially associated to amino acid change in the polypeptidic chain.

The diversity, frequency, and main features of the 25 *G. duodenalis* sequences generated at the *bg* locus are summarized in Table S2. The only sequence assigned to AII was identical to reference sequence AY072723. The other 24 sequences, all belonging to the assemblage B of *G. duodenalis*, presented a comparatively lower degree of genetic diversity than their counterparts at the *gdh* locus. Of them, four sequences were identical to reference sequence AY072727, whereas the remaining 20 sequences differed from it by 1–5 SNPs. Variations involving transitional C↔T or A↔G mutations and double peaks tended to accumulate at positions 183, 309, 519, and 565 of AY072727; some of them (including a transversion C/A mutation) were involved in amino acid substitutions at the protein level.

The diversity, frequency, and main features of the 15 *G. duodenalis* sequences generated at the *tpi* locus were summarized in Table S3. Out of the three sequences identified as AII, two of them varied by 1–2 SNPs (including a transversion C/G mutation at position 287) from reference sequence U57897. The third AII sequence lacked sufficient quality to accurately determine the presence of potential SNPs. The 10 sequences characterised as BIII differed by 1–8 SNPs from reference sequence AF069561. Six of them included only transitional C↔T or A↔G mutations, whereas the remaining four had several heterozygous positions. Detected SNPs tended to accumulate at positions 34, 108, and 141 of AF069561, some of them involved in amino acid substitutions in the polypeptidic chain. Two isolates corresponded to ambiguous BIII/BIV sequences differing by seven SNPs from reference sequence AF069560. Most of these SNPs were the result of transitional C/T or A/G mutations, one of them (T57C) involving an amino acid chain at the protein level. No transversion mutations were detected within sequences generated at the *tpi* locus.

The evolutionary relationships among the *G. duodenalis* sequences generated at the *gdh* locus in the present study were shown in Figure 1. Sequences of human origin generated by our research team in previous studies conducted in geographical areas of high (Ethiopia, Angola, Brazil, and Iran) and low (Spain) endemicity were also included in the analysis for comparative purposes. Assemblage A sequences grouped together in well-defined clusters with appropriate reference sequences. Although assemblage B sequences also formed a well-supported clade (88% of bootstrap), sub-assemblage BIII and BIV sequences could not be segregated in independent clusters. Phylogenetic trees generated at the *bg* (Figure S1) and *tpi* (Figure S2) loci seem to corroborate this finding.

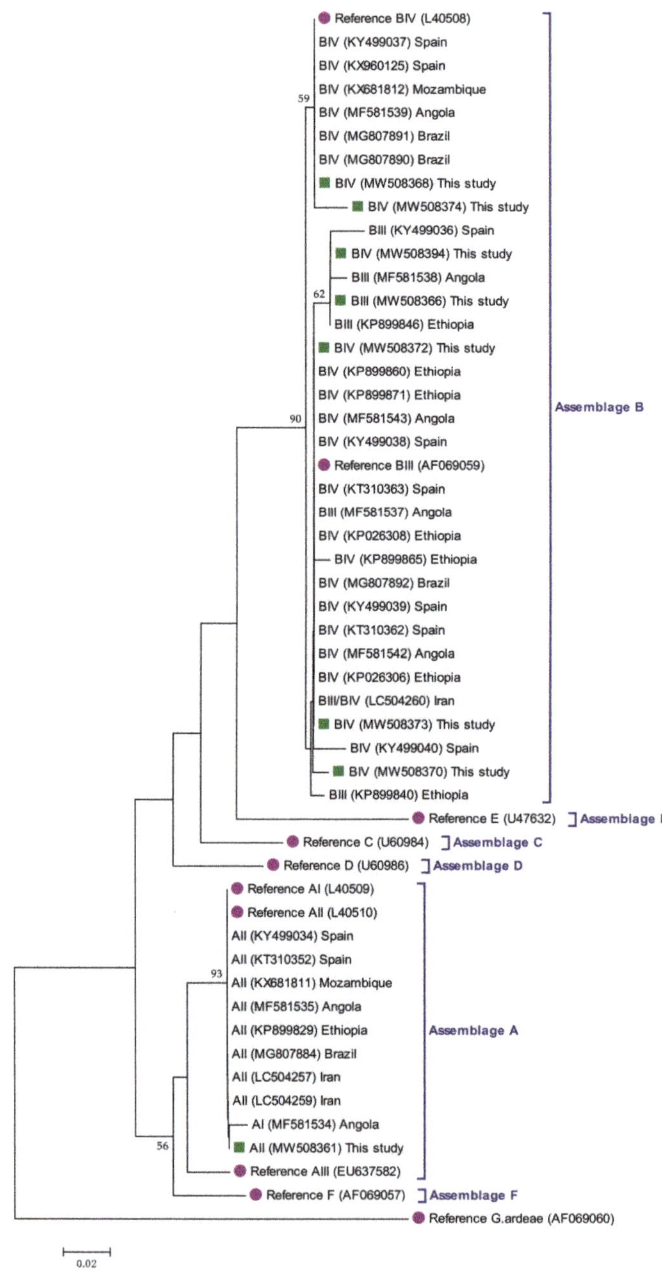

Figure 1. Phylogenetic relationships among *Giardia duodenalis* assemblages and sub-assemblages identified in infected symptomatic and asymptomatic children in the Zambézia province, Mozambique. The analysis was conducted by a neighbor-joining method of a 412-bp fragment (corresponding to position 79–490 of reference sequence L40508) of the *gdh* gene sequence. Genetic distances were calculated using the Kimura two-parameter model. Green filled squares represent sequences generated in the present study. Purple filled dots represent reference sequences. Bootstrap values lower than 50% are not displayed. *Giardia ardeae* was used as outgroup taxon to root the tree.

2.3. Prevalence and Molecular Characterization of Cryptosporidium spp.

Analyses of the 13 *ssu* rDNA sequences generated in the present study revealed the presence of four *Cryptosporidium* species circulating in the paediatric population under study, including *C. hominis* (30.8%; 4/13), *C. parvum* (30.8%, 4/13), *C. felis* (30.8%, 4/13), and *C. viatorum* (7.6%, 1/13) (Table 3). Besides a case of cryptosporidiosis by *C. parvum* detected in a child with diarrhoea, the remaining cases were identified in asymptomatic children. Out of the four sequences characterised as *C. hominis*, two of them were identical to reference sequence AF108865, with the remaining two differing from it by one to two SNPs including a nucleotide deletion and ambiguous positions in the form of double peaks. Three of the sequences assigned to *C. parvum* corresponded to the "bovine genotype" of the parasite, characterised by a four-nucleotide (TAAT) deletion involving positions 686_689 of reference sequence AF112571. These three sequences differed among them by 1–2 SNPs affecting positions 795, 837, and/or 892 of reference sequence AF112571. A fourth *C. parvum* sequence was not investigated further due to insufficient quality to accurately determine the presence of potential SNPs. All *C. hominis* and *C. parvum* isolates failed to be amplified at the *gp60* locus despite repeated attempts to do so. Therefore, their family subtypes remain unknown.

Table 3. Diversity, frequency, and main molecular features of *Cryptosporidium* spp. sequences at the *ssu* rRNA locus in infected symptomatic and asymptomatic children in Zambézia province, Mozambique. GenBank accession numbers are provided.

Species	No. of Isolates	Reference Sequence	Stretch	Single Nucleotide Polymorphisms	GenBank ID
C. hominis	2	AF108865	608–979	None	MW563962
	1	AF108865	540–952	C607T, 697Del_T [1]	MW563963
	1	AF108865	622–956	A808R, G905R	MW563964
C. parvum	1	AF112571	573–997	A646G, T649G, 686_689DelTAAT [1], A691T, C795Y, A892R	MW563965
	1	AF112571	630–997	A646G, T649G, 686_689DelTAAT [1], A691T, C795Y, T837W, A892R	MW563966 [2]
	1	AF112571	539–1025	A646G, T649G, 686_690DelTAATT [1], A691T, A892G	MW563967
	1	AF112571	655–985	Unknown [3]	–
C. felis	3	AF108862	631–980	661InsT [4], T670A, 700DelT [1]	MW563968
	1		648–987	661InsT, T670A,700DelT [1], 824InsT [4]	MW563969
C. viatorum	1	KX174309	290–762	None	MW563970

[1] Nucleotide(s) deletion. [2] Symptomatic child. [3] Sequence of insufficient quality to accurately determine the presence of potential single nucleotide polymorphisms. [4] Nucleotide insertion.

Three of the *C. felis* sequences were identical among them but differed from reference sequence AF108862 by three SNPs including a deletion, an insertion, and a substitution (Table 3). The fourth *C. felis* sequence had an additional T insertion in position 826 of reference sequence AF108862. The only sequence assigned to *C. viatorum* was identical to positions 290–762 of reference sequence KX174309. Sequence analysis at the *gp60* locus revealed that this isolate belonged to the subtype XVaA3a of the parasite, showing 100% identity to positions 1–870 of reference sequence KP115936.

The genetic relationships among *ssu* rRNA gene sequences generated in the present study, as inferred by a neighbor-joining analysis, were shown in Figure 2. All *Cryptosporidium* sequences clustered together (monophyletic groups) with different well-supported clades (58–99% of bootstrap) corresponding to appropriate reference sequences for *Cryptosporidium* species.

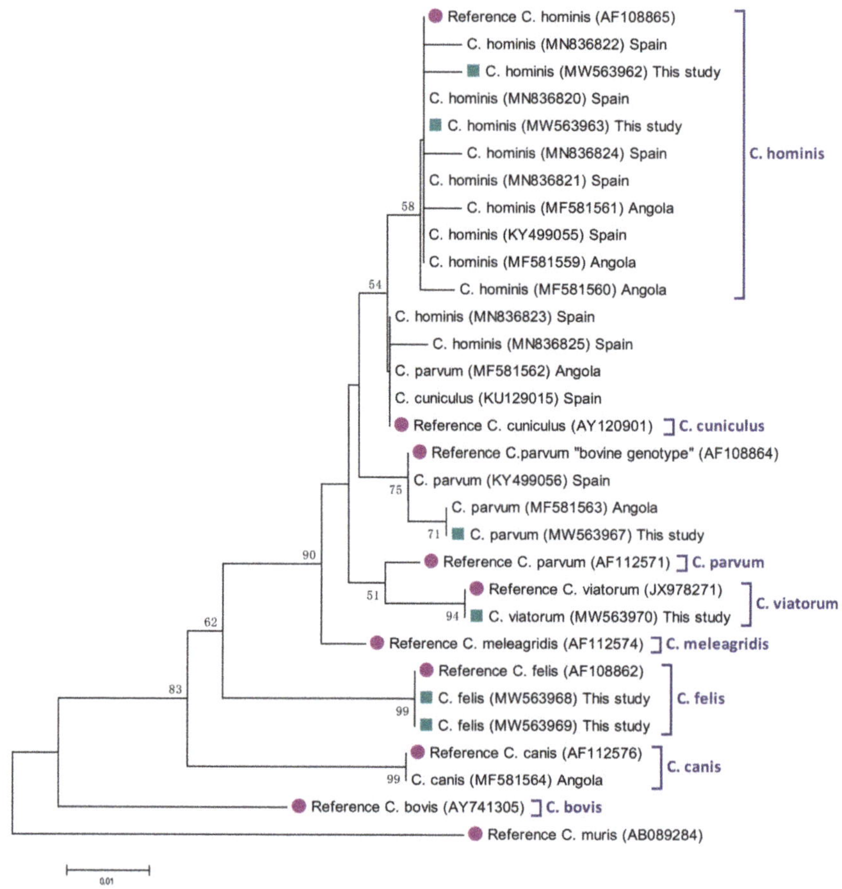

Figure 2. Phylogenetic relationships among *Cryptosporidium* species identified in infected SympTable 490. bp fragment (corresponding to position 538–1027 of reference sequence AF108865) of the *ssu* rRNA gene sequence. Genetic distances were calculated using the Kimura two-parameter model. Green filled squares represent sequences generated in the present study. Purple filled dots represent reference sequences. Bootstrap values lower than 50% are not displayed. *Cryptosporidium muris* was used as outgroup taxon to root the tree.

2.4. Prevalence and Molecular Characterization of Blastocystis sp.

Out of the 227 isolates that yielded amplicons of the expected size for *Blastocystis* sp. by PCR, 67.8% (154/227) were successfully subtyped at the *ssu* rRNA (barcode region) gene of this protist. The remaining 73 isolates produced unreadable or poor-quality sequences usually associated to faint bands on agarose gels. All samples that could not be confirmed by Sanger sequencing were presumptively considered as *Blastocystis*-negative. Overall, 92.9% (143/154) of the subtyped isolates were obtained in asymptomatic children, and 7.1% (11/154) in children with gastrointestinal symptoms. Sequence analyses allowed the identification of four *Blastocystis* subtypes (ST) circulating in the surveyed paediatric population, including ST1 (22.7%; 35/154), ST2 (22.7%; 35/154), ST3 (45.5%; 70/154), and ST4 (9.1%; 14/154) (Figure 1). Neither mixed infection involving different STs of the parasite nor infections caused by animal-specific ST10–ST17, ST21 or ST23–26 were identified. A considerable genetic diversity was observed within ST2 (four different alleles, alone or in combination) and ST3 (six different alleles, alone or in combination). Allele 4 was the most prevalent within ST1 (77.1%, 27/35), allele 12 within ST2 (34.3%, 12/35), and allele 36 within ST3 (54.3%, 38/70). In contrast, allele 42 was the only genetic variant identified within ST4. Several isolates were only genotyped at the subtype level due to insufficient quality sequence data for allele calling (Figure 3).

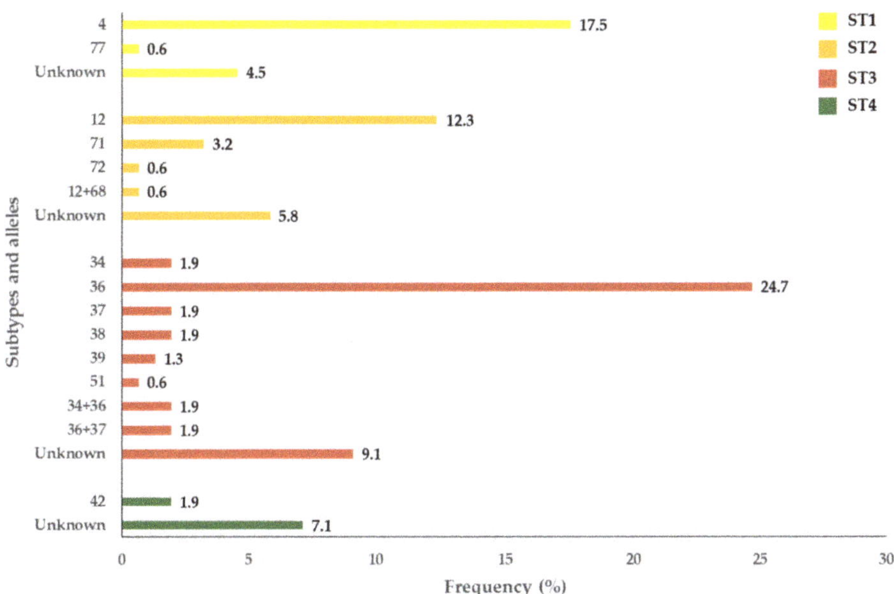

Figure 3. Diversity and frequency of *Blastocystis* subtypes and 18S alleles identified in the symptomatic and asymptomatic children in Zambézia province, Mozambique.

No obvious clusters of the parasite's species/genotypes were found according to the period of sampling or the sex, age group, school of origin or clinical status of the infected children investigated in the present study.

3. Discussion

The molecular diversity of *G. duodenalis* and *Cryptosporidium* spp. in Mozambique has been thoroughly investigated in children younger than five years of age with and without diarrhoea recruited under the GEMS umbrella in the Manhiça district (Maputo province) in two recent studies [23,24]. The present survey expands our current knowledge on the epidemiology of these two protozoan pathogens in the country, exploring their occurrence and genetic diversity in asymptomatic schoolchildren and children with clinical manifestations in Zambézia province. This is also the first report describing the molecular variability of *Blastocystis* sp. in Mozambique. Of note, our research group has previously investigated in this very same paediatric populations the occurrence of the microsporidia *Enterocytozoon bieneusi* and conducted risk association analyses for intestinal parasites [27,30].

In Mozambique, *G. duodenalis* has been described at infection rates of 1–6% by conventional microscopy in paediatric and clinical populations in Maputo province [31–33], and of 8–37% by PCR in patients with HIV and/or tuberculosis in Gaza province and people living in a high endemic area of Sofala province [21,25]. Using enzyme-linked immunosorbent assay (ELISA), *G. duodenalis* was identified in 10–50% of children with clinical manifestations in different provinces of the country [34–36]. The *G. duodenalis* crude prevalence rate found here (42%) is in the upper range limit of those reported in the aforementioned surveys.

Giardia duodenalis genotyping data in Mozambique are far scarcer. In a preliminary study involving a limited number of isolates, *G. duodenalis* sub-assemblages AII and BIV were found at equal proportions in patients with HIV and/or tuberculosis in Gaza province [25]. A much larger molecular survey involving 222 well-characterized *G. duodenalis* isolates obtained during the GEMS project revealed that assemblage B caused 9 out of 10 infections in young children in Maputo province [23]. That study demonstrated that the occurrence of diarrhoea was not linked to a given assemblage of the parasite. Additionally, the elevated level of genetic variability found within assemblage B sequences did not allow the correct differentiation between sub-assemblage BIII and BIV sequences that were, therefore, tentatively identified as ambiguous BIII/BIV results and represented 59% (132/222) of the total sequences analysed. Molecular data showed in the present study were remarkably similar, with assemblages A and B being identified in 7% and 88% of the infections, respectively, and ambiguous BIII/BIV sequences accounting for 49% (21/43) of them. The absence of animal-specific C–F assemblages in both surveys suggest that livestock and companion animals may play a secondary role as sources of human giardiosis, and that most human infections should be of anthropic origin. Taken together, these results indicate that the epidemiology of human giardiosis is very similar in Mozambican regions as separated from each other as Maputo and Zambézia.

Cryptosporidium infections in Mozambique have been previously documented at rates ranging from 6–38% by ELISA in diarrhoeic children in Maputo province [34,35], at 6% in HIV-positive individuals in the same province [33], and of 12% in general population at national scale [36]. Rates lower than 10% have been reported by PCR in patients with HIV and/or tuberculosis in Gaza province and diarrhoeic patients in Maputo province [25,26]. A much lower crude prevalence of 1.6% has been reported in the present study in mostly asymptomatic children. The marked discrepancies observed in *Cryptosporidium* prevalence rates among these studies may be associated with differences not only in the nature of the surveyed populations, but also in the performance of the diagnostic methods used.

Knowledge on the molecular diversity of *Cryptosporidium* sp. in Mozambique is very limited. Early studies revealed the presence of three *Cryptosporidium* species circulating in Mozambican human populations, namely *C. hominis*, *C. parvum*, and *C. felis*. Furthermore, *gp60* subtypes IbA10G2 and IdA22 have been described in patients with HIV and/or tuberculosis [25], and IA23R3, IIcA5G3, and IIeA12G1 in children and adults with diarrhoea [26,37]. More recently, a large panel (*n* = 191) of *Cryptosporidium*-positive faecal samples from young children collected during the GEMS project in the Maputo province revealed the predominance of *C. hominis* (73%) over *C. parvum* (23%) and *C. meleagridis* (4%). Both *C. hominis* and *C. parvum* were more prevalently found in diarrhoeal children than in non-diarrhoeal children. In that survey, a high intra-species genetic variability was observed within *C. hominis* (subtype families Ia, Ib, Id, Ie, and If) and *C. parvum* (subtype families IIb, IIc, IIe, and IIi), but not within *C. meleagridis* (subtype family IIIb). In contrast, in the present study *C. hominis*, *C. parvum*, and *C. felis* were found at equal (31%) proportions mostly in asymptomatic children, whereas the presence of *C. viatorum* subtype XVaA3a represents the first report of this *Cryptosporidium* species in Mozambique. Unfortunately, sequences identified as *C. hominis* or *C. parvum* at the *ssu* rRNA gene did not yield amplicons at the *gp60* locus, so their subtypes remain unknown.

Of interest, three out of four sequences identified as *C. parvum* were associated to the "bovine genotype" of the parasite, which is characterised by a four-base deletion TAAT at positions 686 to 689 of reference sequence AF112571. Indeed, some authors have proposed that this genetic variant should be considered as an independent species named *C. pestis* [38]. The high proportion of infections due to the "bovine genotype" of *C. parvum* and *C. felis* (a *Cryptosporidium* species adapted to infect cats and other felids) reveals that a significant number of cryptosporidiosis cases were the result of zoonotic events through direct contact with infected animals or indirectly through consumption of contaminated water or food with their faecal material. Of interest, *C. felis* has been previously reported in different human populations in other African countries including Ethiopia, Nigeria, and Kenya [39–41]. The presence of *C. viatorum* is also relevant. This *Cryptosporidium* species was initially thought to be a human-specific species [42], but recent epidemiological surveys have demonstrated its presence in rodents from Australia and China and may have, therefore, zoonotic potential [43,44]. In Africa, *C. viatorum* has been reported in primarily asymptomatic children, diarrhoeic patients and HIV+ individuals in Ethiopia [28,39,45], and in the adult population in Kenya [45].

A major contribution of this study is the first thorough description of the molecular diversity of *Blastocystis* sp. conducted in Mozambique to date. The frequency and diversity of the three main STs detected (ST1: 23%, ST2: 23%, ST3: 45%) were in line with those previously published in other African countries such as Angola [29], Ivory Coast [46], and Madagascar [47], among others. In contrast, ST4 was identified at a much lower rate of 9%. Remarkably, most human cases of blastocystosis by ST4 have been documented in Europe [48]. This geographically restricted pattern of ST4, together with its primarily clonal structure, has been interpreted by some authors as the result of a recent entry into the human population, very likely from rodents [49]. In line with these findings, *Blastocystis* ST4 has only been detected in a few African countries at carriage rates of 12–14% in Liberia and Nigeria [48], and of 2% in Senegal and Tunisia [50,51]. Furthermore, ST4 has been proposed as a more virulent *Blastocystis* strain, being linked with diarrhoeic patients in Denmark and Spain [52,53], and with irritable bowel syndrome and chronic diarrhoea in patients in Italy [54]. This does not seem to be the case of the present study, where all the *Blastocystis* isolates characterised as ST4 were identified in apparently healthy children. Absence of STs rarely found in humans (ST5–ST9) or thought to be present only in non-human animal species (ST10–ST17, ST21, ST23–ST26) seem to indicate that transmission of blastocystosis in Zambézia province is mainly of anthropic origin.

This study benefits from the high number of participating children and number of samples analysed, which allowed for a robust estimate of prevalences. Genetic data were strengthened by the adoption of a multi-locus genotyping scheme for the molecular characterization of samples positive to *G. duodenalis* and *Cryptosporidium* spp. However, the survey also presents some limitations that must be taken into consideration when interpreting the obtained results. Perhaps the most relevant is the long period (up to three months) that elapsed between sample collection and sample processing and analysis. During this time, collected stool samples were kept at room temperature in commercial devices intended to preserve the specimens and allow their use in downstream molecular assays. Despite this effort, suboptimal conservation of stool samples may have negatively affected the quality of the purified genomic DNA. This may explain the low proportion of *G. duodenalis*- and *Cryptosporidium*-positive isolates that were successfully amplified at their respective genotyping loci (*gdh*, *bg*, and *tpi* for *G. duodenalis*, *gp60* for *Cryptosporidium*). It is very likely that extraction and purification of DNA from fresh specimens would have significantly improved the genotyping data presented here. Finally, this study focused on human populations only. No attempts were conducted to analyse samples from animal and environmental (e.g., drinking water) sources, so the picture of the epidemiology of giardiosis, cryptosporidiosis and blastocystosis in this geographical area remains incomplete. This task must be accomplished in future molecular surveys.

4. Materials and Methods

4.1. Study Area and Stool Sample Collection

A prospective cross-sectional molecular epidemiological study of diarrhoea-causing enteric parasites including the protozoan *G. duodenalis* and *Cryptosporidium* spp. and the stramenopile *Blastocystis* sp., was conducted with children aged 3–14 from 10 of the 22 districts of Zambézia province, central Mozambique, between October 2017 and February 2019. Stool samples were collected from participating schoolchildren attending 18 primary schools or children seeking medical attention at seven primary healthcare centres (Figure 4). In school settings (range: 35–2111; mean: 651 schoolchildren) informative meetings were held for interested families. Schoolchildren volunteering to participate were given sampling kits to obtain stool samples during school attendance. In primary health clinics, children with gastrointestinal complaints (chronic or acute diarrhoea, bloating, abdominal pain) were invited to participate in the survey. Samples were collected by members of the research team at scheduled times and an aliquot (2–3 g) transferred to REAL Minisystem devices (Durviz, Valencia, Spain) for stool sample conservation and concentration. Preserved samples were maintained at room temperature up to three months before being transported to the Spanish National Centre for Microbiology (Majadahonda, Spain) for processing and analysis. Inclusion and exclusion criteria to participate in the study, school features, and sampling procedures as well as the detailed analysis of potential associations linked with enteric parasite infections in the paediatric populations surveyed here were thoroughly described elsewhere (Muadica et al., 2020) [27].

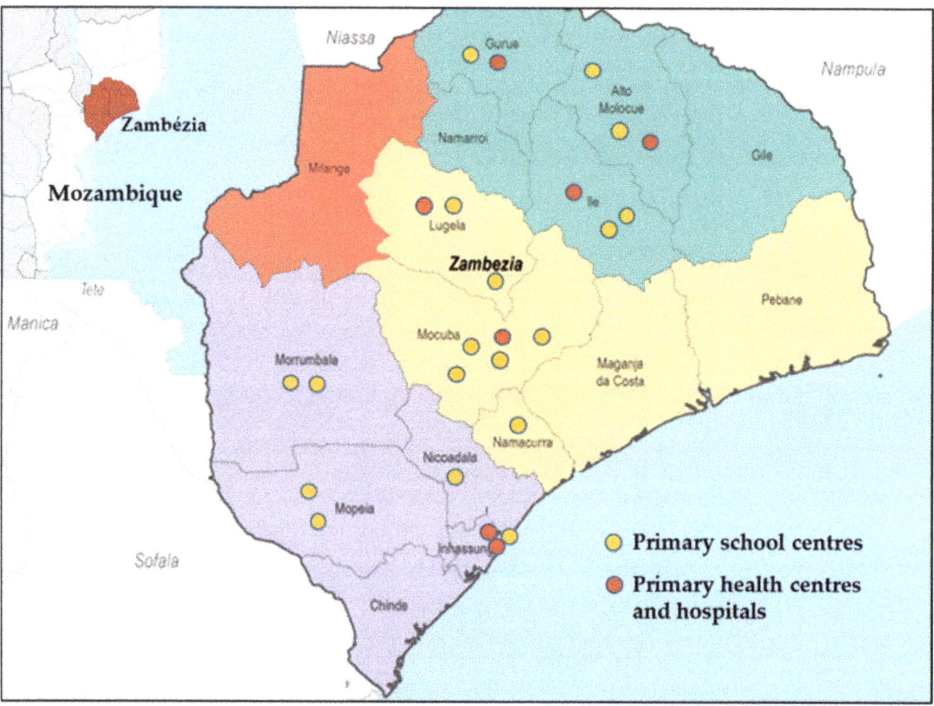

Figure 4. Map showing the geographical location of the Zambézia province in Mozambique (upper left corner) and the primary school and healthcare centres sampled in the present study.

4.2. DNA Extraction and Purification

Genomic DNA was isolated from about 200 mg of each faecal specimen by using the QIAamp DNA Stool Mini Kit (Qiagen, Hilden, Germany) according to the manufacturer's instructions, except that samples mixed with InhibitEX buffer were incubated for 10 min at 95 °C. Extracted and purified DNA samples (200 µL) were kept at −20 °C until further molecular analysis. A water extraction control was included in each sample batch processed.

4.3. Molecular Detection and Characterization of Giardia duodenalis

Giardia duodenalis DNA was detected by qPCR amplification of a 62 bp-fragment of the small subunit ribosomal RNA (*ssu* rRNA) gene of the parasite [55]. Amplification reactions (25 µL) consisted of 3 µL of template DNA, 0.5 µM of primers Gd-80F and Gd-

127R, 0.4 µM of probe (Additional file 1: Table S1), and 12.5 µL TaqMan® Gene Expression Master Mix (Applied Biosystems, CA, USA). Detection of parasitic DNA was performed on a Corbett Rotor Gene™ 6000 real-time PCR system (Qiagen) using an amplification protocol consisting on an initial hold step of 2 min at 55 °C and 15 min at 95 °C followed by 45 cycles of 15 s at 95 °C and 1 min at 60 °C. Water (no-template) and genomic DNA (positive) controls were included in each PCR run.

Giardia duodenalis isolates with a qPCR-positive result were re-assessed by sequence-based multi-locus genotyping of the genes encoding for the glutamate dehydrogenase (*gdh*), ß-giardin (*bg*) and triose phosphate isomerase (*tpi*) proteins of the parasite. A semi-nested PCR was used to amplify a 432-bp fragment of the *gdh* gene [56]. PCR reaction mixtures (25 µL) included 5 µL of template DNA and 0.5 µM of the primer pairs GDHeF/GDHiR in the primary reaction and GDHiF/GDHiR in the secondary reaction (Table S4). Both amplification protocols consisted of an initial denaturation step at 95 °C for 3 min, followed by 35 cycles of 95 °C for 30 s, 55 °C for 30 s and 72 °C for 1 min, with a final extension of 72 °C for 7 min. A nested PCR was used to amplify a 511 bp fragment of the bg gene [57]. PCR reaction mixtures (25 µL) consisted of 3 µL of template DNA and 0.4 µM of the primers sets G7_F/G759_R in the primary reaction and G99_F/G609_R in the secondary reaction (Table S4). The primary PCR reaction was carried out with the following amplification conditions: one step of 95 °C for 7 min, followed by 35 cycles of 95 °C for 30 s, 65 °C for 30 s, and 72 °C for 1 min with a final extension of 72 °C for 7 min. The conditions for the secondary PCR were identical to the primary PCR except that the annealing temperature was 55 °C. Finally, a nested PCR was used to amplify a 530 bp-fragment of the *tpi* gene [58]. PCR reaction mixtures (50 µL) included 2-2.5 µL of template DNA and 0.2 µM of the primer pairs AL3543/AL3546 in the primary reaction and AL3544/AL3545 in the secondary reaction (Table S4). Both amplification protocols consisted of an initial denaturation step at 94 °C for 5 min, followed by 35 cycles of 94 °C for 45 s, 50 °C for 45 s and 72 °C for 1 min, with a final extension of 72 °C for 10 min.

4.4. Molecular Detection and Characterization of Cryptosporidium spp.

The presence of *Cryptosporidium* spp. was assessed using a nested PCR to amplify a 587-bp fragment of the *ssu* rRNA gene of the parasite [59]. Amplification reactions (50 µL) included 3 µL of DNA sample and 0.3 µM of the primer pairs CR-P1/CR-P2 in the primary reaction and CR-P3/CPB-DIAGR in the secondary reaction (Table S4). Both PCR reactions were carried out as follows: one step of 94 °C for 3 min, followed by 35 cycles of 94 °C for 40 s, 50 °C for 40 s and 72 °C for 1 min, concluding with a final extension of 72 °C for 10 min.

Isolates identified as *C. hominis* or *C. parvum* by *ssu*-PCR (and Sanger sequencing, see below) were reanalysed at the 60 kDa glycoprotein (*gp60*) gene for subtyping purposes. Briefly, a nested PCR was conducted to amplify a 870 bp fragment of the *gp60* locus [60]. PCR reaction mixtures (50 µL) included 2-3 µL of template DNA and 0.3 µM of the primer pairs AL-3531/AL-3535 in the primary reaction and AL-3532/AL-3534 in the secondary reaction (Table S4). The primary PCR reaction consisted of an initial denaturation step of 94 °C for 5 min, followed by 35 cycles of 94 °C for 45 s, 59 °C for 45 s, and 72 °C for 1 min with a final extension of 72 °C for 10 min. The conditions for the secondary PCR were identical to the primary PCR except that the annealing temperature was 50 °C. Similarly, samples identified as *C. viatorum* by *ssu*-PCR were reanalysed at the *gp60* locus using a nested PCR protocol specifically developed for this *Cryptosporidium* species [45]. This protocol amplifies a 950 bp fragment of the *gp60* gene. Amplification reactions (50 µL) included 1–2 µL of DNA sample and 0.25 µM of the primer pair CviatF2/CviatR5 in the primary reaction and 0.5 µM of the primer pair CviatF3/CviatR8 in the secondary reaction (Table S4). Both PCR reactions were carried out as follows: one step of 95 °C for 4 min, followed by 35 cycles of 95 °C for 30 s, 58 °C for 30 s and 72 °C for 1 min, concluding with a final extension of 72 °C for 7 min.

4.5. Molecular Detection and Characterization of Blastocystis sp.

Identification of *Blastocystis* sp. was achieved by a direct PCR targeting the *ssu* rRNA gene of the parasite [61]. This protocol uses the pan-*Blastocystis*, barcode primers RD5 and BhRDr (Table S4) to amplify a PCR product of 600 bp. Amplification reactions (25 μL) included 5 μL of template DNA and 0.5 μM of the primer set RD5/BhRDr. Amplification conditions consisted of one step of 95 °C for 3 min, followed by 30 cycles of 1 min each at 94, 59 and 72 °C, with an additional 2 min final extension at 72 °C.

All the direct, semi-nested, and nested PCR protocols described above were conducted on a 2720 thermal cycler (Applied Biosystems). Reaction mixes always included 2.5 units of MyTAQ™ DNA polymerase (Bioline GmbH, Luckenwalde, Germany), and 5× MyTAQ™ Reaction Buffer containing 5 mM dNTPs and 15 mM $MgCl_2$. Laboratory-confirmed positive and negative DNA isolates for each parasitic species investigated were routinely used as controls and included in each round of PCR. PCR amplicons were visualized on 2% D5 agarose gels (Conda, Madrid, Spain) stained with Pronasafe nucleic acid staining solution (Conda). Positive-PCR products were directly sequenced in both directions using the internal primer set described above. DNA sequencing was conducted by capillary electrophoresis using the BigDye® Terminator chemistry (Applied Biosystems) on an ABI PRISM 3130 automated DNA sequencer.

4.6. Sequence and Phylogenetic Analyses

Raw sequencing data in both forward and reverse directions were viewed using the Chromas Lite version 2.1 sequence analysis program (https://technelysium.com.au/wp/chromas/, accessed on 23 February 2021). The BLAST tool (http://blast.ncbi.nlm.nih.gov/Blast.cgi, accessed on 23 February 2021) was used to compare nucleotide sequences with sequences retrieved from the NCBI GenBank database. Generated DNA consensus sequences were aligned to appropriate reference sequences using the MEGA 6 software [62] to identify *G. duodenalis* assemblages/sub-assemblages and *Cryptosporidium* species. *Blastocystis* sequences were submitted at the *Blastocystis* 18S database (http://pubmlst.org/blastocystis/, accessed on 23 February 2021) for sub-type confirmation and allele identification.

The evolutionary relationships among the identified *G. duodenalis* assemblages/sub-assemblages at the three loci investigated and the *Cryptosporidium* species found were inferred by a phylogenetic analysis using the neighbor-joining method in MEGA 6. Only sequences with unambiguous (no double peak) positions were used in the analyses. The evolutionary distances were computed using the Kimura 2-parameter method and modelled with a gamma distribution. The reliability of the phylogenetic analyses at each branch node was estimated by the bootstrap method using 1000 replications. Representative sequences of different *G. duodenalis* assemblages and sub-assemblages and *Cryptosporidium* species were retrieved from the NCBI database and included in the phylogenetic analyses for reference and comparative purposes.

Representative *G. duodenalis* sequences obtained in this study have been deposited in GenBank under accession numbers MW508361–MW508394 (*gdh* locus), MW508394–MW508410 (*bg* locus) and MW556751–MW556764 (*tpi* locus). Representative *Cryptosporidium* spp. sequences were deposited under accession numbers MW563962–MW563970 (*ssu* rRNA locus) and MW574004 (*gp60* locus). Representative *Blastocystis* sp. sequences were deposited under accession numbers MW564221–MW564233.

5. Conclusions

This PCR-based epidemiological study provides novel data on the molecular diversity of *G. duodenalis*, *Cryptosporidium* spp., and *Blastocystis* sp. in children with and without diarrhoea in Zambézia province, a Mozambican region where this information was completely lacking. Generated results complement and expand those obtained previously by the GEMS project in paediatric populations in Maputo province. A high intra-species molecular diversity was observed among the three parasites investigated, a finding compatible with an epidemiological scenario of high endemicity where infections and reinfections were common. No obvious differences in the distribution and frequency of parasite' species/genotypes were observed between apparently healthy children and children with clinical manifestations, suggesting that virulence/pathogenicity was not associated to a given genetic variant. Transmission of giardiosis and blastocystosis was primarily of anthropic nature, but strong molecular evidence indicated that a significant number of cryptosporidiosis cases were the result of zoonotic events. Data presented here highlight the need to conduct new molecular surveys in animal and environmental (drinking water) samples to complete our understanding of the transmission dynamics of these protist species in Mozambique. Measures directed to improve access to safe drinking water, sanitation facilities, and personal hygiene (e.g., hand washing) practices would definitively help in minimizing the transmission of diarrhoea-causing pathogens in highly endemic areas in Mozambique.

Supplementary Materials: The following are available online at https://www.mdpi.com/2076-0817/10/3/255/s1, Table S1: Diversity, frequency, and main molecular features of *Giardia duodenalis* sequences at the *gdh* locus in infected symptomatic and asymptomatic children in Zambézia province, Mozambique. GenBank accession numbers are provided. Superscript numbers identify single nucleotide polymorphisms involving amino acid change, Table S2: Diversity, frequency, and main molecular features of *Giardia duodenalis* sequences at the *bg* locus in infected symptomatic and asymptomatic children in Zambézia province, Mozambique. GenBank accession numbers are provided. Superscript numbers identify single nucleotide polymorphisms involving amino acid change, Table S3: Diversity, frequency, and main molecular features of *Giardia duodenalis* sequences at the *tpi* locus in infected symptomatic and asymptomatic children in Zambézia province, Mozambique. GenBank accession numbers are provided. Superscript numbers identify single nucleotide polymorphisms involving amino acid change, Table S4: Oligonucleotides used for the molecular identification and/or characterization of *Giardia duodenalis*, *Cryptosporidium* spp., and *Blastocystis* sp. in the present study, Figure S1: Phylogenetic relationships among *Giardia duodenalis* assemblages and sub-assemblages identified in infected symptomatic and asymptomatic children in Zambézia province, Mozambique. The analysis was conducted by a neighbor-joining method of a 502-bp fragment (corresponding to position 101–602 of reference sequence AY072727) of the *bg* gene sequence. Genetic distances were calculated using the Kimura two-parameter model. Green-filled squares represent sequences generated in the present study. Purple-filled dots represent reference sequences. Bootstrap values lower than 50% are not displayed. Because *bg* is a *Giardia*-specific protein, no outgroup taxon was used to root the tree, Figure S2: Phylogenetic relationships among *Giardia duodenalis* assemblages and sub-assemblages identified in infected symptomatic and asymptomatic children in Zambézia province, Mozambique. The analysis was conducted by a neighbor-joining method of a 479-bp fragment (corresponding to position 1–479 of reference sequence AF069560) of the *tpi* gene sequence. Genetic distances were calculated using the Kimura two-parameter model. Green-filled squares represent sequences generated in the present study. Purple-filled dots represent reference sequences. Bootstrap values lower than 50% are not displayed. *Giardia muris* was used as outgroup taxon to root the tree.

Author Contributions: Conceptualization, A.S.M., S.B. and D.C.; methodology, S.B. and D.C.; software, P.C.K. and A.D.; validation, S.B. and D.C.; formal analysis, A.S.M., P.C.K., A.D. and D.C.; investigation, A.S.M., P.C.K., A.D., B.B. and M.H.-d.-M.; resources, D.C.; data curation, D.C.; writing—original draft preparation, D.C.; writing—review and editing, A.S.M., P.C.K., A.D., S.B. and D.C.; visualization, S.B. and D.C.; supervision, S.B. and D.C.; project administration, D.C.; funding acquisition, D.C. All authors have read and agreed to the published version of the manuscript.

Funding: This research was funded by the Health Institute Carlos III (ISCIII), Ministry of Economy and Competitiveness (Spain), grant number PI16CIII/00024.

Institutional Review Board Statement: The study was conducted according to the guidelines of the Declaration of Helsinki and approved by the Ethics Committee of the Health Institute Carlos III (CEI PI17_2017-v3; date of approval: 23 October 2017) and the National Bioethics Committee for Health of the Republic of Mozambique (52/CNBS/2017; date of approval: 11 September 2017).

Informed Consent Statement: Informed consent was obtained from all subjects involved in the study.

Data Availability Statement: All relevant data are within the paper and its Supplementary Materials.

Acknowledgments: The authors are grateful to the participating children and their parents/legal guardians, and to school principals and teachers at the participating schools for their logistic assistance.

Conflicts of Interest: The authors declare no conflict of interest. The funders had no role in the design of the study; in the collection, analyses, or interpretation of data; in the writing of the manuscript; or in the decision to publish the results.

References

1. Snyder, J.D.; Merson, M.H. The magnitude of the global problem of acute diarrhoeal disease: A review of active surveillance data. *Bull. World Health Org.* **1982**, *60*, 605–613. [PubMed]
2. GBD 2016 Diarrhoeal Disease Collaborators. Estimates of the global, regional, and national morbidity, mortality, and aetiologies of diarrhoea in 195 countries: A systematic analysis for the Global Burden of Disease Study 2016. *Lancet Infect. Dis.* **2018**, *18*, 1211–1228. [CrossRef]
3. Liu, L.; Johnson, H.L.; Cousens, S.; Perin, J.; Scott, S.; Lawn, J.E.; Rudan, I.; Campbell, H.; Cibulskis, R.; Li, M.; et al. Global, regional, and national causes of child mortality: An updated systematic analysis for 2010 with time trends since 2000. *Lancet* **2012**, *379*, 2151–2161. [CrossRef]
4. Black, R.E.; Morris, S.S.; Bryce, J. Where and why are 10 million children dying every year? *Lancet* **2003**, *361*, 2226–2234. [CrossRef]
5. GBD Diarrhoeal Diseases Collaborators. Estimates of global, regional, and national morbidity, mortality, and aetiologies of diarrhoeal diseases: A systematic analysis for the Global Burden of Disease Study 2015. *Lancet Infect. Dis.* **2017**, *17*, 909–948. [CrossRef]
6. Torgerson, P.R.; Devleesschauwer, B.; Praet, N.; Speybroeck, N.; Willingham, A.L.; Kasuga, F.; Rokni, M.B.; Zhou, X.N.; Fèvre, E.M.; Sripa, B.; et al. World Health Organization estimates of the global and regional disease burden of 11 foodborne parasitic diseases, 2010: A data synthesis. *PLoS Med.* **2015**, *12*, e1001920. [CrossRef]
7. World Health Organization. *The World Health Report—Fighting Disease Fostering Development*; World Health Organization: Geneva, Switzerland, 1996; Available online: http://www.who.int/whr/1996/en/ (accessed on 26 January 2021).
8. Kotloff, K.L.; Nataro, J.P.; Blackwelder, W.C.; Nasrin, D.; Farag, T.H.; Panchalingam, S.; Wu, Y.; Sow, S.O.; Sur, D.; Breiman, R.F.; et al. Burden and aetiology of diarrhoeal disease in infants and young children in developing countries (the Global Enteric Multicenter Study, GEMS): A prospective, case-control study. *Lancet* **2013**, *382*, 209–222. [CrossRef]
9. Tan, K.S.; Mirza, H.; Teo, J.D.; Wu, B.; Macary, P.A. Current views on the clinical relevance of *Blastocystis* spp. *Curr. Infect. Dis. Rep.* **2010**, *12*, 28–35. [CrossRef] [PubMed]
10. Halliez, M.C.; Buret, A.G. Extra-intestinal and long term consequences of *Giardia duodenalis* infections. *World J. Gastroenterol.* **2013**, *19*, 8974–8985. [CrossRef] [PubMed]
11. Korpe, P.S.; Valencia, C.; Haque, R.; Mahfuz, M.; McGrath, M.; Houpt, E.; Kosek, M.; McCormick, B.J.J.; Penataro Yori, P.; Babji, S.; et al. Epidemiology and risk factors for cryptosporidiosis in children from 8 low-income sites: Results from the MAL-ED Study. *Clin. Infect. Dis.* **2018**, *67*, 1660–1669. [CrossRef]
12. Efstratiou, A.; Ongerth, J.E.; Karanis, P. Waterborne transmission of protozoan parasites: Review of worldwide outbreaks—An update 2011–2016. *Water Res.* **2017**, *114*, 14–22. [CrossRef] [PubMed]
13. Ryan, U.; Hijjawi, N.; Xiao, L. Foodborne cryptosporidiosis. *Int. J. Parasitol.* **2018**, *48*, 1–12. [CrossRef] [PubMed]
14. Ryan, U.; Cacciò, S.M. Zoonotic potential of *Giardia*. *Int. J. Parasitol.* **2013**, *43*, 943–956. [CrossRef] [PubMed]
15. Feng, Y.; Ryan, U.; Xiao, L. Genetic diversity and population structure of *Cryptosporidium*. *Trends Parasitol.* **2018**, *34*, 997–1011. [CrossRef]
16. Stensvold, C.R.; Clark, C.G. Pre-empting Pandora's box: *Blastocystis* subtypes revisited. *Trends Parasitol.* **2020**, *36*, 229–232. [CrossRef]
17. Mor, S.M.; Tzipori, S. Cryptosporidiosis in children in Sub-Saharan Africa: A lingering challenge. *Clin. Infect. Dis.* **2008**, *47*, 915–921. [CrossRef]
18. Aldeyarbi, H.M.; El-Ezz, N.M.A.; Karanis, P. *Cryptosporidium* and cryptosporidiosis: The African perspective. *Environ. Sci. Pollut. Res. Int.* **2016**, *23*, 13811–13821. [CrossRef]
19. Squire, S.A.; Ryan, U. *Cryptosporidium* and *Giardia* in Africa: Current and future challenges. *Parasit. Vectors* **2017**, *10*, 195. [CrossRef]
20. Khorshidvand, Z.; Khazaei, S.; Amiri, M.; Taherkhani, H.; Mirzaei, A. Worldwide prevalence of emerging parasite *Blastocystis* in immunocompromised patients: A systematic review and meta-analysis. *Microb. Pathog.* **2021**, in press. [CrossRef]
21. Meurs, L.; Polderman, A.M.; Vinkeles Melchers, N.V.; Brienen, E.A.; Verweij, J.J.; Groosjohan, B.; Mendes, F.; Mechendura, M.; Hepp, D.H.; Langenberg, M.C.; et al. Diagnosing polyparasitism in a high-prevalence setting in Beira, Mozambique: Detection of intestinal parasites in fecal samples by microscopy and real-time PCR. *PLoS Negl. Trop. Dis.* **2017**, *11*, e0005310. [CrossRef]
22. Platts-Mills, J.A.; Babji, S.; Bodhidatta, L.; Gratz, J.; Haque, R.; Havt, A.; McCormick, B.J.; McGrath, M.; Olortegui, M.P.; Samie, A.; et al. MAL-ED Network Investigators. Pathogen-specific burdens of community diarrhoea in developing countries: A multisite birth cohort study (MAL-ED). *Lancet Glob. Health* **2015**, *3*, e564–e575. [CrossRef]

23. Messa, A., Jr.; Köster, P.C.; Garrine, M.; Gilchrist, C.; Bartelt, L.A.; Nhampossa, T.; Massora, S.; Kotloff, K.; Levine, M.M.; Alonso, P.L.; et al. Molecular diversity of *Giardia duodenalis* in children under 5 years from the Manhiça district, Southern Mozambique enrolled in a matched case-control study on the aetiology of diarrhoea. *PLoS Negl. Trop. Dis.* **2021**, *15*, e0008987. [CrossRef]

24. Messa, A., Jr.; Köster, P.C.; Garrine, M.; Nhampossa, T.; Cossa, A.; Kotloff, K.; Levine, M.M.; Alonso, P.L.; Carmena, D.; Mandomando, I. Molecular characterization of *Cryptosporidium* spp. in children younger than 5 years enrolled in a matched case-control study on the aetiology of diarrhoea in the Manhiça district, Southern Mozambique. *Pathogens* **2021**. under review.

25. Irisarri-Gutiérrez, M.J.; Mingo, M.H.; de Lucio, A.; Gil, H.; Morales, L.; Seguí, R.; Nacarapa, E.; Muñoz-Antolí, C.; Bornay-Llinares, F.J.; Esteban, J.G.; et al. Association between enteric protozoan parasites and gastrointestinal illness among HIV- and tuberculosis-infected individuals in the Chowke district, southern Mozambique. *Acta Trop.* **2017**, *170*, 197–203. [CrossRef]

26. Casmo, V.; Lebbad, M.; Maungate, S.; Lindh, J. Occurrence of *Cryptosporidium* spp. and *Cystoisospora belli* among adult patients with diarrhoea in Maputo, Mozambique. *Heliyon* **2018**, *4*, 1–13. [CrossRef]

27. Muadica, A.S.; Balasegaram, S.; Beebeejaun, K.; Köster, P.C.; Bailo, B.; Hernández-de-Mingo, M.; Dashti, A.; Dacal, E.; Saugar, J.M.; Fuentes, I.; et al. Risk associations for intestinal parasites in symptomatic and asymptomatic schoolchildren in central Mozambique. *Clin. Microbiol. Infect.* **2020**, in press. [CrossRef] [PubMed]

28. De Lucio, A.; Amor-Aramendía, A.; Bailo, B.; Saugar, J.M.; Anegagrie, M.; Arroyo, A.; López-Quintana, B.; Zewdie, D.; Ayehubizu, Z.; Yizengaw, E.; et al. Prevalence and genetic diversity of *Giardia duodenalis* and *Cryptosporidium* spp. among school children in a rural area of the Amhara Region, North-West Ethiopia. *PLoS ONE* **2016**, *11*, e0159992. [CrossRef] [PubMed]

29. Dacal, E.; Saugar, J.M.; de Lucio, A.; Hernández-de-Mingo, M.; Robinson, E.; Köster, P.C.; Aznar-Ruiz-de-Alegría, M.L.; Espasa, M.; Ninda, A.; Gandasegui, J.; et al. Prevalence and molecular characterization of *Strongyloides stercoralis*, *Giardia duodenalis*, *Cryptosporidium* spp., and *Blastocystis* spp. isolates in school children in Cubal, Western Angola. *Parasit. Vectors* **2018**, *11*, 67. [CrossRef] [PubMed]

30. Muadica, A.S.; Messa, A.E., Jr.; Dashti, A.; Balasegaram, S.; Santin, M.; Manjate, F.; Chirinda, P.; Garrine, M.; Vubil, D.; Acácio, S.; et al. First identification of genotypes of *Enterocytozoon bieneusi* (Microsporidia) among symptomatic and asymptomatic children in Mozambique. *PLoS Negl. Trop. Dis.* **2020**, *14*, e0008419. [CrossRef] [PubMed]

31. Mandomando, I.M.; Macete, E.V.; Ruiz, J.; Sanz, S.; Abacassamo, F.; Vallès, X.; Sacarlal, J.; Navia, M.M.; Vila, J.; Alonso, P.L.; et al. Etiology of diarrhea in children younger than 5 years of age admitted in a rural hospital of southern Mozambique. *Am. J. Trop. Med. Hyg.* **2007**, *76*, 522–527. [CrossRef] [PubMed]

32. Fonseca, A.M.; Fernandes, N.; Ferreira, F.S.; Gomes, J.; Centeno-Lima, S. Intestinal parasites in children hospitalized at the Central Hospital in Maputo, Mozambique. *J. Infect. Dev. Ctries.* **2014**, *8*, 786–789. [CrossRef]

33. Cerveja, B.Z.; Tucuzo, R.M.; Madureira, A.C.; Nhacupe, N.; Langa, I.A.; Buene, T.; Banze, L.; Funzamo, C.; Noormahomed, E.V. Prevalence of intestinal parasites among HIV Infected and HIV uninfected patients treated at the 1° De Maio Health Centre in Maputo, Mozambique. *EC Microbiol.* **2017**, *9*, 231–240.

34. Nhampossa, T.; Mandomando, I.; Acacio, S.; Quintó, L.; Vubil, D.; Ruiz, J.; Nhalungo, D.; Sacoor, C.; Nhabanga, A.; Nhacolo, A.; et al. Diarrheal disease in rural Mozambique: Burden, risk factors and etiology of diarrheal disease among children aged 0-59 months seeking care at health facilities. *PLoS ONE* **2015**, *10*, e0119824. [CrossRef]

35. Acácio, S.; Mandomando, I.; Nhampossa, T.; Quintó, L.; Vubil, D.; Sacoor, C.; Kotloff, K.; Farag, T.; Nasrin, D.; Macete, E.; et al. Risk factors for death among children 0–59 months of age with moderate-to-severe diarrhea in Manhiça district, southern Mozambique. *BMC Infect. Dis.* **2019**, *19*, 322. [CrossRef] [PubMed]

36. Bauhofer, A.F.L.; Cossa-Moiane, I.; Marques, S.; Guimarães, E.L.; Munlela, B.; Anapakala, E.; Chilaúle, J.J.; Cassocera, M.; Langa, J.S.; Chissaque, A.; et al. Intestinal protozoan infections among children 0–168 months with diarrhea in Mozambique: June 2014–January 2018. *PLoS Negl. Trop. Dis.* **2020**, *14*, e0008195. [CrossRef] [PubMed]

37. Sow, S.O.; Muhsen, K.; Nasrin, D.; Blackwelder, W.C.; Wu, Y.; Farag, T.H.; Panchalingam, S.; Sur, D.; Zaidi, A.K.; Faruque, A.S.; et al. The burden of Cryptosporidium diarrheal disease among children <24 months of age in moderate/high mortality regions of sub-Saharan Africa and South Asia, utilizing data from the Global Enteric Multicenter Study (GEMS). *PLoS Negl. Trop. Dis.* **2016**, *10*, e0004729.

38. Slapeta, J. *Cryptosporidium* species found in cattle: A proposal for a new species. *Trends Parasitol.* **2006**, *22*, 469–474. [CrossRef] [PubMed]

39. Adamu, H.; Petros, B.; Zhang, G.; Kassa, H.; Amer, S.; Ye, J.; Feng, Y.; Xiao, L. Distribution and clinical manifestations of *Cryptosporidium* species and subtypes in HIV/AIDS patients in Ethiopia. *PLoS Negl. Trop. Dis.* **2014**, *8*, e2831. [CrossRef]

40. Akinbo, F.O.; Okaka, C.E.; Omoregie, R.; Dearen, T.; Leon, E.T.; Xiao, L. Molecular characterization of *Cryptosporidium* spp. in HIV-infected persons in Benin City, Edo State, Nigeria. *Fooyin J. Health Sci.* **2010**, *2*, 85–89. [CrossRef]

41. Gatei, W.; Wamae, C.N.; Mbae, C.; Waruru, A.; Mulinge, E.; Waithera, T.; Gatika, S.M.; Kamwati, S.K.; Revathi, G.; Hart, C.A. Cryptosporidiosis: Prevalence, genotype analysis, and symptoms associated with infections in children in Kenya. *Am. J. Trop. Med. Hyg.* **2006**, *75*, 78–82. [CrossRef]

42. Elwin, K.; Hadfield, S.J.; Robinson, G.; Crouch, N.D.; Chalmers, R.M. *Cryptosporidium viatorum* n. sp. (Apicomplexa: Cryptosporidiidae) among travellers returning to Great Britain from the Indian subcontinent, 2007–2011. *Int. J. Parasitol.* **2012**, *42*, 675–682. [CrossRef]

43. Koehler, A.V.; Wang, T.; Haydon, S.R.; Gasser, R.B. *Cryptosporidium viatorum* from the native Australian swamp rat *Rattus lutreolus*—An emerging zoonotic pathogen? *Int. J. Parasitol. Parasites Wildl.* **2018**, *7*, 18–26. [CrossRef] [PubMed]

44. Zhao, W.; Zhou, H.; Huang, Y.; Xu, L.; Rao, L.; Wang, S.; Wang, W.; Yi, Y.; Zhou, X.; Wu, Y.; et al. *Cryptosporidium* spp. in wild rats (*Rattus* spp.) from the Hainan Province, China: Molecular detection, species/genotype identification and implications for public health. *Int. J. Parasitol. Parasites Wildl.* **2019**, *9*, 317–321. [CrossRef] [PubMed]

45. Stensvold, C.R.; Elwin, K.; Winiecka-Krusnell, J.; Chalmers, R.M.; Xiao, L.; Lebbad, M. Development and application of a *gp60*-based typing assay for *Cryptosporidium viatorum*. *J. Clin. Microbiol.* **2015**, *53*, 1891–1897. [CrossRef]

46. D'Alfonso, R.; Santoro, M.; Essi, D.; Monsia, A.; Kaboré, Y.; Glé, C.; Di Cave, D.; Sorge, R.P.; Di Cristanziano, V.; Berrilli, F. *Blastocystis* in Côte d'Ivoire: Molecular identification and epidemiological data. *Eur. J. Clin. Microbiol. Infect. Dis.* **2017**, *36*, 2243–2250. [CrossRef]

47. Greigert, V.; Abou-Bacar, A.; Brunet, J.; Nourrisson, C.; Pfaff, A.W.; Benarbia, L.; Pereira, B.; Randrianarivelojosia, M.; Razafindrakoto, J.L.; Rakotomalala, R.S.; et al. Human intestinal parasites in Mahajanga, Madagascar: The kingdom of the protozoa. *PLoS ONE* **2018**, *13*, e0204576. [CrossRef] [PubMed]

48. Alfellani, M.A.; Stensvold, C.R.; Vidal-Lapiedra, A.; Onuoha, E.S.; Fagbenro-Beyioku, A.F.; Clark, C.G. Variable geographic distribution of *Blastocystis* subtypes and its potential implications. *Acta Trop.* **2013**, *126*, 11–18. [CrossRef]

49. Stensvold, C.R.; Alfellani, M.; Clark, C.G. Levels of genetic diversity vary dramatically between *Blastocystis* subtypes. *Infect. Genet. Evol.* **2012**, *12*, 263–273. [CrossRef]

50. El Safadi, D.; Gaayeb, L.; Meloni, D.; Cian, A.; Poirier, P.; Wawrzyniak, I.; Delbac, F.; Dabboussi, F.; Delhaes, L.; Seck, M.; et al. Children of Senegal River Basin show the highest prevalence of *Blastocystis* sp. ever observed worldwide. *BMC Infect. Dis.* **2014**, *14*, 164. [CrossRef]

51. Abda, I.B.; Maatoug, N.; Romdhane, R.B.; Bouhelmi, N.; Zallegua, N.; Aoun, K.; Viscogliosi, E.; Bouratbine, A. Prevalence and subtype identification of *Blastocystis* sp. in healthy individuals in the Tunis area, Tunisia. *Am. J. Trop. Med. Hyg.* **2017**, *96*, 202–204. [CrossRef]

52. Stensvold, C.R.; Christiansen, D.B.; Olsen, K.E.; Nielsen, H.V. *Blastocystis* sp. subtype 4 is common in Danish *Blastocystis*-positive patients presenting with acute diarrhea. *Am. J. Trop. Med. Hyg.* **2011**, *84*, 883–885. [CrossRef] [PubMed]

53. Domínguez-Márquez, M.V.; Guna, R.; Muñoz, C.; Gómez-Muñoz, M.T.; Borrás, R. High prevalence of subtype 4 among isolates of *Blastocystis hominis* from symptomatic patients of a health district of Valencia (Spain). *Parasitol. Res.* **2009**, *105*, 949–955. [CrossRef] [PubMed]

54. Mattiucci, S.; Crisafi, B.; Gabrielli, S.; Paoletti, M.; Cancrini, G. Molecular epidemiology and genetic diversity of *Blastocystis* infection in humans in Italy. *Epidemiol. Infect.* **2016**, *144*, 635–646. [CrossRef] [PubMed]

55. Verweij, J.J.; Schinkel, J.; Laeijendecker, D.; van Rooyen, M.A.; van Lieshout, L.; Polderman, A.M. Real-time PCR for the detection of *Giardia lamblia*. *Mol. Cell. Probes* **2003**, *17*, 223–225. [CrossRef]

56. Read, C.M.; Monis, P.T.; Thompson, R.C. Discrimination of all genotypes of *Giardia duodenalis* at the glutamate dehydrogenase locus using PCR-RFLP. *Infect. Genet. Evol.* **2004**, *4*, 125–130. [CrossRef]

57. Lalle, M.; Pozio, E.; Capelli, G.; Bruschi, F.; Crotti, D.; Cacciò, S.M. Genetic heterogeneity at the beta-giardin locus among human and animal isolates of *Giardia duodenalis* and identification of potentially zoonotic subgenotypes. *Int. J. Parasitol.* **2005**, *35*, 207–213. [CrossRef]

58. Sulaiman, I.M.; Fayer, R.; Bern, C.; Gilman, R.H.; Trout, J.M.; Schantz, P.M.; Das, P.; Lal, A.A.; Xiao, L. Triosephosphate isomerase gene characterization and potential zoonotic transmission of *Giardia duodenalis*. *Emerg. Infect. Dis.* **2003**, *9*, 1444–1452. [CrossRef] [PubMed]

59. Tiangtip, R.; Jongwutiwes, S. Molecular analysis of *Cryptosporidium* species isolated from HIV-infected patients in Thailand. *Trop. Med. Int. Health* **2002**, *7*, 357–364. [CrossRef]

60. Feltus, D.C.; Giddings, C.W.; Schneck, B.L.; Monson, T.; Warshauer, D.; McEvoy, J.M. Evidence supporting zoonotic transmission of *Cryptosporidium* spp. in Wisconsin. *J. Clin. Microbiol.* **2006**, *44*, 4303–4308. [CrossRef]

61. Scicluna, S.M.; Tawari, B.; Clark, C.G. DNA barcoding of *Blastocystis*. *Protist* **2006**, *157*, 77–85. [CrossRef]

62. Tamura, K.; Stecher, G.; Peterson, D.; Filipski, A.; Kumar, S. MEGA6: Molecular evolutionary genetics analysis version 6.0. *Mol. Biol. Evol.* **2013**, *30*, 2725–2729. [CrossRef] [PubMed]

pathogens

Article

Molecular Characterisation of *Cryptosporidium* spp. in Mozambican Children Younger than 5 Years Enrolled in a Matched Case-Control Study on the Aetiology of Diarrhoeal Disease

Augusto Messa Jr. [1], Pamela C. Köster [2], Marcelino Garrine [1,3], Tacilta Nhampossa [1,4], Sérgio Massora [1], Anélsio Cossa [1], Quique Bassat [1,5,6,7,8], Karen Kotloff [9], Myron M. Levine [9], Pedro L. Alonso [1,5,10], David Carmena [2,*] and Inácio Mandomando [1,4,*]

[1] Centro de Investigação em Saúde de Manhiça, Maputo 1929, Mozambique; augusto.junior@manhica.net (A.M.J.); marcelino.garrine@manhica.net (M.G.); tacilta.nhampossa@manhica.net (T.N.); sergio.massora@manhica.net (S.M.); anelsio.cossa@manhica.net (A.C.); quique.bassat@isglobal.org (Q.B.); alonsop@who.int (P.L.A.)
[2] Parasitology Reference and Research Laboratory, National Centre for Microbiology, Health Institute Carlos III, Majadahonda, 28220 Madrid, Spain; pamkoste@ucm.es
[3] Global Health and Tropical Medicine, Instituto de Higiene e Medicina Tropical, Universidade Nova de Lisboa, 1349-008 Lisbon, Portugal
[4] Instituto Nacional de Saúde, Ministério da Saúde, Marracuene, Maputo 1120, Mozambique
[5] ISGlobal, Hospital Clínic—Universitat de Barcelona, 08036 Barcelona, Spain
[6] ICREA, Pg. Lluís Companys 23, 08010 Barcelona, Spain
[7] Pediatric Infectious Diseases Unit, Pediatrics Department, Hospital Sant Joan de Déu, University of Barcelona, 08950 Barcelona, Spain
[8] Consorcio de Investigación Biomédica en Red de Epidemiología y Salud Pública (CIBERESP), 28029 Madrid, Spain
[9] Center for Vaccine Development, University of Maryland School of Medicine, Baltimore, MD 21201-1509, USA; kkotloff@som.umaryland.edu (K.K.); mlevine@som.umaryland.edu (M.M.L.)
[10] Global Malaria Program, World Health Organization, 1211 Geneva, Switzerland
* Correspondence: dacarmena@isciii.es (D.C.); inacio.mandomando@manhica.net (I.M.)

Citation: Messa, A., Jr.; Köster, P.C.; Garrine, M.; Nhampossa, T.; Massora, S.; Cossa, A.; Bassat, Q.; Kotloff, K.; Levine, M.M.; Alonso, P.L.; et al. Molecular Characterisation of *Cryptosporidium* spp. in Mozambican Children Younger than 5 Years Enrolled in a Matched Case-Control Study on the Aetiology of Diarrhoeal Disease. *Pathogens* **2021**, *10*, 452. https://doi.org/10.3390/pathogens10040452

Academic Editor: Luiz Shozo Ozaki

Received: 22 March 2021
Accepted: 7 April 2021
Published: 9 April 2021

Publisher's Note: MDPI stays neutral with regard to jurisdictional claims in published maps and institutional affiliations.

Abstract: *Cryptosporidium* is a leading cause of childhood diarrhoea and associated physical and cognitive impairment in low-resource settings. *Cryptosporidium*-positive faecal samples (n = 190) from children aged ≤ 5 years enrolled in the Global Enteric Multicenter Study (GEMS) in Mozambique detected by ELISA (11.5%, 430/3754) were successfully PCR-amplified and sequenced at the *gp60* or *ssu* rRNA loci for species determination and genotyping. Three *Cryptosporidium* species including *C. hominis* (72.6%, 138/190), *C. parvum* (22.6%, 43/190), and *C. meleagridis* (4.2%, 8/190) were detected. Children ≤ 23 months were more exposed to *Cryptosporidium* spp. infections than older children. Both *C. hominis* and *C. parvum* were more prevalent among children with diarrhoeal disease compared to those children without it (47.6% vs. 33.3%, p = 0.007 and 23.7% vs. 11.8%, p = 0.014, respectively). A high intra-species genetic variability was observed within *C. hominis* (subtype families Ia, Ib, Id, Ie, and If) and *C. parvum* (subtype families IIb, IIc, IIe, and IIi) but not within *C. meleagridis* (subtype family IIIb). No association between *Cryptosporidium* species/genotypes and child's age was demonstrated. The predominance of *C. hominis* and *C. parvum* IIc suggests that most of the *Cryptosporidium* infections were anthroponotically transmitted, although zoonotic transmission events also occurred at an unknown rate. The role of livestock, poultry, and other domestic animal species as sources of environmental contamination and human cryptosporidiosis should be investigated in further molecular epidemiological studies in Mozambique.

Keywords: *Cryptosporidium*; *gp60*; *ssu* rRNA; genotyping; children; diarrhoea; prevalence; molecular epidemiology; Mozambique; GEMS

1. Introduction

Diarrhoeal diseases remain the second leading cause of mortality after pneumonia in children under 5 years worldwide, accounting for approximately 9% of the 5.8 million deaths associated to this condition reported in 2015 [1,2]. Most of these fatalities disproportionately occur in poor-resource settings where suboptimal hygiene conditions and sanitation prevail [3]. A recent update on diarrhoeal burden from the Global Enteric Multicenter Study (GEMS and GEMS1A) demonstrated that Rotavirus, *Cryptosporidium*, enterotoxigenic *Escherichia coli* producing heat stable toxin (ST_ETEC), and *Shigella* were the main pathogens associated with moderate-to-severe diarrhoea (MSD) and less severe diarrhoea (LSD) in African and Asian children [4,5]. Cryptosporidiosis in children under 5 years presents with watery diarrhoea, abdominal pain, nausea, and vomiting, being often associated to growth faltering and cognitive development impairment [6,7]. In severe cases, the disease can lead to life-threatening sequelae among malnourished and immunocompromised children [8]. *Cryptosporidium* spp. is also a major contributor to the burden of diarrhoeal disease in HIV-positive patients [9].

As other diarrhoea-causing pathogens, *Cryptosporidium* spp. are transmitted through the faecal–oral route. Humans acquire the infection through direct contact with infected hosts (person-to-person or zoonotic transmission) or by ingestion of contaminated food or water (foodborne and waterborne transmission), but the relative importance of these routes is still unclear [10]. At least 40 known *Cryptosporidium* species are currently recognised, and among these, more than 20 species and genotypes have been reported to cause human infections. *Cryptosporidium hominis* and *C. parvum* cause more than 90% of the human cases documented globally [11–13].

Molecular tools for the differentiation of *Cryptosporidium* species and genotypes are currently available, mostly using PCR followed by either restriction length fragment polymorphisms (RFLP) analysis or Sanger sequencing of the small subunit ribosomal ribonucleic acid (*ssu* rRNA) and the 60 kDa glycoprotein (*gp60*) genes of the parasite [10,13]. The *ssu* rRNA gene is largely used for the differential diagnosis of *Cryptosporidium* species due to its multicopy nature and associated high sensitivity. Subtype identification is primarily achieved through DNA sequence analysis of the highly polymorphic *gp60* gene. Subtype assignment is based on the number of TCA, TCG, and TCT repeats in addition to other repetitive sequences, such as the ACATCA, within the *gp60* tandem repeat motif region. Subtype families are named as Ia, Ib, Ic, Id, Ie, If, etc. for *C. hominis* and IIa, IIb, IIc, IId, etc. for *C. parvum*, with further species families named in ascending order [12,14].

In Africa, the epidemiology and genetic diversity of *Cryptosporidium* spp. remains relatively unknown. However, as noted by a recent literature review, at least 13 species and genotypes have been identified in humans, with *C. hominis* followed by *C. parvum* once again dominating the epidemiological landscape [11]. Subtyping studies support the dominance of anthroponotic over zoonotic transmission in African countries, regardless of the close contact with farm and domestic animals. Another interesting observation in Africa is the high level of subtype diversity, where at least six subtype families for *C. hominis* (Ia, Ib, Id, Ie, If, and Ih) have been described. For *C. parvum* the predominant subtypes identified in humans belong to the IIc family, in addition to IIa, IIb, IId, IIg, IIi, IIh, IIm and the rarer anthroponotically transmitted IIe subtype family [11].

In Mozambique, diarrhoea is ranked as the third cause of death in children under 14 years from the capital city Maputo [15] and fourth in children under 5 years from the Manhiça district (Maputo province) [16], being responsible for 20% of hospital paediatric admissions in this district [17]. The GEMS data support that prevention strategies targeting Rotavirus, *Cryptosporidium*, ST_ETEC and *Shigella* could contribute to reduce diarrhoeal cases by approximately 50% in infants, and hence diarrhoeal-associated mortality [18]. However, the genetic diversity of *Cryptosporidium* spp. in GEMS was not investigated. Two previous hospital-based studies carried out in the southern part of the country have identified IaA23R3, IIcA5G3, and IIeA12G1 subtypes among nine isolates from patients with diarrhoea in the capital city Maputo [19], and IbA10G2 and IdA22 subtypes among eight isolates in patients with HIV and tuberculosis in the Chokwe district of Gaza province [20]. However, no extensive molecular epidemiological studies have been conducted to evaluate the genetic diversity within *Cryptosporidium* spp. Herein, we aimed to analyse the diversity and frequency of *Cryptosporidium* species and subtypes detected in stools from children younger than 5 years from the Manhiça district, Mozambique, enrolled in the context of GEMS between 2007 and 2012.

2. Results

2.1. Initial Screening for the Detection of Cryptosporidium spp. by ELISA Immunoassay

During the 5-year study period (December 2007–November 2012) a total of 3754 stool samples were collected. The ELISA positivity rate for *Cryptosporidium* spp. was estimated at 11.5% (430/3754). The prevalence was significantly ($p < 0.001$) higher among diarrhoea cases (MSD and LSD cases) (16.5%, 222/1346) compared to children without diarrhoea (non-cases; 8.5%, 208/2408). Most (91.2%, 392/430) of the *Cryptosporidium*-positive samples by ELISA were available for molecular analyses (Figure 1). Unavailable samples were the result of the depletion of starting material as consequence of testing and analyses in previous studies [4,5,18].

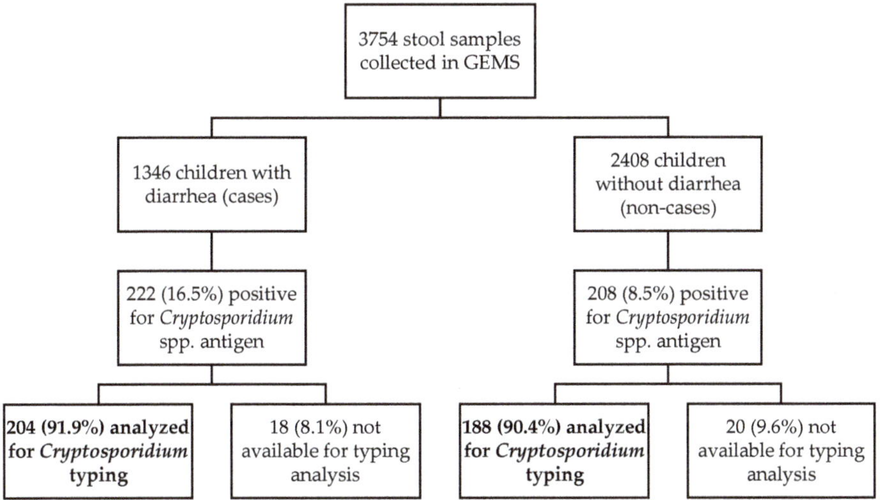

Figure 1. Flow chart summarising the diagnostic and genotyping procedures used in this study.

The distribution of the ELISA-positive *Cryptosporidium* infections in cases and non-cases according to sex, age group, and clinical condition is summarised in Table 1. Approximately one in two (53.6%, 210/392) children with cryptosporidiosis were aged 0–11 months. The male/female ratio was 1.8. Children with MSD and their matched controls were significantly more exposed to *Cryptosporidium* than their counterparts with LSD and corresponding controls ($p < 0.001$). HIV+ patients with diarrhoea were more likely to be infected with *Cryptosporidium* spp. than HIV+ patients without diarrhoea ($x^2 = 9.8758$, $p = 0.001675$). Being undernourished and having diarrhoea were also significantly associated with cryptosporidiosis ($x^2 = 19.769$, $p \leq 0.00001$). Regarding coinfections with other intestinal pathogens, *Cryptosporidium* infection was more likely in children with diarrhoea and rotavirus infection ($p \leq 0.011$). In contrast, coinfections by *Cryptosporidium* spp. and *G. duodenalis* were more frequent in asymptomatic (non-cases) children ($p < 0.001$). The full dataset showing the epidemiological, clinical, diagnostic, and genotyping data used in the analyses conducted in the present survey is presented in Table S1.

Table 1. Main epidemiological and clinical variables of *Cryptosporidium*-positive children under five years of age by ELISA (*n* = 392) with diarrhoea (cases) and without diarrhoea (non-cases) according to age group. Children were recruited during the Global Enteric Multicenter Study at the Manhiça district (Maputo, southern Mozambique), 2007–2012.

	0–11 Months		12–23 Months		24–59 Months	
	Cases	**Non-Cases** [1]	**Cases**	**Non-Cases** [1]	**Cases**	**Non-Cases** [1]
Variable	*n* = 111 (%)	*n* = 99 (%)	*n* = 72 (%)	*n* = 68 (%)	*n* = 21 (%)	*n* = 21 (%)
MSD [2]	87 (78.4)	85 (85.9)	45 (62.5)	57 (83.8)	12 (57.1)	19 (90.5)
LSD [2]	24 (21.6)	14 (14.1)	27 (37.5)	11 (16.2)	9 (42.9)	2 (9.5)
Mean age (months)	7.3	7.1	15.9	16.9	31.1	31
Sex (male)	68 (61.3)	68 (68.7)	48 (66.7)	41 (61.2)	13 (69.9)	14 (66.7)
HIV+ [3]	10/44 (24.7)	2/18 (11.1)	7/37 (18.9)	2/21 (9.5)	2/9 (22.2)	0 (0.0)
Undernutrition	14 (12.6)	2 (2.0)	12 (16.7)	1 (1.5)	1 (4.8)	0 (0.0)
Co-infections						
Rotavirus	33 (29.7)	13 (13.1)	9 (12.5)	10 (14.7)	4 (19.1)	1 (4.7)
Shigella spp.	0 (0.0)	0 (0.0)	6 (8.3)	0 (0.0)	1 (4.8)	0 (0.0)
All ETECs	7 (6.3)	5 (5.1)	13 (18.1)	7 (10.3)	4 (19.1)	4 (19.1)
G. duodenalis	16 (14.4)	30 (30.3)	18 (25.0)	38 (55.9)	4 (19.5)	10 (47.6)
E. histolytica[4]	6 (5.5)[4]	8 (8.1)	7 (9.7)	5 (7.4)	2 (9.5)	4 (19.1)

[1] Non-cases are asymptomatic children without diarrhoea matched by age, sex, and neighbourhood with MSD and LSD cases. [2] Only applicable to cases. [3] Only part of the participants were tested for HIV, and the numbers of participants with known HIV status are specified in the denominator. [4] Missing values: *n* = 1. ETEC: Enterotoxigenic *Escherichia coli*; HIV: Human immunodeficiency virus; LSD: Less severe diarrhoea; MSD: Moderate-to-severe diarrhoea; NA: Not applicable.

2.2. Confirmation of Cryptosporidium spp. by Nested PCR Methods

Out of the 392 samples that tested positive by ELISA, 37.2% (146/392) were successfully sub-genotyped at the *gp60* locus. The remaining 250 isolates with a negative result by *gp60*-PCR were subsequently re-assessed at the *ssu* rRNA marker, allowing the confirmation of *Cryptosporidium* DNA in 44 additional samples. Overall, the presence of the parasite was confirmed by *gp60*-PCR and/or *ssu*-PCR in 48.5% (190/392) of the analysed samples (Table 2). Sequence alignment analyses including appropriate reference sequences allowed the identification of three *Cryptosporidium* species including *C. hominis* (72.6%, 138/190), *C. parvum* (22.6%, 43/190), and *C. meleagridis* (4.2%, 8/190). An additional isolate (0.5%, 1/190) was only identified at the genus level due to poor sequence quality (Table 2). Both *C. hominis* and *C. parvum* were more prevalent among diarrhoeal children (cases) compared to non-diarrhoeal (non-cases) children (47.6% vs. 33.3%, *p* = 0.007 and 23.7% vs. 11.8%, *p* = 0.014, respectively). Infections by *Cryptosporidium* spp. were most common in children younger than 24 months, with *C. hominis* being the *Cryptosporidium* species more prevalent in all age groups investigated (Table 2). Cases of cryptosporidiosis by *C. hominis* and *C. parvum* were consistently detected along the whole study period, peaking during November 2011 and March 2012, particularly in children with LSD (Figure S1).

Table 2. Diagnostic performance of PCR methods and distribution of the *Cryptosporidium* species detected in children under five years of age (*n* = 396) with diarrhoea (cases) and without diarrhoea (non-cases) according to age group. Children were recruited during the Global Enteric Multicenter Study at the Manhiça district (Maputo, southern Mozambique), 2007–2012.

PCR Results	0–11 Months			12–23 Months			24–59 Months		
	Cases	Non-Cases	*p*	Cases	Non-Cases	*p*	Cases	Non-Cases	*p*
	n = 112 (%)	*n* = 101 (%)		*n* = 73 (%)	*n* = 67 (%)		*n* = 21 (%)	*n* = 22 (%)	
gp60 (*n* = 146)	49 (44.1)	24 (24.4)	0.003	37 (51.4)	30 (44.1)	0.389	3 (14.3)	3 (14.3)	1
ssu rRNA [1] (*n* = 44)	12/62 (19.4)	10/75 (13.3)	0.339	9/35 (25.7)	8/38 (21.1)	0.638	4/18 (22.2)	1/18 (5.6)	0.148
Both (*n* = 190)	61 (54.9)	34 (34.3)	0.003	46 (63.9)	38 (55.9)	0.334	7 (33.3)	4 (19.1)	0.292
Species [2]									
C. hominis (*n* = 138)	45 (47.4)	27 (29.4)	0.011	33 (55.9)	26 (46.4)	0.308	3 (22.2)	3 (15.0)	0.687
C. parvum (*n* = 43)	14 (21.9)	6 (8.5)	0.028	12 (31.6)	9 (23.1)	0.402	2 (12.5)	0 (0.0)	0.229
C. meleagridis (*n* = 8)	1 (1.9)	1 (1.5)	1	1 (3.7)	3 (9.1)	0.620	1 (6.7)	1 (5.6)	1
Unknown [3] (*n* = 1)	1 (1.9)	0 (0.0)	0.440	0 (0.0)	0 (0.0)	NA	0 (0.0)	0 (0.0)	NA

[1] Only negative samples by *gp60*-PCR (*n* = 250) were analysed by *ssu*-PCR. [2] Species assigned on the combination of both *gp60*-PCR and *ssu*-PCR results. [3] Poor sequence quality data only allowed subtyping at genus level. NA: Not applicable.

2.3. Genetic Variation within C. hominis and C. parvum

Sequence analysis of the 117 isolates characterised as *C. hominis* at the *gp60* locus revealed the presence of five subtype families including Ia (35.0%, 41/117), Ib (20.5%, 24/117), Id (1.7%, 2/117), Ie (34.2%, 40/117), and If (8.6%, 10/117). The most prevalent subtypes found were IaA23R1 within family Ia, IbA13G2 within family Ib, and IdA20 within family Id. All isolates belonging to families Ie and If were identified as IeA11G3T3 and IfA12G1, respectively. Two genetic variants within IaA24R3 and IbA13G2 corresponded to novel subtypes (Table 3). Sequence analyses of the 29 isolates characterised as *C. parvum* at the same locus revealed the presence of four subtype families including IIb (3.4%, 1/29), IIc (86.2%, 25/29), IIe (6.9%, 2/29), and IIi (3.4%, 1/29). IIbA11 within family IIb, IIcA5G3 within family IIc, IIeA11G1 within family IIe, and IIiA6-like within family IIi were the only subtypes found. Novel genetic variants were found within IIbA11, IIcA5G3, and IIiA6-like subtypes. All four isolates assigned to *C. meleagridis* belonged to subtype IIIbA23G1R1 within family IIIb (Table 3).

Out of the 21 sequences characterised as *C. hominis* at the *ssu* rRNA locus, 81% (17/21) showed 100% identity with reference sequence AF108865. The remaining four sequences differed from AF108865 by 1–3 single nucleotide polymorphisms (SNPs) including a deletion mutation (Table 3). All the 14 sequences assigned to *C. parvum* corresponded to different genetic variations of the "bovine genotype" of this *Cryptosporidium* species, which is characterised by the presence of a four-base deletion TAAT at positions 686 to 689 of reference sequence AF112571. Indeed, some authors have proposed this genetic variant as an independent species named *C. pestis* [22]. *Cryptosporidium hominis* and *C. parvum* sequences differing from reference sequences at the *ssu* rRNA locus included ambiguous (C/T, A/G) positions in the form of double peaks, transition (A↔G, C↔T) and transversion (T↔G, A↔T) mutations, and single- to multiple-base deletions.

Table 3. Diversity, frequency, and main molecular features of *Cryptosporidium*-positive samples at the *gp60* and *ssu* rRNA loci in children under 5 years of age recruited during the Global Enteric Multicenter Study at the Manhiça district (Maputo, southern Mozambique), 2007–2012. GenBank accession numbers of representative sequences were provided.

Locus	Species	Isolates	Family	Subtype	Reference	Stretch	Single Nucleotide Polymorphisms	GenBank ID
gp60	*C. hominis*	37	Ia	IaA23R3	KX579755	3–805	None	MW480826
		1		IaA24R3	KX579755	1–793	84_86InsTCA	MW480827
		3		IaA25R3	JF927194	18–838	None	MW480828
		7	Ib	IbA10G2	AY262031	22–857	None	MW480829
		15		IbA13G2	MT053132	13–896	G85A	MW480830
		2		IbA9G3	DQ665688	14–825	None	MW480831
		1	Id	IdA20	JX088404	48–904	None	MW480832
		1		IdA21	MN904672	47–910	None	MW480833
		39	Ie	IeA11G3T3	AY738184	19–923	None	MW480834
		1		IeA11G3T3	AY738184	48–923	T284Y, A662R	MW480835
		10	If	IfA12G1	EU161655	1–870	None	MW480836
	C. parvum	1	IIb	IIbA11	AY166805	1–782	51_59DelTCATCATCA	MW480837
		11	IIc	IIcA5G3	GU214365	31–851	None	MW480838
		7		IIcA5G3	GU214365	29–854	38 SNPs [2]	MW480839
		5		IIcA5G3	GU214365	29–853	40 SNPs [2]	MW480840
		1		IIcA5G3	GU214365	50–853	C110T	MW480841
		1		IIcA5G3	GU214365	50–853	40 SNPs [2]	MW480842
		1	IIe	IIeA11G1	MN904721	1–813	None	MW480843
		1		IIeA13G1	KU852716	7–795	None	MW480844
		1	IIi	IIiA6-like	AY873782	26–932	85 SNPs [2]	MW480845
	C. meleagridis [1]	4	IIIb	IIIbA23G1R1	MK331716	1–714	None	MW480846
ssu rRNA	*C. hominis*	17	–	–	AF108865	529–954	None	MW487256
		1	–	–	AF108865	587–965	A892R	MW487257
		2	–	–	AF108865	591–969	T795Y, A892R	MW487258
		1	–	–	AF108865	640–956	697delT, T795Y, A892R	MW487259
	C. parvum	1	–	–	AF112571	565–956	A646G, T649G, 686_689DelTAAT, A691T	MW487260
		3	–	–	AF112571	526–1039	A646G, T649G, 686_689DelTAAT, T693A	MW487261
		1	–	–	AF112571	524–1039	A646G, 647_649DelATT, T663C, 686_689DelTAAT, C795T	MW487262
		5	–	–	AF112571	539–1031	A646G, T649G, 686_689DelTAAT, T693A, C795T	MW487263
		1	–	–	AF112571	526–965	A646G, T649G, 686_689DelTAAT, T693A, C795Y	MW487264
		3	–	–	AF112571	539–954	A646G, T649G, 686_689DelTAAT, A691T, C795Y, A892R	MW487265
	Unknown	1	–	–	–	–	–	–
	C. meleagridis	8	–	–	AF112574	524–1034	None	MW487266

[1] Samples initially diagnosed by *ssu*-PCR and subsequently genotyped at the *gp60* locus using the *C. meleagridis*-specific PCR protocol described elsewhere [21]. [2] See details in Table S1. Del: nucleotide(s) deletion(s); NA: Not applicable; Y: C/T; R: A/G. Novel genotypes are shown underlined.

Finally, all eight sequences identified as *C. meleagridis* at the *ssu* rRNA locus showed 100% identity with reference sequence AF112574. Four of these eight isolates were successfully amplified at the *gp60* locus using a specific PCR protocol for this *Cryptosporidium* species (see Section 4.3.2.). Sanger sequencing analyses allowed the identification of subtype IIIbA23G1R1 in all four sequences, which were identical to reference sequence MK331716.

Cryptosporidium hominis was the most prevalent species in children with MSD or LSD (79.6%, 70/88), followed by *C. parvum* (19.3%, 17/88), and *C. meleagridis* (1.1%, 1/88) (Table 4). Within *C. hominis*, nearly three out of every four diarrhoea-associated infections were caused by subtype families Ie (30.7%, 27/88) and Ia (27.3%, 24/88). Subtype family Ie was more frequent in children with MSD (43.1%, 25/58), and subtype family Ia in children with LSD (46.7%, 14/30). No obvious differences in subtype distribution were observed among the age groups considered. Near half of the cryptosporidiosis cases identified in HIV+ patients were caused by the subtype family Ia (55.6%, 5/9) (Table 4).

Table 4. Diversity and frequency of *Cryptosporidium* subtypes families within *C. hominis*, *C. parvum*, and *C. meleagridis* in symptomatic (cases) children under 5 years of age according to severity of the diarrhoea, age group, and HIV coinfection. Children were recruited during the Global Enteric Multicenter Study at the Manhiça district (Maputo, southern Mozambique), 2007–2012.

				C. hominis			*C. parvum*	*C. meleagridis*
Variable	**Total**	**Ia**	**Ib**	**Id**	**Ie**	**If**	**IIc**	**IIIb**
	n = 90 (%)	n = 25 (%)	n = 11 (%)	n = 2 (%)	n = 28 (%)	n = 6 (%)	n = 17 (%)	n = 1 (%)
Diarrhoea								
MSD	60 (66.7)	11 (44.0)	7 (63.6)	1 (50.0)	26 (92.9)	5 (83.3)	9 (52.9)	1 (100)
LSD	30 (33.3)	14 (56.0)	4 (36.4)	1 (50.0)	2 (7.1)	1 (16.7)	8 (47.1)	0 (0)
Age (months)								
0–11	49 (54.4)	16 (64.0)	8 (72.7)	1 (50.0)	14 (50)	2 (33.3)	8 (47.1)	0 (0)
12–23	37 (41.1)	7 (28.0)	3 (27.3)	1 (50.0)	13 (46.4)	4 (66.7)	9 (52.9)	0 (0)
24–59	4 (4.4)	2 (8.0)	0 (0.0)	0 (0.0)	1 (3.6)	0 (0.0)	0 (0.0)	1 (100)
Co-infections								
HIV+ [1]	9/42 (21.4)	5/17 (29.4)	1/4 (25)	0/1 (0.0)	1/5 (20)	0/1 (0.0)	2/14 (14.3)	NA

[1] Frequencies calculated over the total of HIV+ children only. 42 of the 90 children had an HIV test result. HIV: Human immunodeficiency virus; LSD: Less severe diarrhoea; MSD: Moderate-to-severe diarrhoea. NA: Not applicable.

In matched controls without diarrhoea (non-cases), *C. hominis* was also the predominant species found (75.0%, 45/60), followed by *C. parvum* (20.0%, 12/60) and *C. meleagridis* (5.0%, 3/60) (Table S2). Within *C. hominis*, half of the infections were attributed to subtype families Ia (26.7%, 16/60) and Ib (21.7%, 13/60). No obvious differences in subtype distribution were observed among the age groups considered. No *Cryptosporidium* subtype families could be determined in HIV+ patients without diarrhoea (Table S2).

Cryptosporidium hominis subtype families Ib and Ie were more frequently found during study years 1 to 4, whereas *C. hominis* subtype family Ia and *C. parvum* subtype family IIc were observed only in study year 5, suggesting variable seasonal patterns in the frequency of *Cryptosporidium* subtypes (Figure S2).

The genetic relationships among *gp60* gene sequences generated in the present study, as inferred by a neighbor-joining analysis, are shown in Figure 2. All *Cryptosporidium* sequences clustered together (monophyletic groups) with different well-supported clades (≥93% of bootstrap) corresponding to appropriate reference sequences for *Cryptosporidium* subtype families.

3. Discussion

This is the most comprehensive molecular epidemiological study conducted in Mozambique to date investigating the genetic diversity of the diarrhoea-causing enteric protozoan parasite *Cryptosporidium* spp. The analysis took advantage of the large sample repository generated by the GEMS in children younger than five years of age with and without diarrhoea in Maputo province [4,5]. Consequently, a total of 392 stool samples with a positive result by ELISA were available for molecular investigations, of which 190 were successfully genotyped at the *gp60* or *ssu* rRNA loci.

Figure 2. Phylogenetic relationships among *Cryptosporidium hominis* (family I), *C. parvum* (family II), and *C. meleagridis* (family III) genotypes identified in children under 5 years of age recruited during the Global Enteric Multicenter Study at the Manhiça district (Maputo, southern Mozambique), 2007–2012. The analysis was conducted by a neighbor-joining method of the *gp60* gene. Genetic distances were calculated using the Kimura two-parameter model. Green filled dots represent sequences generated in the present study. Purple filled dots represent reference sequences. Bootstrap values lower than 75% are not displayed. *Cryptosporidium cuniculus* was used as outgroup taxon to root the tree.

A preliminary assessment of the *Cryptosporidium*-positive samples by ELISA corroborated results obtained in previous epidemiological studies carried out in sub-Saharan African countries [11]. For instance, cryptosporidiosis was confirmed as a serious public health concern in children younger than 2 years old, particularly if immunocompromised by other infections (e.g., HIV/AIDS) or malnourished. Young children are more susceptible to intestinal parasites and other infectious pathogens due to their low level of immunity [23]. The level of exposure and risk of infections increase in poor settings with limited access to safe drinking water, sanitation, and hygiene [24,25]. A significant association between *Cryptosporidium* infection and malnutrition (stunting, wasting, underweight) has been documented in children from several African countries, including Kenya [26], Mozambique [27], Tanzania [28], and Uganda [29], among others. Similarly, *Cryptosporidium* infection was more frequently found in HIV-positive than in HIV-negative children and patients in Mozambique [20,30] and other African endemic regions [28,31,32].

Our molecular analyses revealed the presence of three *Cryptosporidium* species (*C. hominis*, *C. parvum*, and *C. meleagridis*) in the studied paediatric population. Mostly anthroponotic *C. hominis* and zoonotic *C. parvum* were previously known to be circulating in Mozambique [19,20,33], but this is the first report of *C. meleagridis* in the country. An additional two species, *C. felis* and *C. viatorum*, have also been recently described in adult patients with diarrhoea in the Maputo province and in asymptomatic children in the Zambézia province [19,34]. *Cryptosporidium meleagridis* and *C. felis* are adapted to infect birds and domestic cats, respectively, as primary host species, but they are also responsible for a significant number of human infections globally, particularly the former [10,12]. These data seem to indicate that direct contact with cats, poultry, and other avian species (or their faecal material) may be a risk factor for human cryptosporidiosis in Mozambique. Following the same line of reasoning, the fact that all the *C. parvum* isolates characterised at the *ssu* rRNA gene belonged to the "bovine genotype" of *C. parvum* supported the notion that an unknown number of human cases of cryptosporidiosis are indeed of zoonotic nature. This is without precluding that some of the infections caused by this genetic variant of *C. parvum* may be also transmitted through person-to-person contact. The extent of the exact contribution of each potential transmission pathway (zoonotic, anthropic, direct contact, or indirect through ingestion of contaminated water or food) remains to be elucidated. Finally, *C. viatorum* was initially thought to be a human-adapted species [35], but recent epidemiological surveys conducted in Australia and China have demonstrated that this *Cryptosporidium* species can successfully infect rodents and therefore may have zoonotic potential [36,37]. Overall, these data agree with those reported in the African continent, where *C. hominis* was the most prevalent (2–100%) *Cryptosporidium* species in humans, followed by *C. parvum* (3–100%) and *C. meleagridis* (up to 75%), the latter species found mainly in immunocompromised individuals [11].

Subtyping analyses identified Ia and Ie as the most prevalent subtype families within *C. hominis*, being responsible for nearly 70% of the infections attributable to this *Cryptosporidium* species in both diarrhoeal and non-diarrhoeal children. Similar results have been reported in Kenyan young children with and without HIV infection [38], mostly HIV-positive patients in Nigeria [39], children younger than 10 years in São Tomé and Príncipe [40], and children younger than five years in South Africa [41]. As already described in most previous epidemiological studies conducted in the continent, subtype family Ib was also underrepresented in the paediatric population surveyed here. Indeed, Ib has been shown as the predominant subtype family only in Nigerian children [42]. Subtype families If and Id are typically documented at low frequencies in African human populations. Subtype family If has been identified in Kenyan patients with and without HIV/AIDS [38,43], children of young age in South Africa [41], and individuals from rural areas in Tanzania [44]. Finally, members of the subtype family Id have been found circulating in HIV-positive patients in Ethiopia, Equatorial Guinea, and Mozambique [20,45,46], in children with diarrhoea in Ghana and Madagascar [47,48], in Kenyan children with and without HIV infection [38], and in paediatric populations from Nigeria and South Africa [41,42].

Mainly transmitted anthroponotically, IIc was the predominant (86%) *C. parvum* subtype family circulating in the children investigated here. This is in agreement with previous results obtained in diverse human populations from other sub-Saharan countries including Kenya [38], Madagascar [48], Nigeria [42], South Africa [41], and Uganda [49].

In contrast, subtype families IIb, IIe, and IIi were only sporadically detected, and subtype family IIa was absent. It should be noted that IIa was the most prevalent *C. parvum* subtype family circulating in HIV-positive and diarrhoeal disease patients in Ethiopia and Kenya [43,45,50] and also in Tunisian young children [51]. Taken together, these geographically segregated patterns of *C. parvum* genetic variants may be indicative of differences in sources of infections and transmission pathways.

Very limited information is currently available on the intra-species molecular diversity of *C. meleagridis* in African isolates of human origin. In the present study, all the *C. meleagridis* isolates identified belonged to the subtype family IIIb, and no genetic heterogeneity was observed among their sequences. This very same subtype family has also been reported in an urban population in Tunisia [52], whereas IIId has been described in diarrhoeic paediatric patients in South Africa [41]. Interestingly, a wide range of *C. meleagridis* subtype families including IIIb (but not IIId) has been recently identified in river water and its sediment in South Africa [53]. This finding has important public health implications, as it demonstrated that the consumption of contaminated, non-treated surface waters might lead to waterborne cryptosporidiosis by *C. meleagridis*.

The main strength of this study is the large number of *Cryptosporidium*-positive samples of human origin molecularly characterised by Sanger sequencing. However, certain methodological and study design issues may have hampered its accuracy. For instance, the fact that only half (48.6%, 190/392) of the ELISA-positive samples were amplified at the *gp60* or *ssu* rRNA loci may have biased the actual proportion of *Cryptosporidium* species and genotypes reported here. This may be due to the suboptimal preservation of parasitic DNA through time (stool samples were collected during the period 2007–2012), or to potential false-positive results in the ELISA immunoassay, or to amplification failure associated to suboptimal removal of PCR inhibitors (e.g., proteases, DNases, polysaccharides, bile salts). We cannot completely rule out the possibility that the ELISA immunoassay initially used for screening purposes yielded an unknown number of false-negative results, particularly for *Cryptosporidium* species less frequently found in humans (e.g., *C. felis*, *C. viatorum*, *C. ubiquitum*, among others). Additionally, no attempts were carried out to analyse in depth the potential associations between *Cryptosporidium* species/genotypes and the sociodemographic, epidemiological, and clinical features of the participating children, as this task will be specifically tackled in an independent study.

Overall, the high level of genetic diversity observed within *Cryptosporidium* isolates reveals an epidemiological scenario where infection and re-infection events seem common and environmental contamination high. In this regard, a recent risk association study conducted in the province of Zambézia revealed that drinking untreated water and having regular contact with domestic animals were major risks for acquiring protist infections including cryptosporidiosis [25]. Additionally, a recent quantitative microbial risk assessment analysis has estimated that the consumption of unsafe water causes 2 million cryptosporidiosis cases and 1.6×10^5 disability-adjusted life years in Mozambique annually [54]. These results highlight the relevance of improving access to safe drinking water and sanitary conditions to minimise the risk of environmental contamination and the waterborne and foodborne transmission of diarrhoea-causing enteric pathogens.

4. Materials and Methods

4.1. Study Context

In Mozambique, the GEMS was conducted by the *Centro de Investigação em Saúde de Manhiça* (CISM), in six health facilities in the Manhiça District [55], which is located approximately 80 km north of the capital city Maputo in the country's Southern region. The district covers 2380 km² and has a subtropical climate with two distinct seasons: a warm, rainy season from November to April, and the cool and dry season during the rest of the year [17,56]. Since 1996, CISM has been conducting a continuous Health and Demographic Surveillance System (HDSS) with regular update of demographic events for all surveyed population (current population followed: 203,132 uniquely identified individuals; 46,851 enumerated and geo-positioned households; 27,504 are children under 5 years). During the study period, the HDSS was covering approximately 95,000 inhabitants [17].

The rationale, study design, and methodology of the GEMS have been previously described elsewhere [57]; the study comprised three years of recruitment of acute moderate-to-severe diarrhoea cases (MSD, GEMS1) and one additional year including less-severe diarrhoea (LSD, GEMS1A) [4,5]. In Manhiça, the GEMS collected samples uninterruptedly over 5 years, from December 2007 to October 2011, and GEMS1A from November 2011 to November 2012. The standardised epidemiological and clinical methods for the case-control study as well as the full definitions have been previously described elsewhere [18,55]. Briefly, all children aged 0–59 months (stratified in three age groups: 0–11 months, 12–23 months, and 24–59 months), presenting in the six sentinel health facilities with diarrhoea meeting inclusion criteria for the study were invited to participate. Community controls (up to three for MSD cases and one for LSD cases) matched to the index case by age, sex, and neighbourhood were identified using the HDSS databases and enrolled within 14 days after enrolment of the case, and the stool samples were collected and sent to the laboratory at CISM [58].

4.2. Stool Collection and Initial Testing

The stool samples collection and processing protocols were also standardised across GEMS sites as described elsewhere [55,58]. Samples were collected in sterile flasks and placed in a refrigerator or in a cool-box with a cooler block (2–8 °C) for up to 6 h until transported to the laboratory. Sample aliquots without preservatives were frozen at −80 °C for further testing. The CRYPTOSPORIDIUM II™ ELISA immunoassay (TECHLAB®, Blacksburg, VA, USA) was used as screening method for the specific detection of *Cryptosporidium* spp. following the manufacturer's instructions.

4.3. Molecular Study

4.3.1. DNA Extraction and Purification

Molecular analyses were performed only on stool samples that were *Cryptosporidium*-positive by immunoassay. Genomic DNA was isolated from about 200 mg of faecal material by using the QIAamp DNA Stool Mini Kit (Qiagen, Hilden, Germany) according to the manufacturer's instructions, except that samples mixed with ASL lysis buffer were incubated for 10 min at 95 °C. Resulting eluates (200 µL in PCR-grade water) were stored at −20 °C and shipped to the Spanish National Centre for Microbiology at Majadahonda (Spain) for downstream molecular analysis.

4.3.2. Molecular Detection and Characterisation of *Cryptosporidium* spp.

As this study was based on *Cryptosporidium*-positive samples by ELISA, to optimise time and resources the following diagnostic and genotyping algorithm was implemented. A nested PCR protocol was initially used to amplify an 870-bp fragment of the *gp60* gene of the parasite as previously described [59]. This approach allowed for the differential diagnosis of *C. hominis* and *C. parvum* (the two *Cryptosporidium* species more prevalent in humans), and for the identification of subtype families within these two species. The outer primers were AL-3531_F (5′-ATAGTCTCCGCTGTATTC-3′) and AL-3535_R (5′-GGAAGGAACGATGTATCT-3′), and the inner primers were AL-3532_F (5′-TCCGCTGTATTCTCAGCC-3′) and AL-3534_R (5′-GCAGAGGAACCAGCATC-3′). Reaction mixtures (50 µL) contained 200 nM of each primer and 2–3 µL of template DNA. Cycling conditions included one step of 94 °C for 5 min, followed by 35 cycles of amplification (denaturation at 94 °C for 45 s, annealing at 59 °C for 45 s, and elongation at 72 °C for 1 min), concluding with a final extension of 72 °C for 10 min. The same conditions were used in the secondary reaction, except that the annealing temperature was 50 °C.

Samples with a negative result by *gp60*-PCR were re-analysed by a nested PCR to amplify a 587-bp fragment of the *ssu* rRNA gene of the parasite [60]. This approach allowed for the detection of low burdens of *Cryptosporidium* infections and for the identification of

Cryptosporidium species other than *C. hominis* or *C. parvum*. The outer primers were CR-P1 (5′-CAGGGAGGTAGTGACAAGAA-3′) and CR-P2 (5′-TCAGCCTTGCGACCATACTC-3′), and the inner primers were CR-P3 (5′-ATTGGAGGGCAAGTCTGGTG-3′) and CPB-DIAGR (5′-TAAGGTGCTGAAGGAGTAAGG-3′). In all cases, reaction mixtures (50 µL) contained 300 nM of each primer and 3 µL of template DNA. Cycling conditions consisted of one step of 94°C for 5 min, followed by 35 cycles of amplification (denaturation at 94 °C for 40 s, annealing at 50 °C for 40 s, and elongation at 72 °C for 1 min), finalising with a final extension at 72 °C for 10 min.

Samples that were identified by *ssu*-PCR (and Sanger sequencing, see below) as *C. meleagridis* were re-analysed at the *gp60* locus by a nested PCR specifically developed for this *Cryptosporidium* species [21]. This protocol amplifies a 900 bp fragment of the *gp60* gene. The outer primers were CRSout115F (5′-GATGAGATTGTCGCTCGTTATC-3′) and CRSout1328R (5′-AACCTGCGGAACCTGTG-3′), and the inner primers were ATGFmod (5′-GAGATTGTCGCTCGTTATCG-3′) and GATR2 (5′-GATTGCAAAAACGGAAGG-3′). Reaction mixtures (50 µL) contained 250 nM of each primer and 2–3 µL of template DNA. Cycling conditions included one step of 95 °C for 4 min, followed by 35 cycles of amplification (denaturation at 95 °C for 30 s, annealing at 60 °C for 30 s, and elongation at 72 °C for 1 min), concluding with a final extension of 72 °C for 7 min. The same conditions were used in the secondary reaction, except that the annealing temperature was 58 °C.

Nested PCR protocols described above were conducted on a 2720 Thermal Cycler (Applied Biosystems, CA, USA). Reaction mixes always included 2.5 units of MyTAQ™ DNA polymerase (Bioline GmbH, Luckenwalde, Germany), and 5× MyTAQ™ Reaction Buffer containing 5 mM dNTPs and 15 mM MgCl$_2$. Laboratory-confirmed positive and negative DNA samples of human origin were routinely used as controls and included in each round of PCR. PCR amplicons were visualised on 2% D5 agarose gels (Conda, Madrid, Spain) stained with Pronasafe nucleic acid staining solution (Conda) and recorded using the MiniBIS Pro system controlled by GelCapture version 7.5.2 software (DNR Bio-Imaging Systems, Jerusalem, Israel). A 100 bp DNA ladder (Boehringer Mannheim GmbH, Baden-Wurttemberg, Germany) was used for the sizing of obtained amplicons. Positive-PCR products were directly sequenced in both directions using the internal primer sets described above. DNA sequencing was conducted by capillary electrophoresis using the BigDye® Terminator chemistry (Applied Biosystems) on an on ABI PRISM 3130 automated DNA sequencer at the Core Genomic Facility of the Spanish National Centre for Microbiology, Majadahonda (Spain). Sequencing reactions were repeated on samples for which genotyping was unsuccessful in the first instance.

The *Cryptosporidium* sequences obtained in this study have been deposited in GenBank under accession numbers MW480826–MW480846 (*gp60* locus) and MW487256–MW487266 (*ssu* rRNA locus).

4.4. Data Analysis

4.4.1. Epidemiological Analysis

PCR data were entered in a Microsoft Excel spreadsheet (Redmond, WA, USA) and then checked for accuracy and consistency by independent laboratory personnel. Clinical and demographic data were extracted from the original GEMS dataset. *Cryptosporidium* spp., study groups (diarrhoeal vs. non-diarrhoeal, MSD vs. LSD) age groups, and study years were treated as categorical variables. Differences in frequencies were compared using Chi-squared test or Fisher's exact test as appropriate. A p-value < 0.05 was considered statistically significant. Missing values were excluded from the analyses; thus, denominators for some comparisons may differ. Data analyses were performed in Stata version 14 (StataCorp LP, College Station, TX, USA).

4.4.2. Sequence and Phylogenetic Analysis

Raw sequencing data in both forward and reverse directions were visually inspected using the Chromas Lite version 2.1 sequence analysis program [61]. Special attention was paid to the detection and recording of ambiguous (double peak) positions. The BLAST tool was used to search for identity among sequences deposited in the National Center for Biotechnology Information (NCBI) public repository database [62]. Multiple sequence alignment analyses with appropriate reference sequences were conducted using MEGA 6 to identify *Cryptosporidium* species and to annotate the presence of single nucleotide polymorphisms (SNPs) [63]. *Cryptosporidium hominis* and *C. parvum* subtypes were assigned according to the number of TCA (A), TCG (G), ACATCA/ACATCG (R), and TCTT (T) fragment repeats in the microsatellite region of the *gp60* gene, in accordance with the established nomenclature, as previously described [14].

The evolutionary relationships among the identified *Cryptosporidium* species and subtypes were inferred by a phylogenetic analysis using the neighbor-joining method in MEGA 6 [64]. Only sequences with unambiguous (no double peak) positions were used in the analyses. The evolutionary distances were computed using the Kimura 2-parameter method and modelled with a gamma distribution. The reliability of the phylogenetic analyses at each branch node was estimated by the bootstrap method using 1000 replications. Representative sequences of different *Cryptosporidium* species and subtypes were retrieved from the NCBI database and included in the phylogenetic analysis for reference and comparative purposes.

5. Conclusions

This study provides the most comprehensive description of the molecular diversity of the enteric protozoan parasite *Cryptosporidium* spp. in Mozambique to date. Our findings revealed the circulation of at least three *Cryptosporidium* species in young Mozambican children primarily affected with diarrhoea. A high intra-species genetic variability was observed within *C. hominis* (subtype families Ia, Ib, Id, Ie, and If) and *C. parvum* (subtype families IIb, IIc, IIe, and IIi), but not within *C. meleagridis* (subtype family IIIb). No associations between *Cryptosporidium* species/genetic variants and age-related patterns could be demonstrated. The predominance of mainly anthroponotically transmitted *C. hominis* and *C. parvum* IIc strongly suggests that most of the *Cryptosporidium* infections detected in the surveyed paediatric population are of human origin. However, a significant proportion of the infections were caused by host-adapted *Cryptosporidium* species (e.g., *C. meleagridis*) or genetic variants (e.g., *C. parvum* "bovine genotype") suggesting the occurrence of zoonotic transmission events at an unknown rate. Further molecular epidemiological studies are warranted to assess the actual contribution of livestock, poultry, and other domestic animal species to the environmental (including surface waters intended for human consumption and soils) burden of *Cryptosporidium* oocysts in Mozambique and other African endemic areas.

Supplementary Materials: The following are available online at https://www.mdpi.com/article/10.3390/pathogens10040452/s1, Figure S1: Seasonal distribution and temporal clustering of *Cryptosporidium* species in children under 5 years of age, with and without diarrhoea, recruited during the Global Enteric Multicenter Study at the Manhiça district (Maputo, southern Mozambique), 2007–2012. Figure S2: Seasonal distribution and temporal clustering of *Cryptosporidium* subtype families in children under 5 years of age, with and without diarrhoea, recruited during the Global Enteric Multicenter Study at the Manhiça district (Maputo, southern Mozambique), 2007–2012. Table S1: PCR and sequencing data. Table S2: Diversity and frequency of Cryptosporidium family subtypes within C. hominis (subtype family I), C. parvum (subtype family II) and C. meleagridis (subtype family III) in asymptomatic (non-cases) children under 5 years of age according to severity of clinical manifestations, age group, and HIV coinfection. Children were recruited during the Global Enteric Multicenter Study at the Manhiça district (Maputo, southern Mozambique), 2007–2012. Figures between brackets represent relative frequencies.

Author Contributions: Conceptualisation, A.M.J., K.K., M.M.L., P.L.A., D.C. and I.M.; methodology, K.K., M.M.L., P.L.A., D.C. and I.M.; software, A.M.J. and P.C.K.; validation, D.C. and I.M.; formal analysis, A.M.J., P.C.K., M.G., D.C. and I.M.; investigation, A.M.J., P.C.K., M.G., T.N., S.M., A.C. and Q.B.; resources, K.K., M.M.L., P.L.A., D.C. and I.M.; data curation, D.C. and I.M.; writing—original draft preparation, A.M.J., D.C. and I.M.; writing—review and editing, A.M.J., P.C.K., M.G., T.N., Q.B.,

A.C., K.K., M.M.L., P.L.A., D.C. and I.M.; visualisation, A.M.J., K.K., M.M.L., P.L.A., D.C. and I.M.; supervision, D.C. and I.M.; project administration, D.C. and I.M.; funding acquisition, K.K., M.M.L., P.L.A., D.C. and I.M. All authors have read and agreed to the published version of the manuscript.

Funding: This research was funded by the Bill and Melinda Gates Foundation through the Center for Vaccine Development at the University of Maryland, School of Medicine who coordinated GEMS, grant number 38874 (GEMS) and OPP1033572 (GEMS1A). Additional funding was obtained from the Health Institute Carlos III (ISCIII), Ministry of Economy and Competitiveness (Spain), grant number PI16CIII/00024, from the Fundo Nacional de Investigacão, Ministry of Science and Technology (Mozambique), grant number 245-INV, and from the USAID Country Office of Mozambique, grant number AID-656-F-16-00002.

Institutional Review Board Statement: The study was conducted according to the guidelines of the Declaration of Helsinki and approved by the Review Board of the Mozambican National Bioethics Committee for Health, Mozambique (Ref. 11/CNBS/07), the Ethics Committee of the Hospital Clinic of Barcelona, Spain (Ref. 2006/3260) and the Institutional Review Board for Human Subject Research at University of Maryland Baltimore, USA.

Informed Consent Statement: Informed consent was obtained from all subjects involved in the study.

Data Availability Statement: All relevant data are within the paper and its Supplementary Materials.

Acknowledgments: We thank the children and their caretakers who participated in the study, as well as the clinical, field, and laboratory staff who worked tirelessly to ensure the data collection and laboratory testing was performed according to the standardized protocol. We also thank all the local government authorities (district Administration and Health Directorate) and all community leaders for supporting and collaborating in the study. CISM is supported by the Government of Mozambique and the Spanish Agency for International Development Cooperation (AECID). ISGlobal receives support from the Spanish Ministry of Science and Innovation through the "Centro de Excelencia Severo Ochoa 2019–2023" Program (CEX2018-000806-S), and support from the Generalitat de Catalunya through the CERCA Program.

Conflicts of Interest: The authors declare no conflict of interest. The funders had no role in the design of the study; in the collection, analyses, or interpretation of data; in the writing of the manuscript, or in the decision to publish the results.

References

1. GBD 2015 Mortality and Causes of Death Collaborators. Global, regional, and national life expectancy, all-cause mortality, and cause-specific mortality for 249 causes of death, 1980–2015: A systematic analysis for the Global Burden of Disease Study 2015. *Lancet* **2016**, *388*, 1459–1544. [CrossRef]
2. GBD 2015 Child Mortality Collaborators. Global, regional, national, and selected subnational levels of stillbirths, neonatal, infant, and under-5 mortality, 1980–2015: A systematic analysis for the Global Burden of Disease Study 2015. *Lancet* **2016**, *388*, 1725–1774. [CrossRef]
3. GBD Diarrhoeal Diseases Collaborators. Estimates of global, regional, and national morbidity, mortality, and aetiologies of diarrhoeal diseases: A systematic analysis for the Global Burden of Disease Study 2015. *Lancet Infect. Dis.* **2017**, *17*, 909–948. [CrossRef]
4. Kotloff, K.L.; Nataro, J.P.; Blackwelder, W.C.; Nasrin, D.; Farag, T.H.; Panchalingam, S.; Wu, Y.; Sow, S.O.; Sur, D.; Breiman, R.F.; et al. Burden and aetiology of diarrhoeal disease in infants and young children in developing countries (the Global Enteric Multicenter Study, GEMS): A prospective, case-control study. *Lancet* **2013**, *382*, 209–222. [CrossRef]
5. Kotloff, K.L.; Nasrin, D.; Blackwelder, W.C.; Wu, Y.; Farag, T.; Panchalingham, S.; Sow, S.O.; Sur, D.; Zaidi, A.K.M.; Faruque, A.S.G.; et al. The incidence, aetiology, and adverse clinical consequences of less severe diarrhoeal episodes among infants and children residing in low-income and middle-income countries: A 12-month case-control study as a follow-on to the Global Enteric Multicenter Study (GEMS). *Lancet Glob. Health* **2019**, *7*, e568–e584. [PubMed]
6. Steiner, K.L.; Ahmed, S.; Gilchrist, C.A.; Burkey, C.; Cook, H.; Ma, J.Z.; Korpe, P.S.; Ahmed, E.; Alam, M.; Kabir, M.; et al. Species of Cryptosporidia causing subclinical infection associated with growth faltering in rural and urban Bangladesh: A birth cohort study. *Clin. Infect. Dis.* **2018**, *67*, 1347–1355. [CrossRef] [PubMed]
7. Platts-Mills, J.A.; Babji, S.; Bodhidatta, L.; Gratz, J.; Haque, R.; Havt, A.; McCormick, B.J.; McGrath, M.; Olortegui, M.P.; Samie, A.; et al. MAL-ED Network Investigators. Pathogen-specific burdens of community diarrhoea in developing countries: A multisite birth cohort study (MAL-ED). *Lancet Glob. Health* **2015**, *3*, e564–e575. [CrossRef]
8. Bouzid, M.; Hunter, P.R.; Chalmers, R.M.; Tyler, K.M. *Cryptosporidium* pathogenicity and virulence. *Clin. Microbiol. Rev.* **2013**, *26*, 115–134. [CrossRef]

9. Ehmadpour, E.; Safarpour, H.; Xiao, L.; Zarean, M.; Hatam-Nahavandi, K.; Barac, A.; Picot, S.; Rahimi, M.T.; Rubino, S.; Mahami-Oskouei, M.; et al. Cryptosporidiosis in HIV-positive patients and related risk factors: A systematic review and meta-analysis. *Parasite* **2020**, *27*, 27. [CrossRef]

10. Xiao, L. Molecular epidemiology of cryptosporidiosis: An update. *Exp. Parasitol.* **2010**, *124*, 80–89. [CrossRef] [PubMed]

11. Squire, S.A.; Ryan, U. *Cryptosporidium* and *Giardia* in Africa: Current and future challenges. *Parasit. Vectors* **2017**, *10*, 195. [CrossRef] [PubMed]

12. Ryan, U.; Fayer, R.; Xiao, L. *Cryptosporidium* species in humans and animals: Current understanding and research needs. *Parasitology* **2014**, *141*, 1667–1685. [CrossRef]

13. Feng, Y.; Ryan, U.; Xiao, L. Genetic diversity and population structure of *Cryptosporidium*. *Trends Parasitol.* **2018**, *34*, 997–1011. [CrossRef]

14. Strong, W.B.; Gut, J.; Nelson, R.G. Cloning and sequence analysis of a highly polymorphic *Cryptosporidium parvum* gene encoding a 60-kilodalton glycoprotein and characterization of its 15- and 45-kilodalton zoite surface antigen products. *Infect. Immun.* **2000**, *68*, 4117–4134. [CrossRef]

15. Dgedge, M.; Novoa, A.; Macassa, G.; Sacarlal, J.; Black, J.; Michaud, C.; Cliff, J. The burden of disease in Maputo City, Mozambique: Registered and autopsied deaths in 1994. *Bull. World Health Organ.* **2001**, *79*, 546–552. [PubMed]

16. Sacarlal, J.; Nhacolo, A.Q.; Sigaúque, B.; Nhalungo, D.A.; Abacassamo, F.; Sacoor, C.N.; Aide, P.; Machevo, S.; Nhampossa, T.; Macete, E.V.; et al. A 10 year study of the cause of death in children under 15 years in Manhiça, Mozambique. *BMC Public Health* **2009**, *9*, 67. [CrossRef] [PubMed]

17. Sacoor, C.; Nhacolo, A.; Nhalungo, D.; Aponte, J.J.; Bassat, Q.; Augusto, O.; Mandomando, I.; Sacarlal, J.; Lauchande, N.; Sigaúque, B.; et al. Profile: Manhiça Health Research Centre (Manhiça HDSS). *Int. J. Epidemiol.* **2013**, *42*, 1309–1318. [CrossRef]

18. Nhampossa, T.; Mandomando, I.; Acacio, S.; Quintó, L.; Vubil, D.; Ruiz, J.; Nhalungo, D.; Sacoor, C.; Nhabanga, A.; Nhacolo, A.; et al. Diarrheal disease in rural Mozambique: Burden, risk factors and etiology of diarrheal disease among children aged 0–59 months seeking care at health facilities. *PLoS ONE* **2015**, *10*, e0119824. [CrossRef]

19. Casmo, V.; Lebbad, M.; Maungate, S.; Lindh, J. Occurrence of *Cryptosporidium* spp. and *Cystoisospora belli* among adult patients with diarrhoea in Maputo, Mozambique. *Heliyon* **2018**, *4*, 1–13. [CrossRef]

20. Irisarri-Gutiérrez, M.J.; Mingo, M.H.; de Lucio, A.; Gil, H.; Morales, L.; Seguí, R.; Nacarapa, E.; Muñoz-Antolí, C.; Bornay-Llinares, F.J.; Esteban, J.G.; et al. Association between enteric protozoan parasites and gastrointestinal illness among HIV- and tuberculosis-infected individuals in the Chowke district, southern Mozambique. *Acta Trop.* **2017**, *170*, 197–203. [CrossRef]

21. Stensvold, C.R.; Beser, J.; Axén, C.; Lebbad, M. High applicability of a novel method for *gp60*-based subtyping of *Cryptosporidium meleagridis*. *J. Clin. Microbiol.* **2014**, *52*, 2311–2319. [CrossRef] [PubMed]

22. Slapeta, J. *Cryptosporidium* species found in cattle: A proposal for a new species. *Trends Parasitol.* **2006**, *22*, 469–474. [CrossRef] [PubMed]

23. Valiathan, R.; Ashman, M.; Asthana, D. Effects of ageing on the immune system: Infants to elderly. *Scand. J. Immunol.* **2016**, *83*, 255–266. [CrossRef] [PubMed]

24. Local Burden of Disease Diarrhoea Collaborators. Mapping geographical inequalities in childhood diarrhoeal morbidity and mortality in low-income and middle-income countries, 2000–2017: Analysis for the Global Burden of Disease Study 2017. *Lancet* **2020**, *395*, 1779–1801. [CrossRef]

25. Muadica, A.S.; Balasegaram, S.; Beebeejaun, K.; Köster, P.C.; Bailo, B.; Hernández-de-Mingo, M.; Dashti, A.; Dacal, E.; Saugar, J.M.; Fuentes, I.; et al. Risk associations for intestinal parasites in symptomatic and asymptomatic schoolchildren in central Mozambique. *Clin. Microbiol. Infect.* **2020**, in press.

26. Delahoy, M.J.; Omore, R.; Ayers, T.L.; Schilling, K.A.; Blackstock, A.J.; Ochieng, J.B.; Moke, F.; Jaron, P.; Awuor, A.; Okonji, C.; et al. Clinical, environmental, and behavioral characteristics associated with *Cryptosporidium* infection among children with moderate-to-severe diarrhea in rural western Kenya, 2008-2012: The Global Enteric Multicenter Study (GEMS). *PLoS Negl. Trop. Dis.* **2018**, *12*, e0006640. [CrossRef]

27. Acácio, S.; Mandomando, I.; Nhampossa, T.; Quintó, L.; Vubil, D.; Sacoor, C.; Kotloff, K.; Farag, T.; Nasrin, D.; Macete, E.; et al. Risk factors for death among children 0–59 months of age with moderate-to-severe diarrhea in Manhiça district, southern Mozambique. *BMC Infect. Dis.* **2019**, *19*, 322. [CrossRef]

28. Tellevik, M.G.; Moyo, S.J.; Blomberg, B.; Hjøllo, T.; Maselle, S.Y.; Langeland, N.; Hanevik, K. Prevalence of *Cryptosporidium parvum*/*hominis*, *Entamoeba histolytica* and *Giardia lamblia* among young children with and without diarrhea in Dar es Salaam, Tanzania. *PLoS Negl. Trop. Dis.* **2015**, *9*, e0004125. [CrossRef]

29. Tumwine, J.K.; Kekitiinwa, A.; Nabukeera, N.; Akiyoshi, D.E.; Rich, S.M.; Widmer, G.; Feng, X.; Tzipori, S. *Cryptosporidium parvum* in children with diarrhea in Mulago Hospital, Kampala, Uganda. *Am. J. Trop. Med. Hyg.* **2003**, *68*, 710–715. [CrossRef]

30. Acácio, S.; Nhampossa, T.; Quintó, L.; Vubil, D.; Sacoor, C.; Kotloff, K.; Farag, T.; Dilruba, N.; Macete, E.; Levine, M.M.; et al. The role of HIV infection in the etiology and epidemiology of diarrheal disease among children aged 0–59 months in Manhiça District, Rural Mozambique. *Int. J. Infect. Dis.* **2018**, *73*, 10–17. [CrossRef]

31. Pavlinac, P.B.; John-Stewart, G.C.; Naulikha, J.M.; Onchiri, F.M.; Denno, D.M.; Odundo, E.A.; Singa, B.O.; Richardson, B.A.; Walson, J.L. High-risk enteric pathogens associated with HIV infection and HIV exposure in Kenyan children with acute diarrhoea. *AIDS* **2014**, *28*, 2287–2296. [CrossRef]

32. Mengist, H.M.; Taye, B.; Tsegaye, A. Intestinal parasitosis in relation to CD4+T cells levels and anemia among HAART initiated and HAART naive pediatric HIV patients in a Model ART center in Addis Ababa, Ethiopia. *PLoS ONE* **2015**, *10*, e0117715. [CrossRef]

33. Sow, S.O.; Muhsen, K.; Nasrin, D.; Blackwelder, W.C.; Wu, Y.; Farag, T.H.; Panchalingam, S.; Sur, D.; Zaidi, A.K.; Faruque, A.S.; et al. The burden of *Cryptosporidium* diarrheal disease among children <24 months of age in moderate/high mortality regions of sub-Saharan Africa and South Asia, utilizing data from the Global Enteric Multicenter Study (GEMS). *PLoS Negl. Trop. Dis.* **2016**, *10*, e0004729. [CrossRef] [PubMed]

34. Muadica, A.S.; Köster, P.C.; Dashti, A.; Bailo, B.; Hernández-de-Mingo, M.; Balasegaram, S.; Carmena, D. Molecular diversity of *Giardia duodenalis*, *Cryptosporidium* spp., and *Blastocystis* sp. in symptomatic and asymptomatic school children in Zambézia province (Mozambique). *Pathogens* **2021**, *10*, 255. [CrossRef] [PubMed]

35. Elwin, K.; Hadfield, S.J.; Robinson, G.; Crouch, N.D.; Chalmers, R.M. *Cryptosporidium viatorum* n. sp. (Apicomplexa: Cryptosporidiidae) among travellers returning to Great Britain from the Indian subcontinent, 2007–2011. *Int. J. Parasitol.* **2012**, *42*, 675–682. [CrossRef] [PubMed]

36. Koehler, A.V.; Wang, T.; Haydon, S.R.; Gasser, R.B. *Cryptosporidium viatorum* from the native Australian swamp rat *Rattus lutreolus*—An emerging zoonotic pathogen? *Int. J. Parasitol. Parasites Wildl.* **2018**, *7*, 18–26. [CrossRef] [PubMed]

37. Zhao, W.; Zhou, H.; Huang, Y.; Xu, L.; Rao, L.; Wang, S.; Wang, W.; Yi, Y.; Zhou, X.; Wu, Y.; et al. *Cryptosporidium* spp. in wild rats (*Rattus* spp.) from the Hainan Province, China: Molecular detection, species/genotype identification and implications for public health. *Int. J. Parasitol. Parasites Wildl.* **2019**, *9*, 317–321. [CrossRef]

38. Mbae, C.; Mulinge, E.; Waruru, A.; Ngugi, B.; Wainaina, J.; Kariuki, S. Genetic diversity of *Cryptosporidium* in children in an urban informal settlement of Nairobi, Kenya. *PLoS ONE* **2015**, *10*, e0142055. [CrossRef] [PubMed]

39. Akinbo, F.O.; Okaka, C.E.; Omoregie, R.; Dearen, T.; Leon, E.T.; Xiao, L. Molecular characterization of *Cryptosporidium* spp. in HIV-infected persons in Benin City, Edo State, Nigeria. *Fooyin J. Health Sci.* **2010**, *2*, 85–89. [CrossRef]

40. Lobo, M.L.; Augusto, J.; Antunes, F.; Ceita, J.; Xiao, L.H.; Codices, V.; Matos, O. *Cryptosporidium* spp., *Giardia duodenalis*, *Enterocytozoon bieneusi* and other intestinal parasites in young children in Lobata province, Democratic Republic of São Tomé and Príncipe. *PLoS ONE* **2014**, *9*, e97708. [CrossRef]

41. Abu Samra, N.; Thompson, P.N.; Jori, F.; Frean, J.; Poonsamy, B.; du Plessis, D.; Mogoye, B.; Xiao, L. Genetic characterization of *Cryptosporidium* spp. in diarrhoeic children from four provinces in South Africa. *Zoonoses Public Health* **2013**, *60*, 154–159. [CrossRef] [PubMed]

42. Molloy, S.F.; Smith, H.V.; Kirwan, P.; Nichols, R.A.; Asaolu, S.O.; Connelly, L.; Holland, C.V. Identification of a high diversity of *Cryptosporidium* species genotypes and subtypes in a pediatric population in Nigeria. *Am. J. Trop. Med. Hyg.* **2010**, *82*, 608–613. [CrossRef]

43. Wanyiri, J.W.; Kanyi, H.; Maina, S.; Wang, D.E.; Steen, A.; Ngugi, P.; Kamau, T.; Waithera, T.; O'Connor, R.; Gachuhi, K.; et al. Cryptosporidiosis in HIV/AIDS patients in Kenya: Clinical features, epidemiology, molecular characterization and antibody responses. *Am. J. Trop. Med. Hyg.* **2014**, *91*, 319–328. [CrossRef] [PubMed]

44. Parsons, M.B.; Travis, D.; Lonsdorf, E.V.; Lipende, I.; Roellig, D.M.; Collins, A.; Kamenya, S.; Zhang, H.; Xiao, L.; Gillespie, T.R. Epidemiology and molecular characterization of *Cryptosporidium* spp. in humans, wild primates, and domesticated animals in the Greater Gombe Ecosystem, Tanzania. *PLoS Negl. Trop. Dis.* **2015**, *9*, e0003529. [CrossRef] [PubMed]

45. Adamu, H.; Petros, B.; Zhang, G.; Kassa, H.; Amer, S.; Ye, J.; Feng, Y.; Xiao, L. Distribution and clinical manifestations of *Cryptosporidium* species and subtypes in HIV/AIDS patients in Ethiopia. *PLoS Negl. Trop. Dis.* **2014**, *8*, e2831. [CrossRef] [PubMed]

46. Blanco, M.A.; Montoya, A.; Iborra, A.; Fuentes, I. Identification of *Cryptosporidium* subtype isolates from HIV-seropositive patients in Equatorial Guinea. *Trans. R. Soc. Trop. Med. Hyg.* **2014**, *108*, 594–596. [CrossRef]

47. Eibach, D.; Krumkamp, R.; Al-Emran, H.M.; Sarpong, N.; Hagen, R.M.; Adu-Sarkodie, Y.; Tannich, E.; May, J. Molecular characterization of *Cryptosporidium* spp. among children in rural Ghana. *PLoS Negl. Trop. Dis.* **2015**, *9*, e0003551. [CrossRef]

48. Areeshi, M.; Dove, W.; Papaventsis, D.; Gatei, W.; Combe, P.; Grosjean, P.; Leatherbarrow, H.; Hart, C.A. *Cryptosporidium* species causing acute diarrhoea in children in Antananarivo, Madagascar. *Ann. Trop. Med. Parasitol.* **2008**, *102*, 309–315. [CrossRef]

49. Akiyoshi, D.E.; Tumwine, J.K.; Bakeera-Kitaka, S.; Tzipori, S. Subtype analysis of *Cryptosporidium* isolates from children in Uganda. *J. Parasitol.* **2006**, *92*, 1097–1100. [CrossRef]

50. Adamu, H.; Petros, B.; Hailu, A.; Petry, F. Molecular characterization of *Cryptosporidium* isolates from humans in Ethiopia. *Acta Trop.* **2010**, *115*, 77–83. [CrossRef]

51. Rahmouni, I.; Essid, R.; Aoun, K.; Bouratbine, A. Glycoprotein 60 diversity in *Cryptosporidium parvum* causing human and cattle cryptosporidiosis in the rural region of Northern Tunisia. *Am. J. Trop. Med. Hyg.* **2014**, *90*, 346–350. [CrossRef] [PubMed]

52. Essid, R.; Menotti, J.; Hanen, C.; Aoun, K.; Bouratbine, A. Genetic diversity of *Cryptosporidium* isolates from human populations in an urban area of Northern Tunisia. *Infect. Genet. Evol.* **2018**, *58*, 237–242. [CrossRef] [PubMed]

53. Grace Mphephu, M.; Deogratias Ekwanzala, M.; Ndombo Benteke Momba, M. *Cryptosporidium* species and subtypes in river water and riverbed sediment using next-generation sequencing. *Int. J. Parasitol.* **2021**, in press.

54. Limaheluw, J.; Medema, G.; Hofstra, N. An exploration of the disease burden due to *Cryptosporidium* in consumed surface water for sub-Saharan Africa. *Int. J. Hyg. Environ. Health* **2019**, *222*, 856–863. [CrossRef] [PubMed]

55. Kotloff, K.L.; Blackwelder, W.C.; Nasrin, D.; Nataro, J.P.; Farag, T.H.; van Eijk, A.; Adegbola, R.A.; Alonso, P.L.; Breiman, R.F.; Faruque, A.S.; et al. The Global Enteric Multicenter Study (GEMS) of diarrheal disease in infants and young children in developing countries: Epidemiologic and clinical methods of the case/control study. *Clin. Infect. Dis.* **2012**, *55* (Suppl. 4), S232–S245. [CrossRef]

56. Alonso, P.L.; Saúte, F.; Aponte, J.J.; Gómez-Olivé, F.X.; Nhacolo, A.; Thomson, R.; Macete, E.; Abacassamo, F.; Ventura, P.J.; Bosch, X.; et al. Manhiça DSS, Mozambique. In *Population and Health in Developing Countries*; International Development Research Centre: Otawa, ON, Canada, 2002; Volume 1, pp. 189–195.

57. Levine, M.M.; Kotloff, K.L.; Nataro, J.P.; Muhsen, K. The Global Enteric Multicenter Study (GEMS): Impetus, rationale, and genesis. *Clin. Infect. Dis.* **2012**, *55* (Suppl. 4), S215–S224. [CrossRef]

58. Panchalingam, S.; Antonio, M.; Hossain, A.; Mandomando, I.; Ochieng, B.; Oundo, J.; Ramamurthy, T.; Tamboura, B.; Zaidi, A.K.; Petri, W.; et al. Diagnostic microbiologic methods in the GEMS-1 case/control study. *Clin. Infect. Dis.* **2012**, *55* (Suppl. 4), S294–S302. [CrossRef]

59. Feltus, D.C.; Giddings, C.W.; Schneck, B.L.; Monson, T.; Warshauer, D.; McEvoy, J.M. Evidence supporting zoonotic transmission of *Cryptosporidium* spp. in Wisconsin. *J. Clin. Microbiol.* **2006**, *44*, 4303–4308. [CrossRef]

60. Tiangtip, R.; Jongwutiwes, S. Molecular analysis of *Cryptosporidium* species isolated from HIV-infected patients in Thailand. *Trop. Med. Int. Health* **2002**, *7*, 357–364. [CrossRef]

61. Chromas Lite 2.1—Software Informer. Available online: http://chromaslite.software.informer.com/2.1/ (accessed on 8 April 2021).

62. Agarwala, R.; Barrett, T.; Beck, J.; Benson, D.A.; Bollin, C.; Bolton, E.; Bourexis, D.; Brister, J.R.; Bryant, S.H.; Canese, K.; et al. Database resources of the National Center for Biotechnology Information. *Nucleic Acids Res* **2016**, *44*, D7–D19.

63. Tamura, K.; Stecher, G.; Peterson, D.; Filipski, A.; Kumar, S. MEGA6: Molecular evolutionary genetics analysis version 6.0. *Mol. Biol. Evol.* **2013**, *30*, 2725–2729. [CrossRef] [PubMed]

64. Saitou, N.; Nei, M. The neighbor-joining method: A new method for reconstructing phylogenetic trees. *Mol. Biol. Evol.* **1987**, *4*, 406–425. [PubMed]

Brief Report

High Frequency of *Cryptosporidium hominis* Infecting Infants Points to A Potential Anthroponotic Transmission in Maputo, Mozambique

Idalécia Cossa-Moiane [1,2,*], Hermínio Cossa [3], Adilson Fernando Loforte Bauhofer [1,4], Jorfélia Chilaúle [1], Esperança Lourenço Guimarães [1,4], Diocreciano Matias Bero [1], Marta Cassocera [1,4], Miguel Bambo [1], Elda Anapakala [1], Assucênio Chissaque [1,4], Júlia Sambo [1,4], Jerónimo Souzinho Langa [1], Lena Vânia Manhique-Coutinho [1], Maria Fantinatti [5], Luis António Lopes-Oliveira [5], Alda Maria Da-Cruz [5,6] and Nilsa de Deus [1,7]

[1]   Instituto Nacional de Saúde (INS), EN1, Bairro da Vila–Parcela n° 3943, Distrito de Marracuene, Maputo 264, Mozambique; adilsonbauhofer@gmail.com (A.F.L.B.); jorfeliachilaule@gmail.com (J.C.); espeguima@yahoo.com.br (E.L.G.); dmbero@gmail.com (D.M.B.); marti.life@hotmail.com (M.C.); bambomiguel@gmail.com (M.B.); elda.muianga07@gmail.com (E.A.); assucenyoo@gmail.com (A.C.); juliassiat@gmail.com (J.S.); je_langa@yahoo.com.br (J.S.L.); lenicouto1@gmail.com (L.V.M.-C.); ndeus1@yahoo.com (N.d.D.)
[2]   Institute of Tropical Medicine, 2000 Antwerp, Belgium
[3]   Centro de Investigação em Saúde de Manhiça (CISM), Unidade de Pesquisa Social, Manhiça Foundation (Fundação Manhiça, FM), Manhiça 1929, Mozambique; herminiofernando.cossa@gmail.com
[4]   Instituto de Higiene e Medicina Tropical, Universidade Nova de Lisboa, 1349-008 Lisboa, Portugal
[5]   Laboratório Interdisciplinar de Pesquisas Médicas, Instituto Oswaldo Cruz-FIOCRUZ, Rio de Janeiro 22040-360, Brazil; maria.fantinatti@gmail.com (M.F.); luizpldeoliveira@gmail.com (L.A.L.-O.); alda@ioc.fiocruz.br (A.M.D.-C.)
[6]   Disciplina de Parasitologia, Faculdade de Ciências Médicas, UERJ/RH, Rio de Janeiro 21040-900, Brazil
[7]   Departamento de Ciências Biológicas, Universidade Eduardo Mondlane, Maputo 3453, Mozambique
*   Correspondence: idalecia.moiane@ins.gov.mz; Tel.: +258-84-327-3270

Citation: Cossa-Moiane, I.; Cossa, H.; Bauhofer, A.F.L.; Chilaúle, J.; Guimarães, E.L.; Bero, D.M.; Cassocera, M.; Bambo, M.; Anapakala, E.; Chissaque, A.; et al. High Frequency of *Cryptosporidium hominis* Infecting Infants Points to A Potential Anthroponotic Transmission in Maputo, Mozambique. *Pathogens* **2021**, *10*, 293. https://doi.org/10.3390/pathogens10030293

Academic Editor: David Carmena

Received: 26 January 2021
Accepted: 13 February 2021
Published: 4 March 2021

Abstract: *Cryptosporidium* is one of the most important causes of diarrhea in children less than 2 years of age. In this study, we report the frequency, risk factors and species of *Cryptosporidium* detected by molecular diagnostic methods in children admitted to two public hospitals in Maputo City, Mozambique. We studied 319 patients under the age of five years who were admitted due to diarrhea between April 2015 and February 2016. Single stool samples were examined for the presence of *Cryptosporidium* spp. oocysts, microscopically by using a Modified Ziehl–Neelsen (mZN) staining method and by using Polymerase Chain Reaction and Restriction Fragment Length Polymorphism (PCR-RFLP) technique using 18S ribosomal RNA gene as a target. Overall, 57.7% (184/319) were males, the median age (Interquartile range, IQR) was 11.0 (7–15) months. *Cryptosporidium* spp. oocysts were detected in 11.0% (35/319) by microscopy and in 35.4% (68/192) using PCR-RFLP. The most affected age group were children older than two years, [adjusted odds ratio (aOR): 5.861; 95% confidence interval (CI): 1.532–22.417; *p*-value < 0.05]. Children with illiterate caregivers had higher risk of infection (aOR: 1.688; 95% CI: 1.001–2.845; *p*-value < 0.05). An anthroponotic species *C. hominis* was found in 93.0% (27/29) of samples. Our findings demonstrated that cryptosporidiosis in children with diarrhea might be caused by anthroponomic transmission.

Keywords: acute diarrhea; *Cryptosporidium*; children; risk factor; Mozambique

1. Introduction

Diarrhea is the one main causes of mortality among children less than 5 years old in low and middle-income countries (LMIC) [1,2]. In Mozambique, it was estimated that 11%

of pediatric diseases were due to diarrhea [3]. Rotavirus remains the main etiological agent of diarrhea in children, followed by *Cryptosporidium* spp. [1].

Cryptosporidium spp. is an apicomplexan enteric pathogenic parasite protozoan related to water and foodborne outbreaks worldwide [4–7]. It easily spreads in the environment, through soil, drinking and recreation water (swimming pool, surface waters) or even directly by person-to-person contact and contact with objects surfaces with oocysts [4]. *Cryptosporidium* is also considered an opportunistic parasite that can infect immunocompetent and immunocompromised people [8]. It can also infect wild and domestic animals [4,9,10], which facilitates spread in the environment.

Currently, more than 40 species of *Cryptosporidium* are recognized, some of which were described recently [9,10]. Even with increased efforts and improved laboratory detection, unknown *Cryptosporidium* species remain to be identified [9]. Among humans, the species most frequently associated with cryptosporidiosis are *C. hominis* (anthroponotic species) and *C. parvum* (anthroponotic and zoonotic species) [9,11,12]. Moreover, *C. parvum* is known as having the broadest range of hosts and is otherwise an important zoonotic species [9–12].

The Global Enteric Multicenter Study (GEMS) showed that *Cryptosporidium* spp. is the second most attributable pathogen moderate-to-severe diarrhea (MSD) requiring medical attention among young infants [1,13,14]. In Mozambique, several studies demonstrated the presence of this parasite in different populations and/or regions of the country. Additionally, a study in rural Mozambique indicated that *Cryptosporidium* spp. is one of the two pathogens associated with an increased risk of death in children with MSD [15]. Studies of children under 14 years with diarrhea reported different frequencies of *Cryptosporidium* spp. The highest frequency was observed in children aged 0–11 months at 20.0%, followed by 19.0% in children aged 12–23 months, 9.0% in children aged 24–59 months [16], 12.0% in children aged 0–168 months [17] detected by a commercial immunoassay method, and 3.4% to 34.0% in all individuals with diarrhea using microscopy [18,19]. A community study detected *Cryptosporidium* spp. in less than 5.0% of the study population, aged 0 to 48 months old living in poor environment sanitation [20].

Currently, due to its zoonotic and anthroponotic features, it is becoming more important to determine the molecular epidemiology of *Cryptosporidium* spp. Polymerase Chain Reaction (PCR) is a molecular-based method commonly used to study this parasite [10–12]. Additionally, this method provides information about the occurrence and distribution of *Cryptosporidium* species, contributing to a better understanding of the parasite.

Moreover, risk factors also play an important role in infection dynamics. Infection with human immunodeficiency virus (HIV) and consequently development of diarrhea is one of the risk factors that contribute to the chronic diarrhea profile [4,6]. There are few studies in Mozambique that report the risk factors in children with diarrhea and/or that used molecular tools to characterize *Cryptosporidium* [1,15,21]. Altogether, the current analyses were performed with the aim of determining the frequency, risk factors, and molecular diagnostic of *Cryptosporidium* by using a PCR Restriction Fragment Length Polymorphism (PCR-RFLP) in children hospitalized in Maputo City within the context of National Surveillance of Acute Diarrhea (ViNaDiA).

2. Results

2.1. Characteristics of the Participants

Overall, 319 children were included in the study. A single stool sample was collected from each one. Males composed 57.7% (184/319) of the group. The median age and interquartile interval (in months) were 11 (7–15), with 40.8% (130/319) of children ranging from 7 to 12 months old. Additionally, 38.9% (124/319) of caregivers reported that their children had animal contact, 58.0% (185/319) of the caregivers were literate, and 12.9% (41/319) of the children were HIV-positive (Table 1).

Table 1. Demographic and clinical characteristics of the participants enrolled for the study at Hospital Geral de Mavalane (HGM) and Hospital Geral José Macamo (HGJM), Maputo City.

Characteristics	N = 319	Frequency (%)
Provenience		
HGM	156	48.9
HGJM	163	51.1
Sex		
Female	135	42.3
Male	184	57.7
Age (in months), categorized		
0–6	63	19.7
7–12	130	40.8
13–18	81	25.4
19–24	30	9.4
25–60	15	4.7
Animal contact		
No	195	61.1
Yes	124	38.9
Caregiver literacy status		
Illiterate	133	41.7
Literate	185	58.0
Unknown/missing	1	0.3
Child HIV status		
Negative	168	52.7
Positive	41	12.9
Unknown/missing	110	34.5

N = Sample size.

2.2. Frequency of Cryptosporidium spp. Infection

During the period of the study, 319 stool samples were collected, examined, and tested for presence of *Cryptosporidium* spp. It was possible to test all 319 samples using the modified Ziehl–Neelsen (mZN) staining method and 192 samples using a PCR method. Only samples with a sufficient stool amount were included for PCR analysis, regardless of results from the staining.

Cryptosporidium spp. was detected in 11.0% (35/319) by mZN and in 35.4% (68/192) by PCR (Figure 1).

As shown in Figure 1, it was not possible to perform molecular analyses on all the 319 samples, because 127 samples had not sufficient quantity for further testing (the remaining 192 samples tested using PCR). In our PCR-based protocol, it was not possible to amplify all genetic material, because 4 positives for mZN failed to amplify.

2.3. Molecular Characterization of Cryptosporidium Species

Of the 192 samples tested by the PCR method, 86.5% (166/192) were negative through the mZN technique, and of those, we were able to recover *Cryptosporidium* DNA in 27.1% (46/166). Overall, 29 samples, including 10.4% (20/192) previously positive and 5.0% (9/192) negative by mZN staining method were successfully genotyped (Figure 1).

A molecular analysis of the 18S rRNA locus identified *C. hominis* in 93.0% (27/29), followed by *C. parvum* in 3.5% (1/29), and mixed infection with *C. hominis* and *C. parvum* in (3.5%, 1/29) (Figure 2). Due to insufficient sample amount, the remaining isolates could not be genotyped.

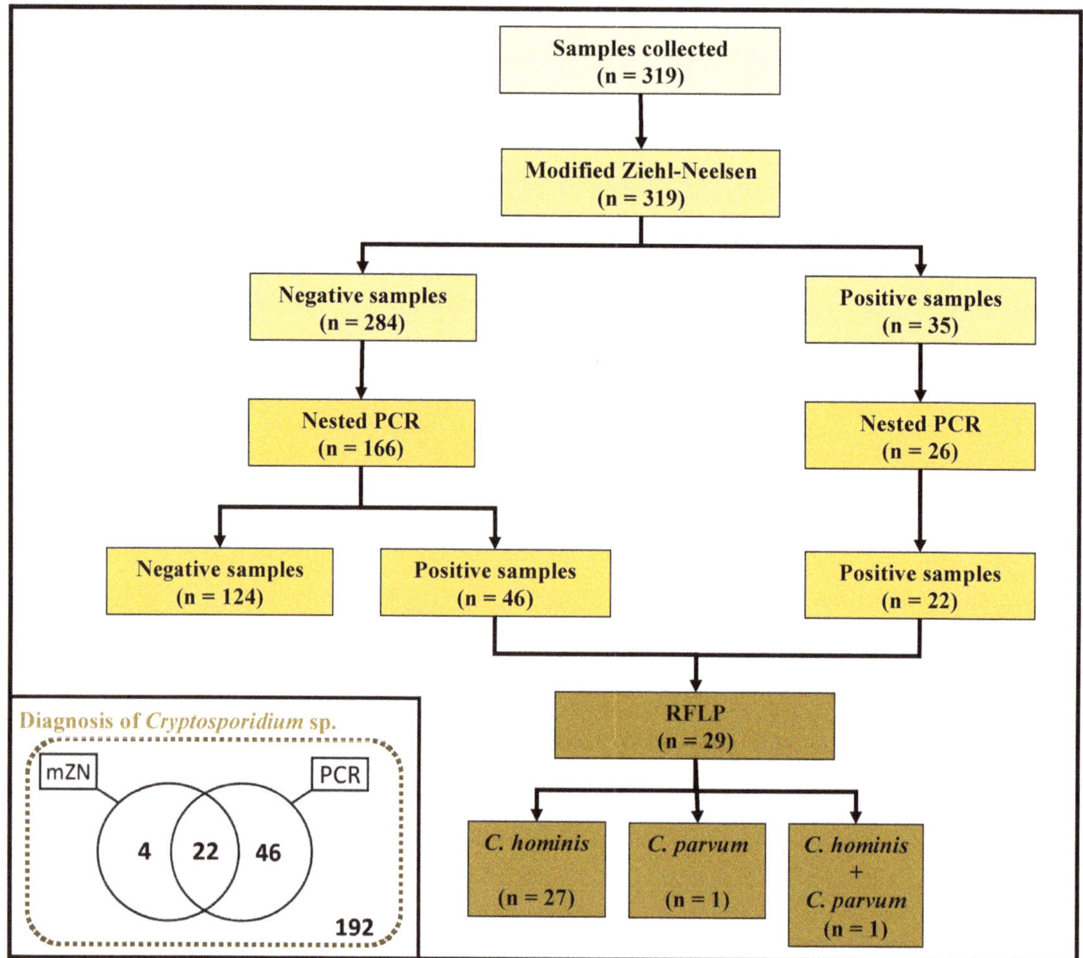

Figure 1. Experimental design. *Cryptosporidium* spp. investigation in samples by microscopy (modified Ziehl–Neelsen staining method) and molecular diagnostic (PCR). The PCR-RFLP was used to *Cryptosporidium* characterization. ZN: modified Ziehl–Neelsen; PCR: polymerase chain reaction; RFLP: restriction fragment length polymorphism.

In the restriction with SspI digestion products for *C. parvum*, *C. hominis* and the mixed infection showed an identical restriction pattern with three visible bands of 111 bp, 254 bp, and 449 bp. A different pattern was seen with AseI digestion products, where *C. parvum* had two visible bands of 104 bp and 268 bp. *C. hominis* also had two visible bands of different molecular sizes, of 104 bp and 561 bp. The mixed infection presented with three bands of 104 bp, 561 bp and 628 bp.

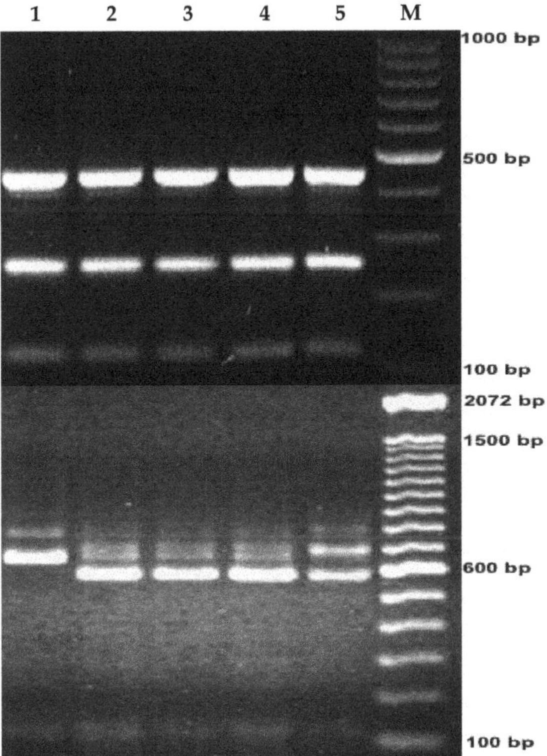

Figure 2. Genotyping of the *Cryptosporidium* parasites by PCR-RFLP targeting 18S rRNA gene. M, molecular size makers (100 bp). Lane 1: *C. parvum*; Lanes 2, 3 and 4: *C. hominis* and Lane 5: mixed infection with *C. hominis* and *C. parvum*. The upper lanes show SspI digestion products showing a molecular size from 111 bp to 449 bp, and the lower lanes show AseI digestion products with molecular size of approximately 104 bp to 628 bp.

2.4. Risk Factors for Cryptosporidium spp. Infection

Older children were more likely to be infected with *Cryptosporidium* spp. The most susceptible group was children older than two years, compared to younger than seven months (p-value < 0.05; adjusted odds ratio (aOR): 5.861, 95% CI: 1.532–22.417) (Table 2).

Children with illiterate caregivers were more likely to be infected by *Cryptosporidium* spp. than ones with literate caregivers (p-value < 0.05; aOR: 1.688, 95% CI: 1.001–2.845) (Table 2).

Being male (crude odds ratio (cOR): 1.252, 95% CI: 0.747–2.097), HIV-positive (cOR: 1.032, 95% CI: 0.466–2.289), and not having animal contact (cOR: 1.280, 95% CI: 0.756–2.166) were not related to infection by *Cryptosporidium* spp. (p-value > 0.05) in this study (Table 2).

Table 2. Demographic characteristics and animal contact information of children frequencies, crude and adjusted odds ratio for children infected by *Cryptosporidium* spp.

Characteristics	N = 319	n = 81	%	Crude OR (95% CI)	Adjusted OR (95% CI)
Provenience					
HGM	156	41	26.3	NA	NA
HGJM	163	40	24.5	NA	NA
Sex					
Female	135	31	23.0	1	
Male	184	50	27.2	1.252 (0.747–2.097)	
Age (in months), categorized					
0–6	63	6	9.5	1	1
7–12	130	37	28.5	3.780 (1.501–9.517) **	3.604 (1.426–9.112) **
13–18	81	25	30.9	4.241 (1.617–11.124) **	4.170 (1.584–10.979) **
19–24	30	7	23.3	2.891 (0.877–9.533)	2.503 (0.727–8.618)
25–60	15	6	40.0	6.333 (1.671–23.999) **	5.861 (1.532–22.417) **
Animal contact					
No	195	53	27.2	1.280 (0.756–2.166)	
Yes	124	28	22.6	1	
Caregiver literacy status					
Illiterate	133	42	31.6	1.785 (1.071–2.976) **	1.688 (1.001–2.845) *
Literate	185	38	20.5	1	1
Unknown/missing	1				
Child HIV status					
Negative	168	40	23.8	1	
Positive	41	10	24.4	1.032 (0.466–2.289)	
Unknown/missing	110				

N: Total number of samples tested; n: number of positive samples; %: percentage/relative frequency; NA: not applicable; * $p < 0.05$; ** $p < 0.01$.

3. Discussion

This study aimed to determine the frequency and risk factors for *Cryptosporidium* spp. infection in children hospitalized in Maputo City through the National Surveillance of Acute Diarrhea (ViNaDiA).

The frequency of *Cryptosporidium* spp. found in our study was higher than those in studies conducted in different areas (urban, peri-urban and/or rural) and regions from Mozambique (north, south and/or central). Those studies indicate lower frequencies of *Cryptosporidium* spp. if we consider our PCR results (35.4%): 12% in children hospitalized with diarrhea by using ELISA [17], 3.4% in children admitted in pediatric ward in one north central hospital by using mZN and rapid test [18] and 34% in children from ViNaDiA tested before the current study by using mZN, between 2013 and 2015 [19]. The differences in frequencies among these studies can be attributed to different study designs, population characteristics [20,22], diagnostic tools (for instance only microscopy, serology or PCR-based methods) and region of the country (north, south and/or central).

Children older than six months showed higher risk of infection. This trend has been reported in other African countries [23–26] and also in one of the studies conducted in Mozambique [22]. Higher frequencies in older children (>six months) may be explained by feeding practices, mobility of the child and/or age at which the child has high contact with other children when playing. In Mozambique, exclusive breastfeeding is recommended for the first six months, and then, complementary food is added to the child's diet. It was previously observed that breastfeeding has a protective effect against any protozoan infection, including *Cryptosporidium* in children from 0 to 48 months old in Maputo, Mozambique [20,27]. The introduction of potentially contaminated complementary foods and the increased mobility of the child expose them to other possible sources of infection, such as soil and animal contact, increasing the risk of infection among older children before mature immunity achieved [4].

Children from illiterate caregivers were more susceptible to infection by *Cryptosporidium* spp. Empirically, literate caregivers suggest that the household is in a higher wealth quintile, compared to those that did not go to school (illiterate). It is reasoned that caregivers who learned better hygiene practices at school are more aware of health risks and practice improved sanitary and hygiene behaviors.

In our analysis, we found no association between *Cryptosporidium* infection and HIV-status, although more than one third of the children had unknown HIV status. Conversely, in Tanzania [28] and Kenya [25], strong associations between *Cryptosporidium* infection and HIV-status have been observed, with higher frequencies of *Cryptosporidium* infection in HIV-positive children.

Animal contact was not a predictor for *Cryptosporidium* spp. infection in children. *Cryptosporidium* infection can be acquired through animal contact but can also be transmitted through an anthroponotic route. In our sample, we observed that the majority of the PCR-tested samples contained *C. hominis*. This is commonly reported in Africa, and its acquisition is related to person-to-person transmission [14,21,23,24,28,29], suggesting that animal contact plays a smaller role in infection.

The increased occurrence of *C. hominis* is corroborated with one hospital-based study of adults with diarrhea in Maputo City which used the 60-kDa glycoprotein gene (gp60) as a target [21]. Occurrence of anthroponomical transmission in children, as described in our study, suggests empiric circulation of the parasite if we consider that adults assist children. An infected adult can easily transmit the parasite to the child, and/or the child can pass it onto another adult or other children.

On the other hand, the PCR-RFLP targeting SSU rRNA (18S rRNA gene) used in this study was the first attempt to molecularly characterize *Cryptosporidium* in children with diarrhea in Mozambique. The target used is the most used among investigators, because this region is less polymorphic, presenting five copies per genome [7,10]. Although gene sequencing could enrich the findings, we did not have the technical conditions at the time of the study. As there was no prior knowledge of the molecular epidemiology of *Cryptosporidium* sp. in our country, we opted for PCR-RFLP to conduct a survey of the circulating species and genotypes.

Furthermore, we analyzed a single stool samples instead of the optimal multiple (at least 3) consecutive approach, which could result in underestimation. We also applied the PCR technique, which is a highly sensitive diagnostic approach [30,31]. Additionally, this was a hospital-based analysis, meaning that our findings can only be extrapolated to the population from the sites included. However, in four samples mZN-positive, the presence of DNA was not identified. This may have been a consequence of DNA degradation due to suboptimal temperature during transportation and storage, due to the presence of inhibitors [32] or due to a different species with a mutation in the primer's region [10].

There are few studies in Mozambique [1,19], reporting the risk factors for *Cryptosporidium* infection in the children with diarrhea and/or using molecular tools. The findings of this study should receive attention, since the high frequency of *C. hominis* in children observed may be a result of anthroponotic transmission. There is a need to expand the analysis to other provinces of the country and complement it with sequencing tools to better characterize the species in circulation. It is also worth noting that other hosts may be participating in the transmission routes, which is corroborated by the identification of *C. parvum* isolates.

4. Materials and Methods

4.1. Ethics Statements

The data used in the present analysis were provided by the ViNaDiA in children. The related protocol was approved by the Mozambique National Bioethics Committee for Health (IRB00002657, reference Nr. 348/CNBS/13). Written informed consent was obtained from children's parents or legal guardians before questionnaire administration and sample collection.

4.2. Study Design, Site and Population

Cross-sectional, hospital-based surveillance was conducted between April 2015 and February 2016 in Hospital Geral de Mavalane (HGM) and Hospital Geral José Macamo (HGJM). These hospitals were selected as sentinel sites because they receive patients from

Maputo City and surrounding areas. Both have pediatric out- and inpatient wards. The HGM and HGJM are referral hospitals for both Mavalane and José Macamo health areas and cover fourteen (14) and nine (9) health centers, respectively.

In each sentinel site, focal points (laboratory technicians, physicians and nurses) were identified and trained to screen diarrhea cases, administer the questionnaire, collect and send stool samples to *Instituto Nacional de Saúde (INS)* where the samples were processed. Children up to 60 months old who presented in the sentinel sites with acute diarrhea, defined as three or more loose or liquid stools within 24 h and less than 14 days, were included [33].

4.3. Sample Size Calculation

Minimum sample size expected was calculated using OpenEpi [34], with 95% confidence interval (CI), desired precision of 3.0% and an estimated frequency of 4.8% for *Cryptosporidium* spp. from a previous study in children aged up to five years with diarrhea [16]. We obtained a minimum sample size of at least 196.

4.4. Data Collection

Demographic, clinical and epidemiological data were assessed by interviewing children's caregivers with a semi-standardized questionnaire. Information regarding sex, age, animal contact (defined as having physical contact with an animal or their excrements) [16] and caregiver education status were collected. HIV status was self-reported by the children's caregivers and confirmed in the children's vaccination cards. If unknown, permission was asked to collect blood samples and tested according to the national testing algorithm. The children newly diagnosed as HIV-positive were followed by the physician at hospital and referred to their neighborhood health facilities for routine assistance after discharge.

4.5. Sample Collection and Management

A single stool sample from each child was collected after inclusion. In cases of liquid diarrhea, non-absorbent diapers were used instead of ordinary diapers. Samples were transferred to sterile polystyrene tubes, kept refrigerated in cooler boxes (approximately 2 °C to 8 °C), without preservative and sent to the Laboratory of Parasitology in INS, Maputo.

A smear was made from fresh stool, and an aliquot was kept in the original tube under 2 to 8 °C for concentration and subsequent microscopic examination for *Cryptosporidium* spp. oocysts. A second aliquot was placed in a vial without preservative, stored under −40 °C and was specifically intended for extracting and purifying genomic deoxyribonucleic acid (DNA) for molecular analysis. A set of previously frozen sub-samples (approximately 0.5 mL) was shipped to *Laboratório Interdisciplinar de Pesquisas Médicas, Instituto Oswaldo Cruz/Fundação Oswaldo Cruz* in Rio de Janeiro, Brazil, under dry ice for DNA extraction, detection and genetic characterization of *Cryptosporidium* species.

4.6. Laboratory Sample Processing
4.6.1. Direct Microscopy

The presence of *Cryptosporidium* oocysts in stools was first established using the modified Ziehl–Neelsen (mZN) staining method as described by the World Health Organization (WHO) [35]. Briefly, thin smears of fresh samples and concentrated pellet from formol-ether concentration technique were prepared on the same glass side and air-dried before mZN staining. Samples were read using a microscope; the results were recorded as a positive for those where oocysts of the parasite were visualized under 100X magnification. All stool samples were collected, labeled, processed and stored following the laboratory standard procedures, including Good Laboratory Practice (GLP) recommendations. For the parasitological assay, each internal quality control was made, and all sample readings were double-checked for concordance by two technicians. In case of discordant results, a third observation was required.

4.6.2. DNA Extraction

DNA was extracted from frozen stool sample using a commercial QIAmp stool Mini Kit (Qiagen, Hilden, Germany) following the manufacturer's protocol with the following modification: the lysis temperature was raised to 95 °C and the DNA was eluted in 100 μL. The pre-treating of the sample and usage of the QIAmp stool Mini Kit was done in a standard procedure as in similar studies, and with this there is no need to freeze or apply thawing cycles [32]. DNA extracts were stored at −20 °C until use.

4.6.3. Molecular Detection by Conventional Polymerase Chain Reaction (PCR)

The 18S rRNA gene was amplified for all samples (positive and negative by mZN staining) by conventional polymerase chain reaction. The forward primer 5′-AACCTGGTTGATC CTGCCAGTAGTC-3′ and reverse primer 5′-TGATCCTTCTGCAGGTTCACCTACG-3′ as described by Xiao et al. [36] were used. Briefly, the PCR contained 10X PCR buffer (MgCl$_2$) at a final concentration of 1X, 5mM MgCl$_2$, 200 mM (each) deoxynucleotides triphosphate (dNTP), 1.0 U of Taq polymerase (Invitrogen Life Technologies, São Paulo, SP-Brasil), 100 nM (each) primer (Extend, SP-Brasil) and 2.5 μL of DNA template in a total 25 μL reaction mixture. The following parameter was adjusted in our study: MgCl$_2$ concentration, Taq polymerase quantity when compared with the study that was used as reference. Each PCR had small adjustments; annealing was set to 61 °C for 45 s and extension to 72 °C for 7 min. The PCR efficacy and the identification of the *Cryptosporidium* genetic material from samples were verified through electrophoresis in agarose gel (1.2%).

Stool samples were considered positive if oocysts with typical characteristics (approximately 4 μm and 6 μm in diameter; stained bright pink within a clear halo under green field) or *Cryptosporidium* DNA with expected base pair (bp) were detected by conventional PCR.

4.6.4. Characterization of *Cryptosporidium* spp. Isolates by Nested PCR and Restriction Fragment Polymorphism (RFLP) Analysis

All samples that were previously positive for 18S rRNA gene amplification were genotyped by a PCR-RFLP technique using genomic DNA as template. Firstly, a PCR product of approximately 1325 bp of the SSU rRNA gene was amplified by a nested PCR using the following primers 5′-TTCTAGAGCTAATACATGCG-3′ and 5′-CCCTAATCCTTCGAAACAG GA-3′ [11,12]. The first PCR contained 1X PCR buffer (MgCl$_2$), 6 mM of MgCl$_2$, 2.5 U of Taq polymerase, 200 μM of each primer concentration (500 nM each) and 2 μL of DNA template in a total volume of the reaction mixture (50 μL). The initial denaturation was 94 °C for 3 min, followed by the amplification performed in 35 total cycles: 94 °C for 45 s for denaturation, 58 °C for 45 s for annealing and 72 °C for 60 s for extension. The final extension was 72 °C for 7 min. For secondary PCR, for a product of 826 to 864 bp, it was done by using the following primers 5′-GGAAGGGTTGTATTTATTAGATAAAG-3′ and 5′-AAGGAGTAAGGAACAACCTC CA-3′ [11]. A total volume of the reaction mixture (50 μL), Taq DNA amount (1 U) and the primary PCR product was optimized (0.5 μL diluted at 1:20 or 1 μL non-diluted product). Amplification condition for secondary PCR was as follows: 94 °C for 45 s, 59 °C for 30 s and 72 °C for 45 s in 25 cycles with an initial hot start at 94 °C for 3 min and a final extension at 72 °C for 7 min. The PCR efficacy and the identification of the *Cryptosporidium* genetic material from samples were verified through electrophoresis in agarose gel (1.2%).

Genotype identification was made through the analysis of pattern of the secondary product after restriction digestion with the enzymes SspI and AseI (New England BioLabs Inc., Beverly, MA, USA) as described by Xiao et al. [11]. The AseI enzyme has the same digestion function and pattern as the enzyme VspI. Each set of experiments included a negative PCR control (laboratory-grade distilled water). For the restriction, 20 μL of the second product of the nested PCR, 10 U of SspI or AseI, 5 μL pf specific enzyme buffer was digested in a 50 μL of the reaction by 37 °C for one hour under conditions recommended by the supplier manufacturer. Aliquots of amplified and digested fragments were separated and visualized under Ultra-Violet light translucent (Bio-Rad, Milan, Italy) after separation in 1.5% to 2% agarose gel (Invitrogen, Aukland, New Zealand) by electrophoresis stained with 3X GelRed (Biotium, San Francisco, CA, USA). The expected band for each enzyme varied according to the species detected were the most observed highlighted in bold (Table 3).

Table 3. Expected band of the restriction digestion using SspI and AseI enzymes.

Specie	Band Expected (in bp)	
	SspI Digestion	**AseI Digestion**
C. hominis	11, 12,111, 254, 449	70, 102, 104, 561
C. parvum	11, 12, 108, 254, 449	102, 104, 628

4.7. Data Management and Statistical Procedures

Data were double entered in Epi Info 3.5.1 (CDC, Atlanta, 2008, Atlanta, GA, USA) to minimize entry errors, followed by data comparison and inconsistencies resolution. Data were analyzed using IBM SPSS software (Statistical Package for the Social Science, Armok, NY: IBM Corp, 2011, version 26.0, Chicago, IL, USA). Categorical variables were summarized as frequencies and continuous variables were summarized as medians and Inter-quartile Range (IQR). Contingency tables were built between dependent and independent variables. Crude and adjusted odds ratio were estimated through simple and multiple logistic regression models. Independent variables with p-values ≤ 0.2 in the simple logistic regression were included in the multiple logistic regression model in order to obtain adjusted odds ratio. Goodness-of-fit was assesses using the Hosmer and Lemeshow test for the multiple logistic regression model. A p-value less than 0.05 was considered evidence of statistical significance and all analyses were performed considering a 95% confidence interval (95% CI).

5. Conclusions

This study showed a high frequency of *Cryptosporidium* spp. infection detected by PCR-RFLP among children admitted to two public hospitals in Maputo City due to acute diarrhea. Child age (25 to 60 months) and caregiver literacy status were predictors for infection. Our findings demonstrated that the infection is mostly due to *C. hominis*. This suggests mainly anthroponomic transmission, with great public health implications. The present study demonstrated the need for improved public health recommendations by including *Cryptosporidium* spp. in routine testing among children with diarrhea in Mozambique. Moreover, the study presented the need for an establishment of more accurate molecular characterization platform of the parasite in a national context, in order to understand which species occur in the country, routes of infection (if anthroponotic and/or zoonotic) and/or regional patterns, informing public health programs.

Author Contributions: Conceptualization, N.d.D. and I.C.-M.; methodology, I.C.-M. and H.C.; validation, I.C.-M., H.C., M.F., L.A.L.-O. and A.M.D.-C.; formal analysis, I.C.-M., H.C. and A.F.L.B.; Investigation, H.C.; resources, N.d.D., M.F., L.A.L.-O. and A.M.D.-C.; data curation, H.C.; writing—original draft preparation, H.C.; writing—review and editing, H.C., I.C.-M. and A.F.L.B.; visualization, M.C., M.B., E.L.G., E.A., J.S., J.S.L., D.M.B., L.V.M.-C. and A.C.; supervision, N.d.D., A.M.D.-C. and I.C.-M.; project administration, J.C.; funding acquisition, N.d.D. All authors have read and agreed to the published version of the manuscript.

Funding: The current analysis was performed as part of the requisites for a master's degree and was supported in part by FIOCRUZ/IOC/RJ-Brazil and Instituto Nacional de Saúde—Moçambique post-graduation program. M.F. received funding from Universal/CNPq at and FAPERJ (Grant E-26/202.078/2020) and fellowship from CAPES (Brasil Sem Miséria/Brazilian governmental program), CNPq (PDJ), INOVA Program (FIOCRUZ), and FAPERJ. The National Surveillance for Acute Diarrhea in Children project was supported by a Senior Fellowship awarded to N.D. by the European Foundation Initiative for African Research into Neglected Tropical Diseases (EFINTD, Grant number: 89539) and World Health Organization (WHO).

Institutional Review Board Statement: The protocol was approved by the Mozambique National Bioethics Committee for Health (IRB00002657, reference Nr. 348/CNBS/13).

Informed Consent Statement: Informed consent was obtained from all subjects involved in the study.

Data Availability Statement: The data presented in this study are available on request from the corresponding author. The data are not publicly available due to the ethical reasons.

Acknowledgments: We would like to thank the caregivers who consented to their children being enrolled in the surveillance. We wish to thank the Sequencing Service DNA Sequencing Platform–Fundação Oswaldo Cruz for their contribution in the study procedures, Direcção Clínica de Pediatria do Hospital Geral de Mavalane and Hospital Geral José Macamo, and Carlos Guilamba and Celina Nhamuave for their amazing work on data entry. Finally, we would like to thank all the focal points in the hospital who helped conduct this study and the members of the surveillance team. We would also like to thank Kristen Heitzinger and Neha Kamat who reviewed the language in the manuscript.

Conflicts of Interest: The authors declare no conflict of interest.

References

1. Kotloff, K.L.; Nataro, J.P.; Blackwelder, W.C.; Nasrin, D.; Farag, T.H.; Panchalingam, S.; Wu, Y.; Sow, S.O.; Sur, D.; Breiman, R.F.; et al. Burden and Aetiology of Diarrhoeal Disease in Infants and Young Children in Developing Countries (the Global Enteric Multicenter Study, GEMS): A Prospective, Case-Control Study. *Lancet Lond. Engl.* **2013**, *382*, 209–222. [CrossRef]
2. GBD 2016 Causes of Death Collaborators. Global, Regional, and National Age-Sex Specific Mortality for 264 Causes of Death, 1980-2016: A Systematic Analysis for the Global Burden of Disease Study 2016. *Lancet Lond. Engl.* **2017**, *390*, 1151–1210. [CrossRef]
3. Ministério da Saúde (MISAU); Instituto Nacional de Estatística (INE); ICF International (ICFI). *Inquérito Demográfico e de Saúde (IDS)*; MISAU: Calverton, MD, USA, 2011.
4. Rey, L. *Bases Da Parasitologia Médica*, 3rd ed.; Guanabara Koogan: Rio de Janeiro, Brazil, 2010.
5. Chalmers, R.M.; Davies, A.P. Minireview: Clinical Cryptosporidiosis. *Exp. Parasitol.* **2010**, *124*, 138–146. [CrossRef] [PubMed]
6. Tzipori, S.; Ward, H. Cryptosporidiosis: Biology, Pathogenesis and Disease. *Microbes Infect.* **2002**, *4*, 1047–1058. [CrossRef]
7. Widmer, G.; Carmena, D.; Kváč, M.; Chalmers, R.M.; Kissinger, J.C.; Xiao, L.; Sateriale, A.; Striepen, B.; Laurent, F.; Lacroix-Lamandé, S.; et al. Update on Cryptosporidium Spp.: Highlights from the Seventh International Giardia and Cryptosporidium Conference. *Parasite Paris Fr.* **2020**, *27*, 14. [CrossRef]
8. Peletz, R.; Mahin, T.; Elliott, M.; Montgomery, M.; Clasen, T. Preventing Cryptosporidiosis: The Need for Safe Drinking Water. *Bull. World Health Organ.* **2013**, *91*, 238A. [CrossRef] [PubMed]
9. Feng, Y.; Ryan, U.M.; Xiao, L. Genetic Diversity and Population Structure of Cryptosporidium. *Trends Parasitol.* **2018**, *34*, 997–1011. [CrossRef]
10. Cunha, F.S.; Peralta, R.H.S.; Peralta, J.M. New Insights into the Detection and Molecular Characterization of Cryptosporidium with Emphasis in Brazilian Studies: A Review. *Rev. Inst. Med. Trop. Sao Paulo* **2019**, *61*, e28. [CrossRef]
11. Xiao, L.; Morgan, U.M.; Limor, J.; Escalante, A.; Arrowood, M.; Shulaw, W.; Thompson, R.C.; Fayer, R.; Lal, A.A. Genetic Diversity within Cryptosporidium Parvum and Related Cryptosporidium Species. *Appl. Environ. Microbiol.* **1999**, *65*, 3386–3391. [CrossRef]
12. Xiao, L.; Bern, C.; Limor, J.; Sulaiman, I.; Roberts, J.; Checkley, W.; Cabrera, L.; Gilman, R.H.; Lal, A.A. Identification of 5 Types of Cryptosporidium Parasites in Children in Lima, Peru. *J. Infect. Dis.* **2001**, *183*, 492–497. [CrossRef]
13. Kotloff, K.L. The Burden and Etiology of Diarrheal Illness in Developing Countries. *Pediatr. Clin. N. Am.* **2017**, *64*, 799–814. [CrossRef] [PubMed]
14. Sow, S.O.; Muhsen, K.; Nasrin, D.; Blackwelder, W.C.; Wu, Y.; Farag, T.H.; Panchalingam, S.; Sur, D.; Zaidi, A.K.M.; Faruque, A.S.G.; et al. The Burden of Cryptosporidium Diarrheal Disease among Children <24 Months of Age in Moderate/High Mortality Regions of Sub-Saharan Africa and South Asia, Utilizing Data from the Global Enteric Multicenter Study (GEMS). *PLoS Negl. Trop. Dis.* **2016**, *10*, e0004729. [CrossRef] [PubMed]
15. Acácio, S.; Mandomando, I.; Nhampossa, T.; Quintó, L.; Vubil, D.; Sacoor, C.; Kotloff, K.; Farag, T.; Nasrin, D.; Macete, E.; et al. Risk Factors for Death among Children 0-59 Months of Age with Moderate-to-Severe Diarrhea in Manhiça District, Southern Mozambique. *BMC Infect. Dis.* **2019**, *19*, 322. [CrossRef]

16. Nhampossa, T.; Mandomando, I.; Acacio, S.; Quintó, L.; Vubil, D.; Ruiz, J.; Nhalungo, D.; Sacoor, C.; Nhabanga, A.; Nhacolo, A.; et al. Diarrheal Disease in Rural Mozambique: Burden, Risk Factors and Etiology of Diarrheal Disease among Children Aged 0-59 Months Seeking Care at Health Facilities. *PLoS ONE* **2015**, *10*, e0119824. [CrossRef]

17. Bauhofer, A.F.L.; Cossa-Moiane, I.; Marques, S.; Guimarães, E.L.; Munlela, B.; Anapakala, E.; Chilaúle, J.J.; Cassocera, M.; Langa, J.S.; Chissaque, A.; et al. Intestinal Protozoan Infections among Children 0-168 Months with Diarrhea in Mozambique: June 2014–January 2018. *PLoS Negl. Trop. Dis.* **2020**, *14*, e0008195. [CrossRef]

18. Ferreira, F.S.; Pereira, F.d.L.M.; Martins, M.d.R.O. Intestinal Parasitic Infections in Children under Five in the Central Hospital of Nampula, Northern Mozambique. *J. Infect. Dev. Ctries.* **2020**, *14*, 532–539. [CrossRef] [PubMed]

19. Cossa-Moiane, I.; Chilaúle, J.J.; Cossa Herminio, H.; Cassocera, M.; Guimarães, E.L.; de Deus, N. Parasitic Infections in Children Presenting with Acute Diarrhea in Mozambique: National Surveillance Data (2013–2015). *Int. J. Infect. Dis.* **2016**, *45*, 356. [CrossRef]

20. Knee, J.; Sumner, T.; Adriano, Z.; Berendes, D.; de Bruijn, E.; Schmidt, W.-P.; Nalá, R.; Cumming, O.; Brown, J. Risk Factors for Childhood Enteric Infection in Urban Maputo, Mozambique: A Cross-Sectional Study. *PLoS Negl. Trop. Dis.* **2018**, *12*, e0006956. [CrossRef] [PubMed]

21. Casmo, V.; Lebbad, M.; Maungate, S.; Lindh, J. Occurrence of Cryptosporidium Spp. and Cystoisospora Belli among Adult Patients with Diarrhoea in Maputo, Mozambique. *Heliyon* **2018**, *4*, e00769. [CrossRef]

22. Muadica, A.S.; Balasegaram, S.; Beebeejaun, K.; Köster, P.C.; Bailo, B.; Hernández-de-Mingo, M.; Dashti, A.; Dacal, E.; Saugar, J.M.; Fuentes, I.; et al. Risk Associations for Intestinal Parasites in Symptomatic and Asymptomatic Schoolchildren in Central Mozambique. *Clin. Microbiol. Infect.* **2020**. [CrossRef]

23. Gatei, W.; Wamae, C.N.; Mbae, C.; Waruru, A.; Mulinge, E.; Waithera, T.; Gatika, S.M.; Kamwati, S.K.; Revathi, G.; Hart, C.A. Cryptosporidiosis: Prevalence, Genotype Analysis, and Symptoms Associated with Infections in Children in Kenya. *Am. J. Trop. Med. Hyg.* **2006**, *75*, 78–82. [CrossRef] [PubMed]

24. Irisarri-Gutiérrez, M.J.; Mingo, M.H.; de Lucio, A.; Gil, H.; Morales, L.; Seguí, R.; Nacarapa, E.; Muñoz-Antolí, C.; Bornay-Llinares, F.J.; Esteban, J.G.; et al. Association between Enteric Protozoan Parasites and Gastrointestinal Illness among HIV- and Tuberculosis-Infected Individuals in the Chowke District, Southern Mozambique. *Acta Trop.* **2017**, *170*, 197–203. [CrossRef]

25. Mbae, C.K.; Nokes, D.J.; Mulinge, E.; Nyambura, J.; Waruru, A.; Kariuki, S. Intestinal Parasitic Infections in Children Presenting with Diarrhoea in Outpatient and Inpatient Settings in an Informal Settlement of Nairobi, Kenya. *BMC Infect. Dis.* **2013**, *13*, 243. [CrossRef]

26. Osman, M.; El Safadi, D.; Cian, A.; Benamrouz, S.; Nourrisson, C.; Poirier, P.; Pereira, B.; Razakandrainibe, R.; Pinon, A.; Lambert, C.; et al. Prevalence and Risk Factors for Intestinal Protozoan Infections with Cryptosporidium, Giardia, Blastocystis and Dientamoeba among Schoolchildren in Tripoli, Lebanon. *PLoS Negl. Trop. Dis.* **2016**, *10*, e0004496. [CrossRef]

27. World Health Organization (WHO). Breastfeeding Support: Close to Mothers. Available online: https://www.afro.who.int/pt/news/breastfeeding-support-close-mothers (accessed on 26 November 2020).

28. Tellevik, M.G.; Moyo, S.J.; Blomberg, B.; Hjøllo, T.; Maselle, S.Y.; Langeland, N.; Hanevik, K. Prevalence of Cryptosporidium Parvum/Hominis, Entamoeba Histolytica and Giardia Lamblia among Young Children with and without Diarrhea in Dar Es Salaam, Tanzania. *PLoS Negl. Trop. Dis.* **2015**, *9*, e0004125. [CrossRef] [PubMed]

29. Checkley, W.; White, A.C.; Jaganath, D.; Arrowood, M.J.; Chalmers, R.M.; Chen, X.-M.; Fayer, R.; Griffiths, J.K.; Guerrant, R.L.; Hedstrom, L.; et al. A Review of the Global Burden, Novel Diagnostics, Therapeutics, and Vaccine Targets for Cryptosporidium. *Lancet Infect. Dis.* **2015**, *15*, 85–94. [CrossRef]

30. Cacciò, S.M.; Pozio, E. Advances in the Epidemiology, Diagnosis and Treatment of Cryptosporidiosis. *Expert Rev. Anti. Infect. Ther.* **2006**, *4*, 429–443. [CrossRef] [PubMed]

31. Fonseca, A.M.; Fernandes, N.; Ferreira, F.S.; Gomes, J.; Centeno-Lima, S. Intestinal Parasites in Children Hospitalized at the Central Hospital in Maputo, Mozambique. *J. Infect. Dev. Ctries.* **2014**, *8*, 786–789. [CrossRef]

32. Paulos, S.; Mateo, M.; de Lucio, A.; Hernández-de Mingo, M.; Bailo, B.; Saugar, J.M.; Cardona, G.A.; Fuentes, I.; Mateo, M.; Carmena, D. Evaluation of Five Commercial Methods for the Extraction and Purification of DNA from Human Faecal Samples for Downstream Molecular Detection of the Enteric Protozoan Parasites *Cryptosporidium* Spp., Giardia Duodenalis, and *Entamoeba* Spp. *J. Microbiol. Methods* **2016**, *127*, 68–73. [CrossRef]

33. World Health Organizatio (WHO). Diarrhoeal Disease. Available online: http://www.who.int/news-room/fact-sheets/detail/diarrhoeal-disease (accessed on 2 December 2020).

34. OpenEpi-Toolkit Shell for Developing New Applications. Available online: https://www.openepi.com/SampleSize/SSPropor.htm (accessed on 10 May 2020).

35. World Health Organizatio (WHO). *Manual of BasicTecnhiques for Health Laboratory*, 2nd ed.; World Health Organizatio: Geneva, Switzerland, 2003.

36. Xiao, L.; Escalante, L.; Yang, C.; Sulaiman, I.; Escalante, A.A.; Montali, R.J.; Fayer, R.; Lal, A.A. Phylogenetic Analysis of Cryptosporidium Parasites Based on the Small-Subunit RRNA Gene Locus. *Appl. Environ. Microbiol.* **1999**, *65*, 1578–1583. [CrossRef]

Article

Multilocus Genotyping of *Giardia duodenalis* in Mostly Asymptomatic Indigenous People from the Tapirapé Tribe, Brazilian Amazon

Pamela Carolina Köster [1,†], Antonio F. Malheiros [2,3,†], Jeffrey J. Shaw [4], Sooria Balasegaram [5], Alexander Prendergast [6], Héloïse Lucaccioni [7], Luciana Melhorança Moreira [3], Larissa M. S. Lemos [8], Alejandro Dashti [1], Begoña Bailo [1], Arlei Marcili [9,10], Herbert Sousa Soares [9], Solange Maria Gennari [9,10], Rafael Calero-Bernal [11], David González-Barrio [1] and David Carmena [1,*]

[1] Parasitology Reference and Research Laboratory, Spanish National Centre for Microbiology, Majadahonda, 28220 Madrid, Spain; pamelakster@yahoo.com (P.C.K.); dashti.alejandro@gmail.com (A.D.); BEGOBB@isciii.es (B.B.); david.gonzalezb@isciii.es (D.G.-B.)

[2] Post-Graduation Program in Environmental Science, Faculty of Agricultural and Biological Sciences, University of State of Mato Grosso, Cáceres, MG 78200-000, Brazil; malheiros@unemat.br

[3] Department of Biological Sciences, Faculty of Agricultural and Biological Sciences, University of State of Mato Grosso, Cáceres, MG 78200-000, Brazil; luciana.melhoranca@unemat.br

[4] Department of Parasitology, Institute of Biomedical Sciences, São Paulo University, São Paulo, SP 05508-000, Brazil; jayusp@hotmail.com

[5] Field Epidemiology Services, National Infection Service, Public Health England, London SE1 8UG, UK; Sooria.Balasegaram@phe.gov.uk

[6] Independent Researcher, Croydon CR0, UK; akprendergast4@gmail.com

[7] European Program for Intervention Epidemiology Training (EPIET), European Centre for Disease Prevention and Control (ECDC), 16973 Stockholm, Sweden; heloiselucaccioni@dgs.min-saude.pt

[8] Department of Nursing, Faculty of Health Sciences, University of State of Mato Grosso, Cáceres, MG 78200-000, Brazil; larissa@unemat.br

[9] Post-Graduation Program in Veterinary Medicine, Santo Amaro University, São Paulo, SP 04829-300, Brazil; amarcili@usp.br (A.M.); hesoares@prof.unisa.br (H.S.S.); sgennari@usp.br (S.M.G.)

[10] Department of Preventive Veterinary Medicine and Animal Health, School of Veterinary Medicine, University of São Paulo, São Paulo, SP 05508-270, Brazil

[11] SALUVET Research Group, Department of Animal Health, Faculty of Veterinary, Complutense University of Madrid, 28040 Madrid, Spain; r.calero@ucm.es

*   Correspondence: dacarmena@isciii.es; Tel.: +34-91-822-3641

†   These two authors contributed equally to this work.

Citation: Köster, P.C.; Malheiros, A.F.; Shaw, J.J.; Balasegaram, S.; Prendergast, A.; Lucaccioni, H.; Moreira, L.M.; Lemos, L.M.S.; Dashti, A.; Bailo, B.; et al. Multilocus Genotyping of *Giardia duodenalis* in Mostly Asymptomatic Indigenous People from the Tapirapé Tribe, Brazilian Amazon. *Pathogens* **2021**, *10*, 206. https://doi.org/10.3390/pathogens10020206

Academic Editor: Siddhartha Das

Received: 14 December 2020
Accepted: 8 February 2021
Published: 14 February 2021

Publisher's Note: MDPI stays neutral with regard to jurisdictional claims in published maps and institutional affiliations.

Abstract: Little information is available on the occurrence and genetic variability of the diarrhoea-causing enteric protozoan parasite *Giardia duodenalis* in indigenous communities in Brazil. This cross-sectional epidemiological survey describes the frequency, genotypes, and risk associations for this pathogen in Tapirapé people (Brazilian Amazon) at four sampling campaigns during 2008–2009. Microscopy was used as a screening test, and molecular (PCR and Sanger sequencing) assays targeting the small subunit ribosomal RNA, the glutamate dehydrogenase, the beta-giardin, and the triosephosphate isomerase genes as confirmatory/genotyping methods. Associations between *G. duodenalis* and sociodemographic and clinical variables were investigated using Chi-squared test and univariable/multivariable logistic regression models. Overall, 574 individuals belonging to six tribes participated in the study, with *G. duodenalis* prevalence rates varying from 13.5–21.7%. The infection was positively linked to younger age and tribe. Infected children <15 years old reported more frequent gastrointestinal symptoms compared to adults. Assemblage B accounted for three out of four *G. duodenalis* infections and showed a high genetic diversity. No association between assemblage and age or occurrence of diarrhoea was demonstrated. These data indicate that the most likely source of infection was anthropic and that different pathways (e.g., drinking water) may be involved in the transmission of the parasite.

Keywords: *Giardia*; Brazil; Amazon; asymptomatic; community; genotyping; indigenous; risk association; Tapirapé; transmission

1. Introduction

The flagellated *Giardia duodenalis* (syn. *G. intestinalis*, *G. lamblia*) is a cosmopolitan protozoan parasite that inhabits the gastrointestinal tract of humans and other vertebrate animals. Giardiasis is the most reported intestinal protozoan infection globally, with an estimated 280 million symptomatic cases every year [1]. Asymptomatic infections are even more frequent, both in developing [2,3] and developed [4] countries. Indeed, large epidemiological case-control studies conducted in high-prevalence settings have demonstrated that *G. duodenalis* infection was significantly more common in asymptomatic controls than in cases with diarrhoea [5–7]. Host immune status and level of nutrition seem to be key factors in the control of the infection or its progression to active disease [8], although the genotype of the parasite may also play a role in the health/disease balance of the host [9]. When present, clinical manifestations associated with *G. duodenalis* infection may include self-limiting acute diarrhoea, persistent diarrhoea, epigastric pain, nausea, and vomiting [10]. Long-term sequelae, including childhood growth retardation and cognitive impairment, have also been recognised [11,12]. Contrary to severe infections by other diarrhoea-causing protozoan parasites such as *Cryptosporidium* spp. or *Entamoeba histolytica*, giardiasis is rarely fatal and is better considered as a debilitating condition.

Transmission of *G. duodenalis* is through the faecal-oral route, either directly via direct contact with infected humans or animals, or indirectly via ingestion of contaminated food or water. Waterborne transmission is likely the most common source of human infections in poor-resource settings with little or no access to safe drinking water and insufficient sanitary facilities [3]. Because of its strong bond with poverty and elevated socioeconomic impact, giardiasis (together with cryptosporidiosis) joined the Neglected Diseases Initiative launched by the World Health Organisation in 2004 [13].

Giardia duodenalis exhibits a considerable degree of genetic heterogeneity, allowing the differentiation of eight (A–H) lineages or assemblages with marked differences in host specificity and range [14]. These genetic variants likely represent cryptic species [15]. Assemblages A and B cause most human infections, but they can also infect other mammalian hosts and are, therefore, considered potentially zoonotic. Assemblages C and D occur mainly in canids, assemblage E in domestic and wild ungulates, assemblage F in cats, assemblage G in rodents, and assemblage H in marine pinnipeds. Human infections by assemblages C–F have been sporadically reported, particularly in children and immunocompromised individuals [14].

A recent review on the epidemiological situation of *G. duodenalis* in Brazil has revealed that this protozoan parasite represents a public health concern in the country, with prevalence rates up to 78% in Minas Gerais State and 70% in São Paulo State in 1998 [16]. Available molecular data in the country have evidenced marked differences in the geographical segregation of *G. duodenalis* assemblages circulating in human populations (Table S1), domestic and wildlife animal species (Table S2), surface waters (Table S3), and fresh produce (Table S4), likely reflecting disparities in infection sources and transmission pathways. Indeed, contaminated surface waters and having contact with domestic (mainly dog) animals were considered as probable sources of human infections [16]. Despite this relative abundance of epidemiological data, giardiasis has been poorly studied in Brazilian indigenous people, partially due to the geographical isolation and difficulty in accessing these fragile communities. Thus, *G. duodenalis* infections have been documented by conventional (microscopy) methods in the range of 7–47% in the Parakanã indigenous people in the eastern Amazon region [17], in indigenous communities in the municipality of São Gabriel da Cachoeira, Amazonas State [18], in native Brazilian children in the Xingu Indian Reservation, Mato Grosso State [19], in the Maxakali and Xukuru-Kariri indigenous communities, Minas Gerais State [20,21], and in the Terena indigenous people, Mato Grosso do Sul State [22]. However, no information is currently available on the *G. duodenalis* assemblages and sub-assemblages circulating in native Brazilian people. This molecular-based epidemiological survey aims at investigating the genetic diversity of *G. duodenalis* and assessing potential risk and/or protective factors associated with the infection in indigenous people from the Tapirapé tribe living in the Brazilian Amazon.

2. Results

2.1. Study Population

In this study, a total of 574 individuals (male/female ratio: 0.96; age range: 0.1–88 years old, median: 14.0 years old) of the Tapirapé ethnicity living in six independent tribes (population range: 40–263 inhabitants, standard deviation: 83.3) were censed and invited to participate in four consecutive sampling campaigns during July 2008 and January 2010, in both dry and wet seasons. Overall, 98% (564/574) of the censed individuals participated at least in one sampling campaign. Participation rates ranged from 40% to 93% depending on the tribe and sampling campaign (Table 1). A total of 141 individuals participated in all four sampling campaigns, 201 in three sampling campaigns, 136 in two sampling campaigns, and 86 in a single sampling campaign. The distribution of the participating individuals according to sex, age group, and tribe of origin is also summarised in Table 1. Females (mean: 53.6%, SD: 1.0) participated in the survey more often than males (mean: 46.4%, SD: 1.0). Adults (>15 years old) were the largest group in the surveyed population (mean: 24.6%, SD: 3.2), with children 5 to 14 years of age (mean: 39.4%, SD: 1.9) and children ≤5 years old accounting, in average, for 14.6% (SD: 2.0) of the investigated individuals.

Table 1. Participation rates and distribution by sex, age group, and tribe of origin of the Tapirapé people (*n* = 564) taking part in the four sampling campaigns conducted in the present survey, Brazilian Amazon.

Sampling Campaign	Participation		Sex		Age group (years)					Tribe *n* (%)					
	Participants (*n*)	Rate (%)	Male	Female	0–4	5–9	10–14	≥15	Unknown	1	2	3	4	5	6
2008 [1]	362	64	171	191	45	73	74	170	0	175 (66.5)	31 (40.3)	46 (65.7)	47 (63.5)	28 (56.0)	35 (87.5)
2009 [2]	374	66	175	199	53	75	66	178	2	174 (66.2)	45 (58.4)	31 (44.3)	41 (55.4)	46 (92.0)	37 (92.5)
2009 [1]	407	72	189	218	59	90	64	192	2	202 (76.8)	49 (63.6)	36 (51.4)	42 (56.8)	45 (90.0)	33 (82.5)
2010 [2]	382	68	172	210	66	94	64	156	2	193 (73.4)	51 (66.2)	38 (54.3)	35 (47.3)	39 (78.0)	26 (65.0)

[1] Dry season. [2] Rainy season.

2.2. Prevalence of G. duodenalis

Microscopy-based prevalence rates for *G. duodenalis* in the Tapirapé community varied from 13.5% (55/407) in the dry season of 2009 to 21.7% (83/382) in the rainy season of 2010 (Table 2). Over the four sampling campaigns, 35.1% (198/564) individuals tested positive at least once. The occurrence of the parasite was influenced by the seasonality (22% rainy versus 17% dry season, Chi-squared test p = 0.022) but not the sampling period (year) (Chi-squared test p = 0.126). Subsequent newly diagnosed infections were also more likely to occur in the rainy season (odds ratio: 2.29, 95% CI 1.46–3.68, p = 0.0001) although this is dependent on the number of samples analysed. *G. duodenalis* infections were more commonly identified in children aged 0-4 years old. During the period of study, *G. duodenalis* prevalence varied greatly within and among the six tribes investigated, but tribe 5 presented the highest infection rates in all sampling campaigns.

A total of 43, 17, and 4 individuals tested positive for *G. duodenalis* in two, three, or all four sampling campaigns, respectively (Table 3), although this was dependent on the number of samples. In all cases, children younger than 15 years of age accounted for 50.0% to 88.2% of the subjects where the parasite was detected in two or more sampling campaigns. When considering observations with repeated samples, 61.7% observations were always negative, 4.2% always positive, and 34.1% discontinuously positive. Repeated *G. duodenalis* infections were more frequently detected in the wet season (odds ratio: 1.60, 95% CI 1.12–2.29, p = 0.0075) in members of tribe 1 (range: 50.0–58.1%) and, to a lesser extent, in tribe 5 (range: 18.6–50.0%).

Table 2. Microscopy-based prevalence of *Giardia duodenalis* by sex, age group, and tribe of origin of the Tapirapé people (*n* = 564) participating in the present survey according to the sampling period, Brazilian Amazon.

	2008 [1]		2009 [2]		2009 [1]		2010 [2]	
Variable	*Giardia* **Positive**	**%**	*Giardia* **Positive**	**%**	*Giardia* **Positive**	**%**	*Giardia* **Positive**	**%**
Sex								
Male	31	19	35	20	30	16	39	23
Female	37	18	46	23	25	12	43	21
Age group (years)								
0–4	13	29	17	32	17	29	28	42
5–9	17	23	20	27	15	17	25	27
10–14	16	22	14	21	9	14	11	17
≥15	22	13	30	17	14	7	19	12
Tribe								
1	41	23	34	20	30	15	48	25
2	2	7	10	22	7	14	6	12
3	4	9	3	10	2	6	6	16
4	4	9	2	5	1	2	5	14
5	10	36	21	46	13	29	14	36
6	7	20	11	30	2	6	4	15
Total	68	19	81	22	55	14	83	22

[1] Dry season. [2] Rainy season.

Table 3. Number of individuals with a positive result to *Giardia duodenalis* by microscopy in two or more of the sampling campaigns conducted in the present study, Brazilian Amazon.

Variable	Positives in 2 Campaigns (*n*)	Frequency (%)	Positives in 3 Campaigns (*n*)	Frequency (%)	Positives in all 4 Campaigns (*n*)	Frequency (%)
Age group (years)						
0–4	14	33	6	35	1	25
5–9	12	28	6	35	1	25
10–14	9	21	3	18	0	0
≥15	8	19	2	12	2	50
Total	43	100	17	100	4	100
Tribe						
1	25	58	7	41	2	50
2	4	9	1	6	0	0
3	1	2	0	0	0	0
4	1	2	0	0	0	0
5	8	19	7	41	2	50
6	4	9	2	12	0	0
Total	43	100	17	100	4	100

2.3. Molecular Characterisation of G. duodenalis

The genetic diversity within *G. duodenalis* was investigated in a subset of 70 stool samples from 65 individuals with a positive result for this parasite by conventional microscopy. Five individuals provided stool samples positive to this parasite at two different sampling periods. The presence of the parasite was confirmed by qPCR in 97% (68/70) of these samples. Generated cycle threshold (Ct) values ranged from 18.2 to 35.4 (median: 27.4; SD: 3.7).

Genotyping/sub-genotyping data were obtained for a total of 63 stool samples belonging to 58 individuals (Table 4). Amplification success rates were 100% (63/63) for glutamate dehydrogenase (*gdh*), and 87% (55/63) for beta-giardin (*bg*) and triosephosphate isomerase (*tpi*), respectively. Multilocus sequence typing (MLST) data at the three loci were obtained from 83% (52/63) of the samples. Sequence analyses revealed the presence of assemblages

A (25%; 16/63) and B (68%; 43/63). Mixed infections A + B were identified in 6% (4/63) of the samples analysed. No mixed infections involving host-specific assemblages C–H were detected.

Table 4. Multilocus genotyping results of the 63 *G. duodenalis*-positive samples of human origin successfully genotyped at any of the three loci investigated in the present survey.

Sample ID	Ct value in qPCR	*gdh*	*bg*	*tpi*	Assigned Genotype
5	27.4	BIII	B	BIII	BIII
13	25.1	AII	AII + AIII	AII	AII + AIII
24	27.2	AII	AII + AIII	AII	AII + AIII
31	26.4	BIII/BIV	B	BIII/BIV	BIII/BIV
33	28.2	BIII/BIV	B	BIII/BIV	BIII/BIV
41	24.0	BIII	B	BIII/BIV	BIII/BIV
49	25.5	BIII/BIV	B	BIII	BIII/BIV
50a	21.2	AII	AIII	AII	AII/AIII
50b	23.8	BIII	AIII + B	BIII	AIII + BIII
55	27.3	BIII	B	BIV	BIII/BIV
58	29.6	BIII	B	BIII/BIV	BIII/BIV
60	29.1	BIII	B	BIII/BIV	BIII/BIV
70	34.0	BIII	-	-	BIII
71	30.0	BIII	B	BIII/BIV	BIII/BIV
72	21.7	BIII	B	BIII/BIV	BIII/BIV
78	18.2	BIII	B	BIII	BIII
79	29.4	BIII/BIV	B	BIII	BIII/BIV
82	30.5	BIII	B	AII + BIII	AII + BIII
85	29.7	BIII/BIV	B	BIII	BIII/BIV
86a	27.0	BIII	B	BIII/BIV	BIII/BIV
86b	26.3	BIII	B	BIII/BIV	BIII/BIV
93a	30.4	BIII/BIV	B	BIII/BIV	BIII/BIV
93b	24.9	BIII/BIV	B	BIII/BIV	BIII/BIV
94	22.0	AII	-	AII	AII
106	21.3	AII	AIII	AII	AII/AIII
111	26.2	AII	AIII	AII	AII/AIII
123	34.9	BIII	-	-	BIII
128	30.5	AII	AIII	AII	AII/AIII
131	23.5	BIII/BIV	B	BIII/BIV	BIII/BIV
132	29.5	BIII	B	BIII	BIII
149	23.6	BIII	B	-	BIII
157	28.2	BIII/BIV	B	BIII	BIII/BIV
168	24.0	BIII/BIV	B	BIII	BIII/BIV
172	28.3	BIII/BIV	B	BIII	BIII/BIV
176b	29.3	BIII	-	-	BIII
179	27.5	AII	B	AII	AII + B
180	23.3	BIII/BIV	B	BIII/BIV	BIII/BIV
182	27.5	BIII/BIV	B	BIII	BIII/BIV
184	31.3	AII	-	AII	AII
186	26.3	BIII/BIV	B	BIII	BIII/BIV
198	22.6	BIII/BIV	B	BIII	BIII/BIV
200	25.2	BIII/BIV	B	BIII	BIII/BIV
216	25.8	BIII	B	-	BIII
227	19.5	AII	AII + AIII	AII	AII + AIII
242	28.3	BIII/BIV	B	BIII	BIII/BIV
245	22.4	BIII	B	BIII	BIII
269a	31.2	BIII	-	BIII	BIII
269b	25.1	AII	AII	AII	AII
282a	27.3	BIII/BIV	B	BIII	BIII/BIV
282b	23.4	BIII/BIV	B	BIII	BIII/BIV
328	32.0	AII	-	-	AII
330	30.3	BIII/BIV	B	-	BIII/BIV
353	25.3	BIII	B	BIII	BIII
374	21.2	BIII/BIV	B	BIII	BIII/BIV
379	23.7	AII	AII + AIII	AII	AII + AIII
420	19.2	AII	AII + AIII	AII	AII + AIII
570	24.7	BIII/BIV	B	BIII	BIII/BIV
572	28.3	AII	AII + B	AII	AII + B
585	24.9	BIII	B	BIII	BIII
595	25.1	AII	AIII	-	AII/AIII
596	35.4	AII	-	-	AII
604	21.7	BIII	B	BIII/BIV	BIII/BIV
607	28.1	BIII	B	BIII	BIII

bg, beta-giardin; Ct, cycle threshold; *gdh*, glutamate dehydrogenase; qPCR: real-time polymerase chain reaction; *tpi*, triosephosphate isomerase.

Subtyping analyses revealed that sub-assemblage AII (8%, 5/63), mixed AII + AIII infections (8%, 5/63), and ambiguous AII/AIII results (8%, 5/63) were equally distributed within assemblage A. No isolates were identified as sub-assemblage AIII. Within assemblage B, most (50%, 31/63) of the sequences corresponded to ambiguous BIII/BIV results.

BIII was identified in 21% (13/63) of the sequences, whereas no isolates belonging to BIV were detected. Out of the four A + B mixed infections detected, one (2%, 1/63) involved sub-assemblages AII + BIII, one (2%, 1/63) sub-assemblages AIII + BIII, and two (3%, 2/63) sub-assemblage AII + B (unknown sub-assemblage). Out of the five individuals with giardiasis at two consecutive sampling periods, three of them (ID: 86, 93, and 282) were infected by BIII/BIV at both sampling periods, indicative of prolonged *G. duodenalis* infection, or re-infection by that very same genotype. In contrast, one individual (ID: 50) was first infected by AII/AIII, and by AIII + B at the following sampling campaign. The remaining individual (ID: 269) was first infected by BIII and then by AII at the following sampling campaign. Both cases were strongly suggestive of re-infection events by different genotypes of the parasite.

2.4. Intra-Assemblage Genetic Diversity

Tables 5–7, Table S5 show the genetic diversity of the *gdh*, *bg*, and *tpi* representative, partial sequences generated in the present study. These Tables provide information for each sequence including stretch, single nucleotide polymorphisms (SNPs), and GenBank accession number. Assemblage/sub-assemblage assignment was conducted by direct comparison of the sequencing results obtained at the three loci investigated. Sequences presenting double peak positions that could not be unequivocally assigned to a given assemblage/sub-assemblage were reported as ambiguous sequences.

A total of 63 sequences were successfully characterised at the *gdh* locus (Table 5). All 17 assemblage A sequences were unequivocally identified as sub-assemblage AII. Of them, seven sequences were 100% identical to reference sequence L40510. The remaining 10 sequences differed by 1–6 SNPs from L40510. BIII sequences showed a high degree of genetic diversity among them, explaining that 21/24 of the sequences assigned to this sub-assemblage corresponded to distinct genotypes (genetic variants) of the parasite. These sequences differed by 4–13 SNPs from reference sequence AF069059, most of them associated with ambiguous (double peak) positions. Similarly, most (20/22) sequences identified as ambiguous BIII/BIV sequences were different among them, differing by 9–17 SNPs from reference sequence L40508. Virtually all SNPs detected in BIII/IV sequences corresponded to double peaks at single nucleotide positions.

At the *bg* locus, a total of 55 sequences were fully characterised (Table 6). Out of the 14 assemblage A sequences, two belonged to AII and five to AIII. All AII and AIII sequences were identical to reference sequences AY072723 and AY072724, respectively. Five sequences were considered mixed AII + AIII infections based on the presence of two double peak (C415Y and T423Y) positions and taking sequence AY072723 as reference. Two additional sequences corresponded to AII + B and AIII + B mixed infections, differing by 32 and 38 SNPs from reference sequence AY072727, respectively. Except one, all the detected SNPs corresponded to clear double peak positions. Compared to the *gdh* locus, a lower (but still substantial) degree of genetic variability was observed within the 41 sequences assigned to assemblage B at the *bg* locus. All of them differed by 1–6 SNPs from reference sequence AY072727. A genetic variant showing two transitional mutations at positions C165T and A183G was the genotype most frequently detected.

A total of 55 sequences were fully characterised at the *tpi* locus (Table 7). Within assemblage A, 14 sequences were assigned to the sub-assemblage AII. Of them, eight had 100% homology with reference sequence U57897, whereas the remaining six sequences differed by 1–2 SNPs from the latter. Two additional sequences were identified as AII + BIII sequences and presented 94–95 SNPs when aligned with reference sequence U57897. Out of the 25 sequences assigned to BIII, seven showed 100% identity with reference sequence AF069561. The remaining 18 sequences grouped in 16 distinct genotypes that differed by 1–6 SNPs from reference sequence A AF069561. Only a single sequence was confirmed ad BIV, differing by 3 SNPs with reference sequence AF069560. Finally, virtually all (12/13) sequences with a BIII/BIV ambiguous result were different among them, differing from reference sequence AF069560 by 7–11 SNPs. As in the case of the BIII/BIV sequences identified at the *gdh* locus, most of the SNPs identified at the *tpi* locus were associated with ambiguous nucleotide positions.

Table 5. Diversity, frequency, and main molecular features of *G. duodenalis* sequences at the *gdh* locus generated in the present study. GenBank accession numbers are provided.

Assemblage	Sub-Assemblage	No. of Isolates	Reference Sequence	Stretch	Single Nucleotide Polymorphisms	GenBank ID
A	AII	7	L40510	64–491	None	MT542718
		1	L40510	66–486	T88C, C128T, C179T, T385C, A410G, T418A	MT542719
		1	L40510	64–491	T132Y, C179T, A224M, G332R	MT542720
		1	L40510	64–491	C150Y, C179T	MT542721
		6	L40510	64–491	C179T	MT542722
		1	L40510	68–491	C179T, T355Y	MT542723
		1	L40510	64–491	C179Y	MT542724
B	BIII	1	AF069059	28–455	C39T, C87Y, C99Y, C123Y, T147Y, G150R, C204Y, C309Y	MT542725
		1	AF069059	30–455	C39T, T219Y, C309T, C351Y	MT542726
		1	AF069059	28–455	C39T, C87Y, C135Y, T147Y, G150R, C204Y, C309T, C336Y	MT542727
		1	AF069059	28–455	C39Y, C99Y, T147Y, G150R, C204Y, C360Y, C387Y, G402R	MT542728
		1	AF069059	28–455	C39T, C309T, G372A, C375T	MT542729
		2	AF069059	40–455	C87Y, C99Y, C123Y, T147Y, G150R, C204Y, C309T	MT542730
		1	AF069059	40–455	C87Y, C99Y, C123Y, T147Y, G150R, C204Y, C309T, G402R	MT542731
		1	AF069059	47–455	C87Y, C99Y, T147Y, G150R, C204Y, T237Y, C309Y, G354R, G372R, C375Y	MT542732
		1	AF069059	47–455	C87Y, C132Y, T147Y, G150R, C174Y, C204Y, T237Y, C309Y, A324R, G354R, G372R, C375Y, G402R	MT542733
		2	AF069059	49–455	C87Y, C132Y, T147Y, G150R, C174Y, C204Y, C309Y, A324R, G372R, C375Y, G402R	MT542734
		1	AF069059	41–460	C87T, C132T, T147C, G150A, C174T, C204T, A324G, G402A	MT542735
		1	AF069059	40–455	C87T, C132Y, T147C, G150A, C174T, C204T, A324R, G402A	MT542736
		1	AF069059	44–443	C87T, C132T, T147C, G150A, C174Y, C204T, G402A, G406A	MT542737
		1	AF069059	46–454	C87Y, C135Y, T147Y, G150R, T209Y, C309T, C336Y	MT542738
		1	AF069059	40–455	C87Y, T147Y, G150R, T237Y, C309Y, G354R, C360Y, C387Y, G402R	MT542739
		2	AF069059	40–455	C87Y, C204Y, C309T, G372R, C375Y	MT542740
		1	AF069059	48–455	C99Y, C135Y, T147Y, G150R, C204Y, C309Y	MT542741
		1	AF069059	82–443	C99T, T147C, G150A, C204T	MT542742
		1	AF069059	44–389	C99T, T147C, G150A, T308Y	MT542743
		1	AF069059	46–455	C87Y, C99Y, C123Y, C132Y, T147Y, G150R, C174Y, C204Y, C309Y, A324R, G402R	MT542744
		1	AF069059	40–455	C87T, T147C, G150A, C309T	MT542745
	BIII/BIV	1	L40508	76–491	G84R, C159Y, T183Y, G186R, C255Y, C273Y, C345Y, T366Y, T387C, G408R, A438G	MT542746
		1	L40508	85–491	C123Y, T135Y, C159Y, T183Y, G186R, C240Y, C255Y, C273Y, C345Y, T366Y, T387Y, A438R	MT542747
		2	L40508	76–491	C123Y, T135Y, C168Y, G180R, T183Y, G186R, C210Y, C240Y, C255Y, C273Y, A360R, T366Y, T387Y	MT542748
		1	L40508	76–491	C123Y, T135Y, C168Y, T183Y, G186R, C210Y, C240Y, C255Y, C273Y, C291Y, C345Y, A360R, T366Y, C372Y, T387Y, C423Y, A438R	MT542749
		1	L40508	76–491	C123Y, T135Y, C168Y, T183Y, G186R, C210Y, C240Y, C255Y, C273Y, C345Y, A360R, T366Y, C372Y, T387Y, C423Y, A438R	MT542750
		1	L40508	101–491	C123Y, T135Y, C168Y, T183Y, G186R, C210Y, C240Y, C255Y, C273Y, C345Y, T366C, T387Y, A438R	MT542751
		2	L40508	77–491	C123Y, T135Y, C171Y, G180R, C240Y, C255Y, C273Y, C345Y, T366Y, T387Y, C423Y, A438R	MT542752
		1	L40508	76–491	C123Y, T135Y, T183Y, C240Y, C255Y, C273Y, C291Y, C345Y, T366Y, C372Y, T387Y, G408R, C411Y, A438R	MT542753
		1	L40508	76–491	C123Y, T135Y, T183Y, C240Y, C255Y, C273Y, C345Y, T366Y, T387Y, G408R, C411Y, A438R	MT542754
		1	L40508	76–491	C123Y, T135Y, C240Y, C255Y, C273Y, C291Y, C345T, T366C, C372Y, T387Y, A438G	MT542755
		1	L40508	76–491	T135Y, C159Y, T183Y, G186R, C240Y, C255Y, C273Y, C345Y, T366Y, T387Y, G408R, C423Y, A438R	MT542756
		1	L40508	84–491	T135Y, C159Y, T183Y, G186R, C240Y, C255Y, C273Y, C345Y, T366Y, T387Y, C423Y, A438R	MT542757
		1	L40508	82–491	T135Y, C159Y, T183Y, G186R, C255Y, C273Y, C345Y, T366Y, T387C, G408R, C411Y, A438R	MT542758
		1	L40508	76–491	T135Y, C159Y, T183Y, G186R, C255Y, C273Y, C345Y, T366Y, T387Y, C396Y, C423Y, A438R	MT542759
		1	L40508	84–491	T135Y, C177Y, T183Y, C255Y, C273Y, C345Y, T366Y, T387Y, G390R, G408R, C411Y, A438R	MT542760
		1	L40508	76–491	T135Y, T183Y, C255Y, C273Y, T366Y, T387C, G408R, A438R	MT542761
		1	L40508	76–491	C159Y, T183Y, G186R, C255Y, C273Y, C345Y, T366Y, T387Y, A438R	MT542762

Table 5. *Cont.*

Assemblage	Sub-Assemblage	No. of Isolates	Reference Sequence	Stretch	Single Nucleotide Polymorphisms	GenBank ID
		1	L40508	76–491	C159Y, T183C, G186R, C255Y, C273Y, C345Y, T366Y, T387Y, C423Y, A438R	MT542763
		1	L40508	76–491	C159Y, T183Y, G186R, C255Y, C273Y, C345Y, T366Y, T387Y, A438R, A476R	MT542764
		1	L40508	76–491	T183C, G186A, C240Y, C255Y, C273Y, T366Y, T387C, A438R	MT542765

M, A/C; R, A/G; Y, C/T.

Table 6. Diversity, frequency, and main molecular features of *G. duodenalis* sequences at the *bg* locus generated in the present study. GenBank accession numbers are provided.

Assemblage	Sub-Assemblage	No. of Isolates	Reference Sequence	Stretch	Single Nucleotide Polymorphisms	GenBank ID
A	AII	2	AY072723	97–719	None	MT542766
	AIII	5	AY072724	1–753	None	MT542767
	AII + AIII	5	AY072723	1–729	C415Y, T423Y	MT542768
	AIII + B	1	AY072727	98–604	T132Y, G159R, A171R, A183R, A228R, T240Y, G258R, C288Y, T306Y, C309Y, T312K, G327R, C339Y, G345R, C348Y, C372Y, A387R, C390Y, C415Y, A432R, C438Y, C453M, G456R, T471Y, A489R, T519Y, C522R, C558Y, C561Y, C564Y, G576R, C579S	MT542769
	AII + B	1	AY072727	96–603	T132Y, G159R, C165Y, A171R, A183R, A228R, T240Y, G258R, C288Y, T306Y, C309T, T312K, G318R, G327R, C339Y, G345R, C348Y, C372Y, A387R, C390Y, C415Y, A432R, C438Y, C453M, G456R, T471Y, A489R, T519Y, C522S, C558Y, C561Y, C564Y, G576R, C579S	MT542770
B	–	10	AY072727	1–729	C165T, A183G	MT542771
		3	AY072727	93–753	C165Y, A183R	MT542772
		1	AY072727	97–593	C165Y, A183R, C204M, C309Y, C366Y, T519Y	MT542773
		1	AY072727	96–592	C165T, A183G, A272R, C406Y	MT542774
		1	AY072727	93–709	C165Y, A183R, C309Y	MT542775
		4	AY072727	97–753	C165Y, A183R, C309Y, T519Y	MT542776
		1	AY072727	93–600	C165Y, A183R, C309Y, T519Y, C543Y	MT542777
		2	AY072727	93–753	C165T, A183R, C309Y, C543Y	MT542778
		2	AY072727	97–753	C165Y, A183R, T519Y	MT542779
		1	AY072727	93–604	C165Y, C309Y, G354R, T519C	MT542780
		1	AY072727	97–719	C165Y, C309Y, G354R, T519Y, C543Y	MT542781
		1	AY072727	93–753	C165Y, C309T, G354R, T519Y, C543Y	MT542782
		1	AY072727	97–601	C165Y, C309T, T519Y	MT542783
		1	AY072727	93–711	C165Y, C309T, T519Y, C543Y	MT542784
		1	AY072727	97–701	C165Y, C309Y, C543Y	MT542785
		1	AY072727	97–719	C165Y, C309T, C543Y	MT542786
		1	AY072727	97–753	G180R, C309Y, T519C	MT542787
		1	AY072727	103–604	A183R	MT542788
		1	AY072727	98–714	A183R, C204M, T519Y	MT542789
		1	AY072727	97–706	C309Y	MT542790
		1	AY072727	102–604	C309T, T519C	MT542791
		1	AY072727	93–753	C309T, T519Y	MT542792
		1	AY072727	97–719	T519C	MT542793
		2	AY072727	1–753	T519Y	MT542794

K, G/T; M, A/C; R, A/G; Y, C/T.

Table 7. Diversity, frequency, and main molecular features of *G. duodenalis* sequences at the *tpi* locus generated in the present study. GenBank accession numbers are provided.

Assemblage	Sub-Assemblage	No. of Isolates	Reference Sequence	Stretch	Single Nucleotide Polymorphisms	GenBank ID
A	AII	8	U57897	294–805	None	MT542795
		5	U57897	275–805	C287G	MT542796
		1	U57897	275–805	C287G, A381M	MT542797
	AII + BIII	1	U57897	289–754	94 SNPs	MT542798
		1	U57897	313–720	95 SNPs	MT542799
B	BIII	7	AF069561	1–456	None	MT542800
		1	AF069561	1–456	A8R, C108T, C111Y, C201Y	MT542801
		1	AF069561	1–456	A8R, C108Y, C201Y	MT542802
		2	AF069561	1–456	A10R	MT542803
		1	AF069561	1–456	A10R, C104Y, A372R	MT542804
		1	AF069561	1–456	A10R, C108Y, C111Y	MT542805
		1	AF069561	1–456	C15Y, C33Y, C34Y, G105R, C111Y, G254R	MT542806
		1	AF069561	1–456	G21A, C34T, C108T	MT542807
		1	AF069561	26–456	C34Y, G105R, T314Y	MT542808
		1	AF069561	1–456	G48A, G391A	MT542809
		1	AF069561	17–456	G105R, C108Y, C201Y, G391R,	MT542810
		1	AF069561	1–456	C108Y	MT542811
		1	AF069561	1–456	C108Y, C111Y	MT542812
		1	AF069561	38–456	C108Y, C111Y, C201Y	MT542813
		1	AF069561	1–456	C108Y, C201Y	MT542814
		1	AF069561	1–456	C108Y, G258R	MT542815
		2	AF069561	1–456	G198R, G207R	MT542816
	BIV	1	AF069560	1–479	A176G, A395G, C470T	MT542817
	BIII/BIV	1	AF069560	1–479	A5R, A33R, T57Y, C112Y, T131Y, T134Y, A176G, G281R, T314Y, A395G, C470Y	MT542818
		1	AF069560	1–479	A5R, A33R, T57Y, T131Y, T134Y, A176G, G281R, A395G, C470Y	MT542819
		1	AF069560	1–479	A5R, T57Y, T131Y, T134Y, A176G, A181R, A395G, C470T	MT542820
		1	AF069560	1–479	A5R, T57Y, T131Y, T134Y, A176G, G281R, A395G, A421M, C470Y	MT542821
		1	AF069560	1–479	A5R, A33R, T57Y, G128R, T131Y, T134Y, A176G, A395G, C470Y	MT542822
		2	AF069560	1–479	A5R, T57Y, G128R, T131Y, T134Y, A176G, A395G, C470Y	MT542823
		1	AF069560	1–479	A5R, T57Y, G128R, T131Y, T134Y, A176G, A395G	MT542824
		1	AF069560	1–479	A5R, T57Y, T131Y, T134Y, A176G, C237Y, A395G, C470Y,	MT542825
		1	AF069560	1–479	A5R, G44R, T57Y, C127Y, T131Y, T134C, A176G, A395R	MT542826
		1	AF069560	1–479	T57Y, T131Y, T134Y, A176G, G221R, G230R, A395G, C470Y	MT542827
		1	AF069560	44–479	T57Y, T131Y, T134Y, A176G, T317Y, A395G, C470Y	MT542828
		1	AF069560	1–479	T57Y, T131Y, T134Y, A176G, A395G, A437R, A449R, C470Y, G476R	MT542829

M, A/C; R, A/G; S, G/C; Y, C/T.

Figure 1 shows the phylogenetic tree obtained for the *gdh* gene by maximum parsimony and Bayesian methods. All *G. duodenalis* sequences clustered together (monophyletic groups) with different well-supported clades (100% of bootstrap and 1.0 posterior probability). Two major branches were formed and included all (A–F) *G. duodenalis* assemblages. The sequences of indigenous people from the Brazilian Amazon clustered in branches for assemblage A (97% of bootstrap and 1.0 posterior probability) and B (100% of bootstrap and 1.0 posterior probability). In assemblage B, the sequences obtained in this study clustered with sub-assemblages BIII and BIV reference strains. Similar phylogenetic trees for the *bg* and *tpi* sequences generated in the present study are shown in Figures S1 and S2, respectively.

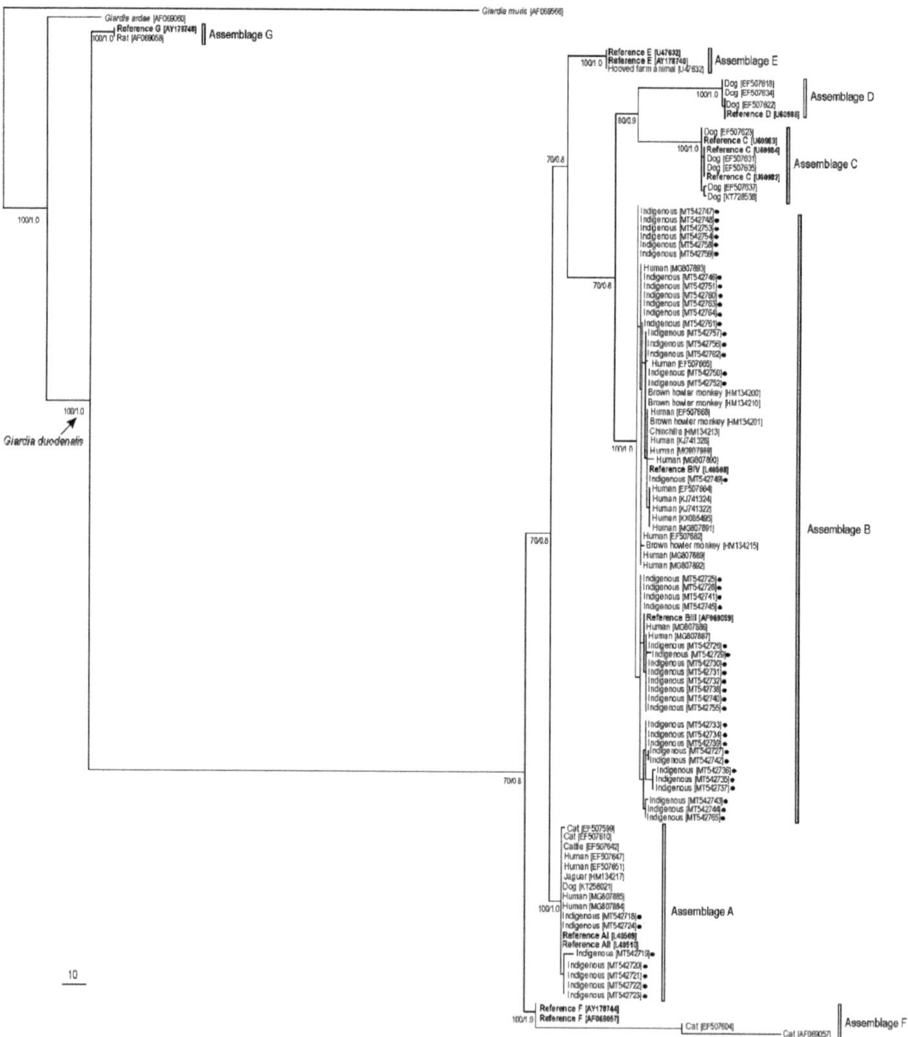

Figure 1. Maximum parsimony phylogenetic tree based on *gdh* sequences of *G. duodenalis*. Numbers on nodes indicate the bootstrap/posterior probability values. Black filled circles represent sequences generated in the present study. GenBank accession numbers for all sequences used for the phylogenetic analysis were embedded in the tree. *Giardia muris* was used as the outgroup.

2.5. Intra-Assemblage B Genetic Diversity Analysis

Genetic diversity was far higher within assemblage B than within assemblage A sequences regardless of the molecular marker used. Multiple sequence alignments of BIII, BIV, and ambiguous BIII/BIV sequences at the *gdh*, *bg*, and *tpi* loci revealed the presence of SNPs in multiple sites across used reference sequences, varying from 11 (for B sequences at the *bg* locus) to 32 (for BIII/BIV sequences at the *tpi* locus) sites (Table S6). Overall, 611 SNPs were identified among assemblage B sequences in all three loci. Of them, 16.9% (103/611) corresponded to single point mutations, and 83.1% (508/611) to double peaks. Defined positions (hotspots) at each investigated locus tended to accumulate the bulk of these SNPs (66.9%; 409/611). Within *gdh*, C87, T147, G150, C204, C309, and G402 were the main hotspots for BIII sequences (reference sequence: AF069059), and C123, T135, T183, G186, C255, C273, C345, T366, T387, and A438 for BIII/BIV sequences (reference sequence: L40508). Within *tpi*, C108 was the only hotspot for BIII sequences (reference sequence: AF069561), and A5, T57, T131, T134, A176, A395, and C470 for BIII/BIV sequences (reference sequence: AF069560). Within *bg*, the main hotspots identified for B sequences were C165, A183, C309, and T519 (reference sequence: AY072727).

The distribution of single point mutations and double peaks differed substantially among sub-assemblages and loci. At the *gdh* locus, hotspot sites accumulated 57.7% of all SNPs detected in BIII sequences, but this figure increased to 72.9% in BIII/BIV sequences. Double peaks accounted for 37.8% of the SNPs detected in BIII sequences, but for 67.8% of the ambiguous BIII/BIV sequences (Figure 2A). At the *tpi* locus, hotspot sites accumulated 18.2% of all SNPs detected in BIII sequences, but this figure increased to 54.5% in BIII/BIV sequences (Figure 2B). Finally, at the *bg* locus, hotspot sites clustered 78.4% of the SNPs detected in assemblage B sequences, of which 58.1% corresponded to double peaks (Figure 2C).

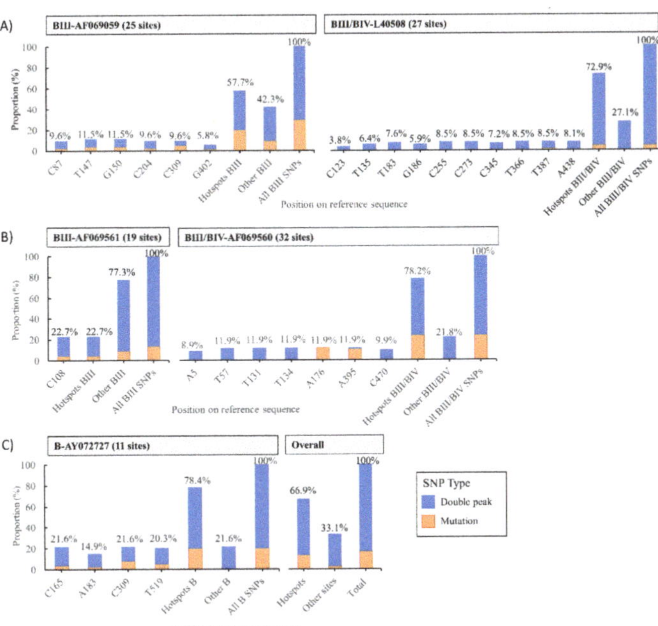

Figure 2. Distribution of single nucleotide polymorphisms segregated by mutations and double peaks, in *G. duodenalis* assemblage B sequences. (**A**) Single nucleotide polymorphisms (SNPs) at the glutamate dehydrogenase (*gdh*) locus; (**B**) SNPs at the triosephosphate isomerase (*tpi*); (**C**) SNPs at the beta-giardin (*bg*) locus, and overall figures for all assemblage B sequences.

2.6. Risk Association Analysis

2.6.1. Comparing *G. duodenalis* Negative/Ever Positive

Overall, 55% of individuals who tested positive for *G. duodenalis* were female, the median age was 10 years old, with 49% <10 years old. A total of 55% were from tribe 1, followed by 15% from tribe 5. The most frequent clinical signs were normal stool appearance (92%), abdominal pain (53%), and stool consistency type 2 (45%). Overall, 58% did not report hand washing, 84% reported eating with hands, and 18% did not report washing fresh produce. Sanitation was predominantly defecation in the woods (70%) and open defecation near households (21.7%). Microscopy examination also revealed that 53% of individuals were coinfected with *Endolimax nana*, 48% with *Entamoeba coli*, 18% with *Chilomastix mesnili*, 16% with *Ancylostoma* spp., and 11% with *Blastocystis* sp.

Children under <15 years old reported more frequently vomiting, abdominal pain, and abnormal (mucous, bloody, mucous-bloody) faecal appearance compared to adults. However, only differences in abdominal pain appeared significant (Chi-squared test, $p = 0.016$). There were no differences in age or symptoms between the two assemblages A and B (Chi-squared test: age group, $p = 0.552$; faecal consistency, $p = 0.732$; abdominal pain, $p = 1$; vomit, $p = 0,953$).

In univariable analysis, hand washing, older age, tribes 2–4, coinfection with *E. coli*, *E. nana*, and *Iodamoeba* were negatively associated with *G. duodenalis*, while tribe 5, faecal consistency 4, open defecation, and the number of samples were positively associated with *G. duodenalis* (Table 8). Regarding the public health features and symptoms, the multivariable model retained the number of samples, age group, tribe, faecal consistency, faecal appearance, and washing fresh produce. Older age groups had a protective effect (adjusted odds ratio (aOR) = 0.40, 95% CI: 0.20–0.81 in 10–14 years old, and aOR = 0.20 95%, CI: 0.11–0.39 in ≥15 years old, respectively), as tribes 2–3 compared to tribe 1 (aOR = 0.46, 95% CI: 0.24–0.85, aOR = 0.43, 95% CI: 0.21–0.85, aOR = 0.31, 95% CI: 0.14–0.63, respectively). The number of samples was positively associated with higher odds of a *G. duodenalis*-positive result (aOR = 1.46, 95% CI: 1.18–1.81), as washing fresh produce (aOR = 1.95, 95% CI: 1.12–3.44) and faecal consistency type 4 (aOR = 1.84, 95% CI: 1.10–3.37) (Table 9). None of the other pathogens considered was found significantly associated with *G. duodenalis* in the multivariable model with coinfections.

Table 8. Univariable analysis. Crude association between *G. duodenalis* infections and variables of interest. *p*-values marked in bold indicate numbers that are significant on the 95% confidence limit (CI).

	G. duodenalis (Negative vs. Ever Positive)			Crude Association		
Variable	0, *n* = 366 [1]	1, *n* = 198 [1]	*n*	OR [2]	95% CI [2]	*p*-Value
Sex			564			
Female	177 (62%)	109 (38%)		—	—	
Male	189 (68%)	89 (32%)		0.76	0.54–1.08	0.13
Age group (years)			562			
0–4	38 (45%)	46 (55%)		—	—	
5–9	67 (57%)	50 (43%)		0.62	0.35–1.08	0.093
10–14	58 (62%)	35 (38%)		0.5	0.27–0.91	**0.023**
≥15	201 (75%)	67 (25%)		0.28	0.16–0.46	**<0.001**
Unknown	2	0				
Tribe			564			
1	155 (59%)	108 (41%)		—	—	
2	50 (73%)	19 (27%)		0.55	0.30–0.96	**0.041**
3	55 (80%)	14 (20%)		0.37	0.19–0.67	**0.002**
4	62 (85%)	11 (15%)		0.25	0.12–0.49	**<0.001**

Table 8. *Cont.*

	G. duodenalis (Negative vs. Ever Positive)			Crude Association		
Variable	0, *n* = 366 [1]	1, *n* = 198 [1]	*n*	OR [2]	95% CI [2]	*p*-Value
5	20 (40%)	30 (60%)		2.15	1.17–4.04	**0.015**
6	24 (60%)	16 (40%)		0.96	0.48–1.87	0.9
Faecal consistency			564			
1	101 (68%)	48 (32%)		—	—	
2	200 (69%)	89 (31%)		0.94	0.61–1.44	0.8
3	9 (56%)	7 (44%)		1.64	0.56–4.65	0.4
4	56 (51%)	54 (49%)		2.03	1.22–3.38	**0.006**
Faecal appearance			564			
Normal	336 (65%)	182 (35%)		—	—	
Other (mucus, bloody)	30 (65%)	16 (35%)		0.98	0.51–1.83	>0.9
Abdominal pain			563			
No	175 (65%)	93 (35%)		—	—	
Yes	190 (64%)	105 (36%)		1.04	0.74–1.47	0.8
Unknown	1	0				
Vomit			564			
No	353 (65%)	192 (35%)		—	—	
Yes	13 (68%)	6 (32%)		0.85	0.29–2.18	0.7
Drinking water source			564			
River	1 (100%)	0 (0%)		—	—	
Well	1 (100%)	0 (0%)		1.00	0.00–36,409	>0.9
Piped	364 (65%)	198 (35%)		1,152,197	0.00–NA	>0.9
Treated water			564			
No	350 (65%)	190 (35%)		—	—	
Yes	16 (67%)	8 (33%)		0.92	0.37–2.13	0.9
Hand washing			564			
No	166 (59%)	115 (41%)		—	—	
Yes	200 (71%)	83 (29%)		0.60	0.42–0.85	**0.004**
Washing fresh produce			563			
No	77 (68%)	36 (32%)		—	—	
Yes	289 (64%)	161 (36%)		1.19	0.77–1.87	0.4
Unknown	0	1				
Eating with			563			
Hand	301 (64%)	167 (36%)		—	—	
Flatware	64 (67%)	31 (33%)		0.87	0.54–1.38	0.6
Unknown	1	0				
Defecation place			564			
Toilets	35 (69%)	16 (31%)		—	—	
Woods	294 (68%)	139 (32%)		1.03	0.56–1.98	>0.9
Yard	37 (46%)	43 (54%)		2.54	1.23–5.41	**0.013**
Contact with animals			564			
No	55 (63%)	32 (37%)		—	—	
Yes	311 (65%)	166 (35%)		0.92	0.57–1.49	0.7

Table 8. *Cont.*

| Variable | *G. duodenalis* (Negative vs. Ever Positive) | | | Crude Association | | |
	0, *n* = 366 [1]	1, *n* = 198 [1]	*n*	OR [2]	95% CI [2]	*p*-Value
Rotavirus			564			
No	365 (65%)	198 (35%)		—	—	
Yes	1 (100%)	0 (0%)		0.00		>0.9
Ancylostoma			564			
No	283 (64%)	162 (36%)		—	—	
Yes	83 (70%)	36 (30%)		0.76	0.49–.16	0.2
Ascaris			564			
No	364 (65%)	195 (35%)		—	—	
Yes	2 (40%)	3 (60%)		2.80	0.46–21.4	0.3
Blastocystis			564			
No	305 (64%)	173 (36%)		—	—	
Yes	61 (71%)	25 (29%)		0.72	0.43–1.18	0.2
Chilomastix			564			
No	307 (66%)	162 (34%)		—	—	
Yes	59 (62%)	36 (38%)		1.16	0.73–1.82	0.5
E. coli			564			
No	141 (59%)	97 (41%)		—	—	
Yes	225 (69%)	101 (31%)		0.65	0.46–0.93	**0.017**
E. histolytica			564			
No	216 (63%)	125 (37%)		—	—	
Yes	150 (67%)	73 (33%)		0.84	0.59–1.20	0.3
E. nana			564			
No	129 (59%)	88 (41%)		—	—	
Yes	237 (68%)	110 (32%)		0.68	0.48–0.97	**0.032**
Hymenolepis			564			
No	333 (65%)	180 (35%)		—	—	
Yes	33 (65%)	18 (35%)		1.01	0.54–1.82	>0.9
Iodamoeba			564			
No	336 (64%)	191 (36%)		—	—	
Yes	30 (81%)	7 (19%)		0.41	0.16–0.90	**0.038**
Isospora			564			
No	366 (65%)	197 (35%)		—	—	
Yes	0 (0%)	1 (100%)		1,447,714	0.00, NA	>0.9
Sarcocystis			564			
No	359 (65%)	196 (35%)		—	—	
Yes	7 (78%)	2 (22%)		0.52	0.08–2.19	0.4
Strongyloides			564			
No	345 (64%)	191 (36%)		—	—	
Yes	21 (75%)	7 (25%)		0.60	0.23–1.38	0.3
Taenia			564			
No	364 (65%)	198 (35%)		—	—	
Yes	2 (100%)	0 (0%)		0.00		>0.9

Table 8. *Cont.*

	G. *duodenalis* (Negative vs. Ever Positive)			Crude Association		
Variable	0, *n* = 366 [1]	1, *n* = 198 [1]	*n*	OR [2]	95% CI [2]	*p*-Value
Trichuris			564			
No	365 (65%)	197 (35%)		—	—	
Yes	1 (50%)	1 (50%)		1.85	0.07–47.0	0.7
Cyclospora			564			
No	351 (65%)	188 (35%)		—	—	
Yes	15 (60%)	10 (40%)		1.24	0.53–2.80	0.6
No. of samples			564	1.72	1.43–2.08	**<0.001**
1	71 (83%)	15 (17%)				
2	99 (73%)	37 (27%)				
3	129 (64%)	72 (36%)				
4	67 (48%)	74 (53%)				

[1] Statistics presented: *n* (%). [2] OR, crude odds ratio; CI, confidence interval; NA, not applicable.

Table 9. Multivariable analysis comparing G. *duodenalis*-negative results versus G. *duodenalis* ever positive results. *p*-values marked in bold indicate numbers that are significant on the 95% confidence limit (CI).

Variable	aOR [1]	95% CI [1]	*p*-Value
Age group (years)			
0–4	—	—	
5–9	0.58	0.30–1.12	0.11
10–14	0.40	0.20–0.81	**0.011**
≥15	0.20	0.11–0.39	**<0.001**
Number_samples	1.46	1.18–1.81	**<0.001**
Tribe			
1	—	—	
2	0.46	0.24–0.85	**0.016**
3	0.43	0.21–0.85	**0.018**
4	0.31	0.14–0.63	**0.002**
5	1.83	0.94–3.60	0.075
6	0.94	0.45–1.95	0.9
Washing fresh produce			
No	—	—	
Yes	1.95	1.12–3.44	**0.020**
Faecal consistency			
1	—	—	
2	0.93	0.59–1.49	0.8
3	2.37	0.73–7.51	0.14
4	1.84	1.01–3.37	**0.046**
Faecal appearance			
Normal	—	—	
Other (mucus, bloody)	0.51	0.23–1.08	0.087

[1] aOR, adjusted odds ratio; CI, confidence interval. *n* = 559 (removing five observations with missing values).

2.6.2. Comparing *G. duodenalis* Serial Results

Out of the 478 observations with more than one sample, 62% (295/478) always tested negative for *G. duodenalis*, 4% (20/478) always tested positive, and 34% (163/478) were discontinuously positive. As such, among observations ever positive for *G. duodenalis*, 11% (20/183) were continuously positive. Of those, 70% were aged 0–9 years old, 55% were from tribe 1, and 35% from tribe 5. Overall, 85% did not report hand washing, 30% not washing fresh produce, 50% reported open defecation near the households, and 45% open defecation in the woods. However, in the multivariable analysis, discontinuous positivity was strongly associated with the number of samples (aOR = 0.30, 95% CI: 0.12–0.66), and tribe 5 (aOR = 4.76, 95% CI: 1.30–18.6), but no further significant association was found (Tables S7 and S8).

Finally, comparing observations that were always negative versus always positive, the multivariable model only retained age group and tribe, showing evidence of a protective effect of older age groups (aOR = 0.24, 95% CI: 0.06–0.89 for children 5–9 years old; aOR = 0.17, 95% CI: 0.03–0.68 for children 10–14 years old; aOR = 0.05, 95% CI: 0.01–0.20 for children ≥15 years old), and a strong positive association with tribe 5 (aOR = 5.55, 95% CI: 1.61–19.4) (Tables S9 and S10). However, when adjusting for the effect of coinfections; the best fit model suggested a protective effect of *E. nana* (aOR = 0.25, 95% CI: 0.08–73), oldest age group ≥15 years old (aOR = 0.07, 95% CI: 0.01–0.28), and a positive effect of tribe 5 (aOR = 5.87, 95% CI: 1.60–22.1) (Table S11).

3. Discussion

This survey presents new insights into the epidemiology of *G. duodenalis* in Amazonian indigenous communities. The main contributions of the study include the demonstration that (i) giardiasis is a common finding (13–22%) in apparently healthy Tapirapé people, mainly affecting children in the age group of 0–9 years old; (ii) assemblage B was responsible for near 70% of the mostly asymptomatic infections detected; and (iii) a high degree of genetic heterogeneity was observed within assemblage B (but not assemblage A) sequences, regardless of the molecular marker used.

Several epidemiological studies conducted in endemic areas worldwide have shown that *G. duodenalis* infections do not seem to correlate positively with diarrhoea [23,24], demonstrating that asymptomatic giardiasis is the rule rather than the exception in these settings. This fact would explain why giardiasis is systematically absent in global burden estimations of diarrhoeal disease [25]. This seems to be also the case of the present study, where *G. duodenalis* infections were detected similarly in asymptomatic individuals (33.8%) and individuals presenting with diarrhoea or other gastrointestinal manifestations (35.3%). Taken together, this information supports the hypothesis that some enteric protist species (e.g., *Blastocystis* sp., *Dientamoeba fragilis*, *G. duodenalis*) might in fact be protective against disease [26]. This is an attractive possibility implying that these agents are indeed acting as pathobionts (that is, microorganisms that normally live as harmless symbionts but under certain circumstances can be pathogenic) forming part of the host eukaryome.

We have shown in our study that *G. duodenalis* infection was strongly related to younger age and tribe (with tribes 1 and 5 having a higher association) and to seasonality. This may be due to external factors associated with indirect transmission pathways of the infection (e.g., source of drinking water, consumption of contaminated fresh produce, swimming in contaminated surface waters, defecation on the open ground near households, and high density of companion or domestic animals) or increased risk of reinfection within the tribe from other infected members through direct person-to-person contact. Contact with faecally contaminated water and produce may be more likely in the rainy season. Children <15 years old with giardiasis reported more frequently vomiting, abdominal pain, and presence of mucus/blood in faeces compared to adults, although observed differences did not reach statistical significance. Young children with an immature immune system may be at higher risk of infections and probably more severe disease episodes. Thus, older adults may have acquired immunity after a previous infection. Indeed, it has been shown that levels of intestinal inflammation caused by *G. duodenalis* infection decrease with subsequent infections [27,28]. This implies that there is acquired protection against the severity of giardiasis but not from reinfection [29]. In this regard, it should be noted that the composition and abundance of the host's microbiota have also been suggested to play an important role in the outcome of the infection [30].

Giardiasis was also strongly dependent on the number of samples taken, even considering that conventional microscopy (a method that is largely known to be of limited diagnostic sensitivity) was the screening method for the initial detection of *G. duodenalis* in the present survey. This suggests that possible reinfections or chronic infections with intermittent positivity may be more common than initially anticipated. Reinfection may be more pronounced in the rainy season. In addition, no evident differences between individuals continuously positive/discontinuously positive to *G. duodenalis* were found. However, we should exclude a bias in those presenting for sampling. This is unlikely to be a major factor due to the lack of symptoms in most cases.

Regarding coinfections, the presence of *G. duodenalis* was not associated with any other enteric parasite species, except possibly *E. nana*. These results may be biased by the relatively small number of positive samples detected for certain pathogens and should, therefore, be interpreted with caution. Similarly, a counter-intuitive positive association between *G. duodenalis* with washing fresh produce was found. This result may be the consequence of the potential confounder effect of other variables no considered here such as the manipulation of fresh produce or the use of contaminated washing water. The latter possibility would support the relevance of waterborne transmission for human giardiasis.

Molecular sequence analyses of the three loci used here for genotyping purposes also revealed interesting data. There were no differences in age between individuals infected either by the assemblage A or the assemblage B of *G. duodenalis*. Regarding age-related patterns in the distribution of *G. duodenalis* assemblages, our results are in contrast with those previously obtained in surveys targeting clinical populations. For instance, children have been shown to be more commonly infected by assemblage B (83%, 44/53) than adults (52%, 22/42) in patients of all age groups in Spain [31]. Moreover, in that country, assemblage B was significantly more prevalent than assemblage A in asymptomatic outpatient children, but not in individuals of older age [32].

Remarkably, no association between the occurrence of diarrhoea (or any other gastrointestinal manifestation) and the *G. duodenalis* assemblage involved in the infection was found in the investigated population. This result corroborates that observed in children under 5 years of age (*n* = 222) recruited under the Global Enteric Multicentre Study (GEMS) in Mozambique [33]. However, it should be noted that other surveys have shown different, even contradictory, results. For instance, assemblage A was more prevalent than assemblage B in Bangladeshi people (*n* = 343) [34], in Turkish clinical patients (*n* = 44) [35], and in Spanish outpatient children (*n* = 43) [32]. The opposite trend was reported in asymptomatic infected individuals (*n* = 18) in the Netherlands [36].

Genotyping data generated here demonstrated that assemblage B was responsible for three out of four *G. duodenalis* infections in the Tapirapé people, a similar proportion of that (78%) described in paediatric populations in the Amazonas State [37]. Of note, assemblage A tends to be the predominant *G. duodenalis* genetic variant circulating in humans in Brazil (Table S1). These facts may be indicative of differences in sources of infection, transmission pathways, or even geographical segregation patterns of the parasite in the country. Lack of non-human, host-specific assemblages C–F seem to suggest that companion, production, and free-living animal species are no significant contributors of giardiasis in the surveyed population. This is in spite of the fact that swine and poultry were reared in all seven tribes, and that domestic dog and cat densities were also high. In addition, cattle (but not sheep) farming was also frequent in the proximity of them. Taking together, these data indicate that human giardiasis is mainly of anthropic nature among the Tapirapé people. The extent and accuracy of this statement should be corroborated in future molecular epidemiological studies including animal and environmental (water) samples.

This study also confirms the high genetic variability within *G. duodenalis* assemblage B (but not assemblage A) reported frequently in similar molecular epidemiological surveys conducted in endemic areas globally [38,39] including Brazil [40,41]. This finding was particularly evident at the *gdh* and *tpi* loci, for which most of the generated BIII (78–87%), BIV (100%), and BIII/BIV (90–92%) sequences corresponded to distinct genotypes of the parasite. Sequences unmistakably assigned to BIII and BIV at the *gdh*/*tpi* loci tended to vary only in one to six positions (hotspots) either as mutations or ambiguous (double peak) sites. In these sets of hotspots, the proportion of sites involving double peaks in BIII sequences varied from 38% at the *gdh* locus to 18% at the *tpi* locus. Interestingly, these percentages increased in both cases to 55–68% in ambiguous BIII/BIV sequences, explaining why these isolates were difficult to allocate to a given sub-assemblage. Two independent mechanisms have been proposed to explain the presence of ambiguous (double peak) positions. The first one involves the occurrence of true mixed infections (e.g., BIII + BIV) and would fit well with an epidemiological scenario characterised by high infection and reinfection rates as the one described in the present study. The second one would be associated with the occurrence of genetic recombination. Evidence for the latter possibility comes from independent investigations demonstrating low levels of allelic sequence heterozygosity (implying a genetic homogenisation mechanism) within assemblage A [42] and, to a lesser extent, within assemblage B [43]. Additional evidence of genetic recombination events has been demonstrated within assemblage B in single (trophozoite and cyst) cells [44] and within sub-assemblages BIII and BIV at the genetic population level [45].

The results obtained in the present study may be biased by certain design and methodological constricts. For instance, the initial screening of *G. duodenalis* was based on conventional microscopy, so the true prevalence of the infection is likely to be underestimated. In addition, there may be a response bias as people may be more or less inclined to return to the study if they had a negative or positive test result. Interestingly, the positivity rate was increased by the number of tests performed, suggesting that over time people were likely to have had a giardiasis episode, that they may have had a false-negative result at microscopy examination, or an inherent response bias in that people who were likely to be positive would return for testing. Limitations associated with the main dataset may arise from the combination of period-specific data, although most of the independent variables considered (e.g., demographics, access to safe drinking water, and sanitary conditions) were not expected to change over time. As our analyses used the first negative test result, we could not further explore the effect of seasonality in the multivariable analysis. However, we have already shown in the descriptive data that seasonality is associated with infections and repeated infections. Lack of association between *G. duodenalis* genetic variants and occurrence of clinical symptoms may be influenced by the fact that other diarrhoea-causing agents (including viral and bacterial pathogens) were not assessed. In addition, suspected mixed infections were not further investigated by cloning of PCR amplicons or next-generation sequencing, methods with high sensitivity able to detect genetic variants of the parasite that are underrepresented in the population pool, and that are otherwise undetectable using conventional PCR methods and Sanger sequencing. Finally, the typing scheme used in the present study may lack enough phylogenetic resolution to correctly differentiate between sub-assemblage BIII and BIV sequences. This issue has been highlighted in recent molecular studies for assemblage B and assemblage A sequences [46,47]. This important point emphasises the need of identifying new markers and of developing novel methods for MLST purposes.

4. Materials and Methods

4.1. Study Area

Brazil extends over 8,511,965 km^2 and includes 724 indigenous lands (ILs) covering a total area of 1,173,770 km^2 and accounting for 14% of the country's territory [48]. Most ILs are concentrated in the Legal Amazon, representing 23% of the Amazon territory) [49]. The indigenous people from the Tapirapé ethnicity live in the Serra do Urubu Branco region,

Mato Grosso State, a region of tropical forest with typical Amazonian flora and fauna interspersed with clean and closed fields. The Tapirapé exploit this environment alternating agriculture, hunting, gathering, and fishing according to the time of year [49,50]. Farmers´villages have traditionally been in the vicinity of dense forests on high, non-flooding lands. Currently, the Tapirapé ethnic group is made up of approximately 700 individuals living in six tribes with maximum and minimum distances from the main tribe of 70 km and 10 km, respectively. The main tribe is in the municipality of Confresa, Mato Grosso State (Figure 3). Tapirapé people interact frequently with individuals from other ethnic tribes at social events, hunt parties, and other activities.

Figure 3. Map showing the exact geographical location of the sites sampled in the present study.

4.2. Sampling and Data Collection

This is a prospective, cross-sectional epidemiological study including four sampling periods covering two dry (July 2008 and July 2009) and two rainy (January 2009 and January 2010) seasons. After obtaining the chief's ('cacique') consent for permission to survey, all members of the tribe were informed about the aim of the project and invited to provide a single stool sample at each of the four scheduled sampling periods. Designated persons in each household were given polystyrene plastic flasks for each member of the household and stool samples were collected on the following day.

Individual standardised questionnaires were completed by a member of our research team in face-to-face interviews with designated persons at sample collection, who provided the requested information for each member of his/her household. Questions included demographics (gender, age, village of origin), clinical manifestations (vomit, abdominal pain), hand and vegetable washing, source of drinking water, use of water treatment, defecation place, and contact with domestic animals and livestock. Provided stool samples were visually inspected for consistency and the presence of mucus or blood. Each participant was assigned a unique distinctive code through the whole period of study, which was used to identify his/her stool sample(s) and associated epidemiological questionnaire(s).

4.3. Microscopy Examination

Stool samples were kept at 4 °C before microscopy examination, usually within 48 h of collection. A conventional flotation method using sucrose solution (specific gravity: 1.2 g/cm^3) was conducted in all stool samples as previously described [51]. Two additional techniques were performed—spontaneous sedimentation [52] and centrifugal-sedimentation in formalin-ether [53]. A sample was considered *G. duodenalis*-positive if cysts of the parasite were detected by at least one of the three methods used. Aliquots of faecal positive samples were stored at –20 °C for downstream molecular analyses. Any other enteric parasite (including helminthic and protist) species found during microscopy observation were also identified and recorded.

4.4. DNA Extraction and Purification

Positive stool samples were defrosted and *G. duodenalis* cysts concentrated and purified using the Faust method [54]. Obtained supernatants were subjected to three freeze–thaw cycles to facilitate the mechanical breakage of the cyst wall [55]. Genomic DNA was extracted from the processed supernatants (ca 200 µL) using the PureLink Genomic DNA Mini Kit (Thermo Fisher Scientific, Waltham, MA, USA) according to the manufacturer's instructions. Extracted and purified DNA samples in molecular grade water (200 µL) were kept at −20 °C and shipped to the Spanish National Centre for Microbiology (Health Institute Carlos III) in Majadahonda (Spain) for downstream genotyping analyses.

4.5. Molecular Confirmation of G. duodenalis

Confirmation of *G. duodenalis* infection was achieved using a real-time PCR (qPCR) method targeting a 62-bp region of the gene codifying the small subunit ribosomal RNA (SSU rRNA) of the parasite [56]. Amplification reactions (25 µL) consisted of 3 µL of template DNA, 0.5 µM of each primer Gd-80F and Gd-127R, 0.4 µM of the probe (Table S12), and 12.5 µL TaqMan® Gene Expression Master Mix (Applied Biosystems, Foster City, CA, USA). Detection of parasitic DNA was performed on a Corbett Rotor GeneTM 6000 real-time PCR system (Qiagen, Hilden, Germany) using an amplification protocol consisting of an initial hold step of 2 min at 55 °C and 15 min at 95 °C, followed by 45 cycles of 15 s at 95 °C and 1 min at 60 °C. Water (no template) and genomic DNA (positive) controls were included in each PCR run.

4.6. Molecular Characterisation of G. duodenalis

Giardia duodenalis isolates with a qPCR-positive result were re-assessed by sequence-based multi-locus genotyping of the genes encoding for the glutamate dehydrogenase (gdh), beta-giardin (bg), and triosephosphate isomerase (tpi) proteins of the parasite. A semi-nested PCR was used to amplify a ~432-bp fragment of the gdh gene [57]. PCR reaction mixtures (25 µL) included 5 µL of template DNA and 0.5 µM of the primer pairs GDHeF/GDHiR in the primary reaction and GDHiF/GDHiR in the secondary reaction (Table S12). Both amplification protocols consisted of an initial denaturation step at 95 °C for 3 min, followed by 35 cycles of 95 °C for 30 s, 55 °C for 30 s, and 72 °C for 1 min, with a final extension of 72 °C for 7 min.

A nested PCR was used to amplify a ~511 bp-fragment of the *bg* gene [58]. PCR reaction mixtures (25 µL) consisted of 3 µL of template DNA and 0.4 µM of the primers sets G7_F/G759_R in the primary reaction and G99_F/G609_R in the secondary reaction (Table S12). The primary PCR reaction was carried out with the following amplification conditions: one step of 95 °C for 7 min, followed by 35 cycles of 95 °C for 30 s, 65 °C for 30 s, and 72 °C for 1 min, with a final extension of 72 °C for 7 min. The conditions for the secondary PCR were identical to the primary PCR except that the annealing temperature was 55 °C.

A nested PCR was used to amplify a ~530 bp-fragment of the *tpi* gene [59]. PCR reaction mixtures (50 µL) included 2–2.5 µL of template DNA and 0.2 µM of the primer pairs AL3543/AL3546 in the primary reaction and AL3544/AL3545 in the secondary reaction (Table S12). Both amplification protocols consisted of an initial denaturation step at 94 °C for 5 min, followed by 35 cycles of 94 °C for 45 s, 50 °C for 45 s, and 72 °C for 1 min, with a final extension of 72 °C for 10 min.

The semi-nested and nested PCR protocols described above were conducted on a 2720 Thermal Cycler (Applied Biosystems). Reaction mixes always included 2.5 units of MyTAQ^{TM} DNA polymerase (Bioline GmbH, Luckenwalde, Germany), and 5× MyTAQTM reaction buffer containing 5 mM dNTPs and 15 mM MgCl$_2$. Laboratory-confirmed positive and negative DNA samples for each parasite species investigated were routinely used as controls and included in each round of PCR. PCR amplicons were visualised on 2% D5 agarose gels (Conda, Madrid, Spain) stained with Pronasafe nucleic acid staining solution (Conda). Positive PCR products were directly sequenced in both directions using appropriate internal primer sets (Table S12). DNA sequencing was conducted by capillary electrophoresis using BigDye® Terminator chemistry (Applied Biosystems) on an on ABI PRISM 3130 Genetic Analyser.

4.7. Sequence and Phylogenetic Analyses

Raw sequencing data in both forward and reverse directions were viewed using the Chromas Lite version 2.1 sequence analysis program (https://technelysium.com.au/wp/chromas/ (accessed on 1 February 2021)). The Basic Local Alignment Search Tool (BLAST) (http://blast.ncbi.nlm.nih.gov/Blast.cgi (accessed on 1 February 2021)) was used to compare nucleotide sequences with sequences retrieved from the NCBI GenBank database. Generated DNA consensus sequences were aligned to appropriate reference sequences using the MEGA 6free software [60] for species confirmation and assemblage/sub-assemblage identification.

For the estimation of the phylogenetic relationships among the identified Giardia-positive samples, gdh sequences generated in this study and human- and animal-derived homologue sequences mostly from Brazil retrieved from GenBank were aligned using Clustal X and adjusted manually with GeneDoc [61,62]. Inferences by maximum parsimony (MP) were constructed by PAUP version 4.0b10 using a heuristic search in 1000 replicates, 500 bootstrap replicates, random stepwise addition starting trees (with random addition sequences), and tree bisection and reconnection branch swapping [63]. MrBayes v3.1.2 was used to perform Bayesian analyses with four independent Markov chain runs for 1,000,000 metropolis-coupled MCMC generations, sampling a tree every 100th generation [64]. References [65–109] are cited in the supplementary materials. The first 25% of trees represented burn-in and the remaining trees were used to calculate Bayesian posterior probability. The GTR +I + G substitution model was used. The gdh sequence of G. ardeae was used as the outgroup.

The sequences obtained in this study have been deposited in GenBank under accession numbers MT542718-MT542765 (gdh), MT542766-MT542794 (bg), and MT542795-MT542829 (tpi).

4.8. Statistical Analysis

We investigated factors (public health features, clinical symptoms, coinfection with other pathogens) associated with a positive *G. duodenalis* result. The main dataset was constructed with data from one of the four sampling points—if the observation ever tested positive for *G. duodenalis*, we used data from the sampling point of the first positive *G. duodenalis* result; otherwise, we used data from the first sampling point in order.

We conducted Chi-squared tests ($p < 0.05$) to compare characteristics of cases and non-cases, and we calculated crude odds ratios (OR) with 95% confidence intervals (CI) to investigate the crude association between independent variables and a *G. duodenalis*-positive result. We constructed multivariable logistic regression models to assess the association between *G. duodenalis* and (i) public health features and clinical signs or (ii) coinfection with other intestinal pathogens, adjusted by age, tribe, and the number of samples. Additionally, we considered the serial results of *G. duodenalis* for observations with at least two samples. We conducted similar analyses by comparing those continuously negative versus continuously positive, and those discontinuously positive versus continuously positive.

Univariable analyses were conducted on all available observations, but observations with missing values were removed from multivariable analyses. All the independent variables were included in the analyses and we used the stepwise backward selection method, removing successively the least significant variable and using Akaike information criterion (AIC) and Bayesian information criterion (BIC) to construct the best fit model. Analyses were performed in R (package stats).

4.9. Ethics Approval

This study has been approved by the National Research Ethics Commission (CONEP), Ministry of Health (Brazil), under reference number 120/2008.

5. Conclusions

This microscopy-based survey demonstrates that symptomatic and asymptomatic giardiasis are common in indigenous people from the Brazilian Amazon. Children under 15 years of age were particularly exposed to the infection, suggesting that acquired immunity plays a role in modulating the frequency and virulence of the disease. *G. duodenalis* infection rates varied largely among the surveyed tribes and sampling periods, suggesting that different pathways may be involved in the transmission of the parasite. Molecular sequence data indicated that the most likely source of infection was anthropic. The distribution of assemblages was independent of the occurrence of clinical manifestations, indicating that the genotype of the parasite was not associated with the outcome of the infection. Assemblage B accounted for near 75% of the infections detected and showed a high genetic diversity that impaired the correct identification of sub-assemblages BIII and BIV. This diversity was mainly associated with the presence of ambiguous positions (double peaks) at the chromatogram level, suggesting that coinfections and/or genetic recombination events were taking place, at unknown rates, in the investigated population. Further molecular epidemiological studies targeting animal (including domestic and wildlife) and environmental (drinking water) samples are needed to elucidate the transmission dynamics of *G. duodenalis* in this Brazilian geographical region.

Supplementary Materials: The following are available online at https://www.mdpi.com/2076-0817/10/2/206/s1, Table S1: Prevalence and molecular diversity of *Giardia duodenalis* in humans in Brazil, Table S2: Prevalence and molecular diversity of *Giardia duodenalis* in domestic and wildlife animal species in Brazil, Table S3: Prevalence and molecular diversity of *Giardia duodenalis* in water samples in Brazil, Table S4: Prevalence and molecular diversity of *Giardia duodenalis* in fresh produce in Brazil, Table S5: Full dataset showing the molecular diversity of *G. duodenalis* at the *gdh*, *bg*, and *tpi* molecular markers, Table S6: Intra-assemblage B single nucleotide polymorphisms distribution and classification among *G. duodenalis* sequences at the *gdh*, *bg* and *tpi* loci. Hotspots for SNPs are identified and the summarised comparisons of frequencies between the hotspot and non-hotspot sites are highlighted with darker shades, Table S7: Univariable analysis comparing discontinuously *G. duodenalis*-positive results versus continuously *G. duodenalis*-positive results. *p*-values marked in bold indicate numbers that are significant on the 95% confidence limit, Table S8: Multivariable analysis comparing discontinuously *G. duodenalis*-positive results versus always *G. duodenalis*-positive results. *p*-values marked in bold indicate numbers that are significant on the 95% confidence limit, Table S9: Univariable analysis comparing always *G. duodenalis*-negative results versus always *G. duodenalis*-positive results. *p*-values marked in bold indicate numbers that are significant on the 95% confidence limit, Table S10: Multivariable analysis always comparing *G. duodenalis*-negative results versus always *G. duodenalis*-positive results. *p*-values marked in bold indicate numbers that are significant on the 95% confidence limit, Table S11: Multivariable analysis comparing always *G. duodenalis*-negative results versus always *G. duodenalis*-positive results and considering the presence of coinfections. *p*-values marked in bold indicate numbers that are significant on the 95% confidence limit, Table S12: Oligonucleotides used for the molecular identification and characterisation of *G. duodenalis* in the present study, Figure S1: Maximum parsimony phylogenetic dendogram based on *bg* sequences of *G. duodenalis*. Numbers on nodes indicate the bootstrap/posterior probability values. GenBank accession numbers for all sequences used for the phylogenetic analysis were embedded in the tree, Figure S2: Maximum parsimony phylogenetic dendogram based on *tpi* sequences of *G. duodenalis*. Numbers on nodes indicate the bootstrap/posterior probability values. GenBank accession numbers for all sequences used for the phylogenetic analysis were embedded in the tree.

Author Contributions: Conceptualisation, A.F.M., J.J.S., S.B., S.M.G., R.C.-B., and D.C.; methodology, A.F.M., S.B., R.C.-B., and D.C.; software, P.C.K., H.L., A.P., and A.M.; validation, A.F.M., A.P., S.B., S.M.G., and D.C.; formal analysis, P.C.K., S.B., H.L., A.P., and A.M.; investigation, L.M.M., L.M.S.L., A.D., B.B., and H.S.S.; resources, A.F.M., J.J.S., S.B., S.M.G., and D.C.; data curation, A.F.M., S.B., S.M.G., and D.C.; writing—original draft preparation, D.C.; writing—review and editing, P.C.K., A.F.M., S.B., H.L., A.D., S.M.G., R.C.-B., D.G.-B., and D.C.; supervision, A.F.M., S.M.G., S.B., R.C.-B., D.G.-B., and D.C.; project administration, A.F.M., S.M.G., S.B., and D.C.; funding acquisition, A.F.M. and D.C. All authors have read and agreed to the published version of the manuscript.

Funding: This research was funded by the São Paulo State Research Support Foundation (FAFESP, Brazil), the National Health Foundation (FUNASA, Brazil), and the Mato Grosso State Research Support Foundation (FAPEMAT, Brazil), grant number 0839/2006. Additional funding was obtained from the Health Institute Carlos III (ISCIII), Ministry of Economy and Competitiveness (Spain), grant number PI16CIII/00024.

Institutional Review Board Statement: The study was conducted according to the guidelines of the Declaration of Helsinki and approved by the Institutional Review Board of Instituto de Ciências Biomédicas da Universidade de São Paulo (protocol code 14485—CONEP—on 17/06/2008).

Informed Consent Statement: Informed consent was obtained from all subjects involved in the study.

Data Availability Statement: All relevant data are within the article and its additional files. The sequences obtained in this study have been deposited in GenBank under accession numbers MT542718-MT542765 (*gdh*), MT542766-MT542794 (*bg*), and MT542795-MT542829 (*tpi*).

Acknowledgments: S.M. Gennari, A. Marcili, and J. J. Shaw are in receipt of a Research Productivity Fellowship from the National Council for Scientific and Technological Development (CNPq, Brazil).

Conflicts of Interest: The authors declare no conflict of interest. The funders had no role in the design of the study; in the collection, analyses, or interpretation of data; in the writing of the manuscript, or in the decision to publish the results.

References

1. Einarsson, E.; Ma'Ayeh, S.; Svärd, S.G. An up-date on Giardia and giardiasis. *Curr. Opin. Microbiol.* **2016**, *34*, 47–52. [CrossRef]
2. Ankarklev, J.; Hestvik, E.; Lebbad, M.; Lindh, J.; Kaddu-Mulindwa, D.H.; Andersson, J.O.; Tylleskär, T.; Tumwine, J.K.; Svärd, S.G. Common coinfections of Giardia intestinalis and Helicobacter pylori in non-symptomatic Ugandan Children. *PLoS Negl. Trop. Dis.* **2012**, *6*, 2–9. [CrossRef]
3. Muadica, A.S.; Balasegaram, S.; Beebeejaun, K.; Köster, P.C.; Bailo, B.; Hernández-De-Mingo, M.; Dashti, A.; DaCal, E.; Saugar, J.M.; Fuentes, I.; et al. Risk associations for intestinal parasites in symptomatic and asymptomatic schoolchildren in central Mozambique. *Clin. Microbiol. Infect.* **2020**. [CrossRef]
4. Reh, L.; Muadica, A.S.; Köster, P.C.; Balasegaram, S.; Verlander, N.Q.; Chércoles, E.R.; Carmena, D. Substantial prevalence of enteroparasites Cryptosporidium spp., Giardia duodenalis and Blastocystis sp. in asymptomatic schoolchildren in Madrid, Spain, November 2017 to June 2018. *Eurosurveillance* **2019**, *24*, 1900241. [CrossRef]
5. Kotloff, K.L.; Nataro, J.P.; Blackwelder, W.C.; Nasrin, D.; Farag, T.H.; Panchalingam, S.; Wu, Y.; Sow, O.S.; Sur, D.; Breiman, R.F.; et al. Burden and aetiology of diarrhoeal disease in infants and young children in developing countries (the Global Enteric Multicenter Study, GEMS): A prospective, case-control study. *Lancet* **2013**, *382*, 209–222. [CrossRef]
6. Tellevik, M.G.; Moyo, S.J.; Blomberg, B.; Hjøllo, T.; Maselle, S.Y.; Langeland, N.; Hanevik, K. Prevalence of Cryptosporidium parvum/hominis, Entamoeba histolytica and Giardia lamblia among young children with and without Diarrhea in Dar es Salaam, Tanzania. *PLoS Negl. Trop. Dis.* **2015**, *9*, e0004125. [CrossRef] [PubMed]
7. Breurec, S.; Vanel, N.; Onambélé, M.; Rafaï, C.; Razakandrainibe, R.; Tondeur, L.; Tricou, V.; Sansonetti, P.J.; Vray, M.; Bata, P.; et al. Etiology and Epidemiology of Diarrhea in Hospitalized Children from Low Income Country: A Matched Case-Control Study in Central African Republic. *PLoS Negl. Trop. Dis.* **2016**, *10*, e0004283. [CrossRef] [PubMed]

8.	Dupont, H.L. Giardia: Both a harmless commensal and a devastating pathogen. *J. Clin. Investig.* **2013**, *123*, 2352–2354. [CrossRef] [PubMed]

9.	Xiao, L.; Feng, Y. Molecular epidemiologic tools for waterborne pathogens *Cryptosporidium* spp. and *Giardia duodenalis*. *Food Waterborne Parasitol.* **2017**, *8–9*, 14–32. [CrossRef]

10.	Certad, G.; Viscogliosi, E.; Chabé, M.; Cacciò, S.M. Pathogenic mechanisms of cryptosporidium and Giardia. *Trends Parasitol.* **2017**, *33*, 561–576. [CrossRef] [PubMed]

11.	Berkman, D.S.; Lescano, A.G.; Gilman, R.H.; Lopez, S.L.; Black, M.M. Effects of stunting, diarrhoeal disease, and parasitic infection during infancy on cognition in late childhood: A follow-up study. *Lancet* **2002**, *359*, 564–571. [CrossRef]

12.	Al-Mekhlafi, H.M.; Al-Maktari, M.T.; Jani, R.; Ahmed, A.; Anuar, T.S.; Moktar, N.; Mahdy, M.A.K.; Lim, Y.A.L.; Mahmud, R.; Surin, J. Burden of Giardia duodenalis infection and its adverse effects on growth of schoolchildren in rural Malaysia. *PLoS Negl. Trop. Dis.* **2013**, *7*, e2516. [CrossRef] [PubMed]

13.	Savioli, L.; Smith, H.; Thompson, A. Giardia and Cryptosporidium join the 'Neglected Diseases Initiative'. *Trends Parasitol.* **2006**, *22*, 203–208. [CrossRef]

14.	Cacciò, S.M.; Lalle, M.; Svärd, S.G. Host specificity in the Giardia duodenalis species complex. *Infect. Genet. Evol.* **2018**, *66*, 335–345. [CrossRef] [PubMed]

15.	Ryan, U.; Cacciò, S.M. Zoonotic potential of Giardia. *Int. J. Parasitol.* **2013**, *43*, 943–956. [CrossRef] [PubMed]

16.	Coelho, C.H.; Durigan, M.; Leal, D.A.G.; Schneider, A.D.B.; Franco, R.M.B.; Singer, S.M. Giardiasis as a neglected disease in Brazil: Systematic review of 20 years of publications. *PLoS Negl. Trop. Dis.* **2017**, *11*, e0006005. [CrossRef] [PubMed]

17.	Miranda, R.A.; Xavier, F.B.; Menezes, R.C. Intestinal parasitism in a Parakanã indigenous community in southwestern Pará State, Brazil. *Cad Saude Publica* **1998**, *14*, 507–511. [CrossRef] [PubMed]

18.	Bóia, M.N.; Carvalho-Costa, F.A.; Sodré, F.C.; Porras-Pedroza, B.E.; Faria, E.C.; Magalhães, G.A.; Silva, I.M. Tuberculosis and intestinal parasitism among indigenous people in the Brazilian Amazon region. *Rev. Saude Publica* **2009**, *43*, 176–178. [CrossRef] [PubMed]

19.	Escobar-Pardo, M.L.; De Godoy, A.P.O.; Machado, R.S.; Rodrigues, D.; Neto, U.F.; Kawakami, E. Prevalence of Helicobacter pylori infection and intestinal parasitosis in children of the Xingu Indian Reservation. *J. Pediatr.* **2011**, *87*, 393–398. [CrossRef] [PubMed]

20.	Assis, E.M.; Olivieria, R.C.; Moreira, L.E.; Pena, J.L.; Rodrigues, L.C.; Machado-Coelho, G.L. Prevalence of intestinal parasites in the Maxakali indigenous community in Minas Gerais, Brazil, 2009. *Cad Saude Publica* **2013**, *29*, 681–690. [CrossRef] [PubMed]

21.	Simões Bdos, S.; Machado-Coelho, G.L.; Pena, J.L.; de Freitas, S.N. Environmental conditions and prevalence of parasitic infection in Xukuru-Kariri indigenous people, Caldas, Brazil. *Rev. Panam. Salud Publica* **2015**, *38*, 42–48. [PubMed]

22.	Neres Norberg, A.; Guerra-Sanches, F.; Moreira-Norberg, P.R.B.; Madeira-Oliveira, J.T.; Santa-Helena, A.A.; Serra-Freire, N.M. Intestinal parasitism in Terena indigenous people of the Province of Mato Grosso do Sul, Brazil. *Rev. Salud Pública* **2015**, *16*, 859–870.

23.	Platts-Mills, A.J.; Babji, S.; Bodhidatta, L.; Gratz, J.; Haque, R.; Havt, A.; McCormick, B.J.; McGrath, M.; Olortegui, M.P.; Samie, A.; et al. Pathogen-specific burdens of community diarrhoea in developing countries: A multisite birth cohort study (MAL-ED). *Lancet Glob. Health* **2015**, *3*, e564–e575. [CrossRef]

24.	Bartelt, L.A.; Platts-Mills, J.A. Giardia: A pathogen or commensal for children in high-prevalence settings? *Curr. Opin. Infect. Dis.* **2016**, *29*, 502–507. [CrossRef] [PubMed]

25.	GBD 2013 Mortality and Causes of Death Collaborators. Global, regional, and national age-sex specific all-cause and cause-specific mortality for 240 causes of death, 1990-2013: A systematic analysis for the Global Burden of Disease Study 2013. *Lancet* **2015**, *385*, 117–171. [CrossRef]

26.	Stensvold, C.R. Pinning down the role of common luminal intestinal parasitic protists in human health and disease–status and challenges. *Parasitology* **2019**, *146*, 695–701. [CrossRef]

27.	Kohli, A.; Bushen, O.Y.; Pinkerton, R.C.; Houpt, E.; Newman, R.D.; Sears, C.L.; Lima, A.A.M.; Guerrant, R.L. Giardia duodenalis assemblage, clinical presentation and markers of intestinal inflammation in Brazilian children. *Trans. R. Soc. Trop. Med. Hyg.* **2008**, *102*, 718–725. [CrossRef]

28.	Hanevik, K.; Hausken, T.; Morken, M.H.; Strand, E.A.; Mørch, K.; Coll, P.; Helgeland, L.; Langeland, N. Persisting symptoms and duodenal inflammation related to Giardia duodenalis infection. *J. Infect.* **2007**, *55*, 524–530. [CrossRef] [PubMed]

29.	Donowitz, J.R.; Alam, M.; Kabir, M.; Ma, J.Z.; Nazib, F.; Platts-Mills, J.A.; Bartelt, L.A.; Haque, R.; Petri, W.A. A prospective longitudinal cohort to investigate the effects of early life giardiasis on growth and all cause diarrhea. *Clin. Infect. Dis.* **2016**, *63*, 792–797. [CrossRef] [PubMed]

30.	Fink, M.Y.; Singer, S.M. The intersection of immune responses, microbiota, and pathogenesis in giardiasis. *Trends Parasitol.* **2017**, *33*, 901–913. [CrossRef] [PubMed]

31.	Wang, Y.; Gonzalez-Moreno, O.; Roellig, D.M.; Oliver, L.; Huguet, J.; Guo, Y.; Feng, Y.; Xiao, L. Epidemiological distribution of genotypes of Giardia duodenalis in humans in Spain. *Parasites Vectors* **2019**, *12*, 1–10. [CrossRef] [PubMed]

32.	Sahagún, J.; Clavel, A.; Goñi, P.; Seral, C.; Llorente, M.T.; Castillo, F.J.; Capilla, S.; Arias, A.; Gómez-Lus, R. Correlation between the presence of symptoms and the Giardia duodenalis genotype. *Eur. J. Clin. Microbiol. Infect. Dis.* **2007**, *27*, 81–83. [CrossRef]

33.	Messa, A., Jr.; Köster, P.C.; Garrine, M.; Gilchrist, C.; Bartelt, L.A.; Nhampossa, T.; Massora, S.; Kotloff, K.; Levine, M.M.; Alonso, P.L.; et al. Molecular diversity of Giardia duodenalis in children under 5 years from the Manhiça district, Southern Mozambique enrolled in a matched case-control study on the aetiology of diarrhoea. *PLoS Negl. Trop. Dis.* **2021**, *15*, e0008987. [CrossRef]

34. Haque, R.; Mondal, D.; Petri, J.W.A.; Karim, A.; Molla, I.H.; Rahim, A.; Faruque, A.S.G.; Ahmad, N.; Kirkpatrick, B.D.; Houpt, E.; et al. Prospective case-control study of the association between common enteric protozoal parasites and diarrhea in Bangladesh. *Clin. Infect. Dis.* **2009**, *48*, 1191–1197. [CrossRef]
35. Aydin, A.F.; Besirbellioglu, B.A.; Avci, I.Y.; Tanyuksel, M.; Araz, E.; Pahsa, A. Classification of Giardia duodenalis parasites in Turkey into groups A and B using restriction fragment length polymorphism. *Diagn. Microbiol. Infect. Dis.* **2004**, *50*, 147–151. [CrossRef]
36. Homan, W.L.; Mank, T.G. Human giardiasis: Genotype linked differences in clinical symptomatology. *Int. J. Parasitol.* **2001**, *31*, 822–826. [CrossRef]
37. Nunes, B.C.; Calegar, D.A.; Pavan, M.G.; Jaeger, L.H.; Monteiro, K.J.L.; Dos Reis, E.R.C.; Lima, M.M.; Neves Bóia, M.; Carvalho-Costa, F.A. Genetic diversity of Giardia duodenalis circulating in three Brazilian biomes. *Infect. Genet. Evol.* **2018**, *59*, 107–112. [CrossRef]
38. De Lucio, A.; Amor-Aramendía, A.; Bailo, B.; Saugar, J.M.; Anegagrie, M.; Arroyo, A.; López-Quintana, B.; Zewdie, D.; Ayehubizu, Z.; Yizengaw, E.; et al. Prevalence and genetic diversity of Giardia duodenalis and Cryptosporidium spp. among school children in a Rural Area of the Amhara Region, North-West Ethiopia. *PLoS ONE* **2016**, *11*, e0159992. [CrossRef]
39. Dacal, E.; Saugar, J.M.; De Lucio, A.; Hernández-De-Mingo, M.; Robinson, E.; Köster, P.C.; Aznar-Ruiz-De-Alegría, M.L.; Espasa, M.; Ninda, A.; Gandasegui, J.; et al. Prevalence and molecular characterization of Strongyloides stercoralis, Giardia duodenalis, Cryptosporidium spp., and Blastocystis spp. isolates in school children in Cubal, Western Angola. *Parasites Vectors* **2018**, *11*, 1–18. [CrossRef] [PubMed]
40. Oliveira-Arbex, A.P.; David, E.B.; Oliveira-Sequeira, T.C.G.; Bittencourt, G.N.; Guimaraes, S.T.D.L. Genotyping of Giardia duodenalis isolates in asymptomatic children attending daycare centre: Evidence of high risk for anthroponotic transmission. *Epidemiol. Infect.* **2015**, *144*, 1418–1428. [CrossRef] [PubMed]
41. Seguí, R.; Muñoz-Antoli, C.; Klisiowicz, D.R.; Oishi, C.Y.; Köster, P.C.; De Lucio, A.; Hernández-De-Mingo, M.; Puente, P.; Toledo, R.; Esteban, J.G.; et al. Prevalence of intestinal parasites, with emphasis on the molecular epidemiology of Giardia duodenalis and Blastocystis sp., in the Paranaguá Bay, Brazil: A community survey. *Parasites Vectors* **2018**, *11*, 1–19. [CrossRef]
42. Prabhu, A.; Morrison, H.G.; Martinez, C.R.; Adam, R.D. Characterisation of the subtelomeric regions of Giardia lamblia genome isolate WBC6. *Int. J. Parasitol.* **2007**, *37*, 503–513. [CrossRef] [PubMed]
43. Teodorovic, S.; Braverman, J.M.; Elmendorf, H.G. Unusually Low Levels of Genetic Variation among Giardia lamblia Isolates. *Eukaryot. Cell* **2007**, *6*, 1421–1430. [CrossRef]
44. Ankarklev, J.; Svärd, S.G.; Lebbad, M. Allelic sequence heterozygosity in single Giardia parasites. *BMC Microbiol.* **2012**, *12*, 65. [CrossRef]
45. Siripattanapipong, S.; Leelayoova, S.; Mungthin, M.; Thompson, R.A.; Boontanom, P.; Saksirisampant, W.; Tan-Ariya, P. Clonal diversity of the glutamate dehydrogenase gene in Giardia duodenalis from Thai Isolates: Evidence of genetic exchange or mixed infections? *BMC Microbiol.* **2011**, *11*, 206. [CrossRef]
46. Seabolt, M.H.; Konstantinidis, K.T.; Roellig, D.M. Hidden diversity within common protozoan parasites revealed by a novel genomotyping scheme. *Appl. Environ. Microbiol.* **2021**. [CrossRef] [PubMed]
47. Ankarklev, J.; Lebbad, M.; Einarsson, E.; Franzén, O.; Ahola, H.; Troell, K.; Svärd, S.G. A novel high-resolution multilocus sequence typing of Giardia intestinalis Assemblage A isolates reveals zoonotic transmission, clonal outbreaks and recombination. *Infect. Genet. Evol.* **2018**, *60*, 7–16. [CrossRef]
48. Instituto Brasileiro de Geografia e Estatística. Indígenas. Available online: https://indigenas.ibge.gov.br/ (accessed on 4 November 2020).
49. Fundação Nacional do Índio. Índios do Brasil. Available online: http://www.funai.gov.br/funai/ (accessed on 4 November 2020).
50. Instituto Sócio-Ambiental. Povos Indígenas no Brasil. Available online: http://www.socioambiental.org/ (accessed on 4 November 2020).
51. Sheather, A.L. The detection of intestinal protozoa and mange parasites by a flotation technique. *J Comp. Pathol.* **1936**, *36*, 266–275. [CrossRef]
52. Hoffman, W.A.; Pons, J.A.; Janer, J.L. The sedimentation concentration method in Schistosomiasis mansoni. *J. Public Health Trop. Med.* **1934**, *9*, 283–291.
53. Allen, A.V.; Ridley, D.S. Further observations on the formol-ether concentration technique for faecal parasites. *J. Clin. Pathol.* **1970**, *23*, 545–546. [CrossRef] [PubMed]
54. Faust, E.C.; D'Antoni, J.S.; Odom, V.; Miller, M.J.; Peres, C.; Sawitz, W.; Thomen, L.F.; Tobie, J.; Walker, J.H. A critical study of clinical laboratory technics for the diagnosis of protozoan cysts and helminth eggs in feces 1. *Am. J. Trop. Med. Hyg.* **1938**, *18*, 169–183. [CrossRef]
55. Weiss, J.B. PCR detection of Giardia lamblia. In *Diagnostic Molecular Microbiology: Principles and Applications*; Persing, D.H., Smith, T.F., Tenover, F.C., White, T.J., Eds.; American Society for Microbiology: Washington, WA, USA, 1993; pp. 480–485.
56. Verweij, J.J.; Schinkel, J.; Laeijendecker, D.; Van Rooyen, M.A.A.; Van Lieshout, L.; Polderman, A.M. Real-time PCR for the detection of Giardia lamblia. *Mol. Cell. Probes* **2003**, *17*, 223–225. [CrossRef]
57. Read, C.M.; Monis, P.T.; Thompson, R.C. Discrimination of all genotypes of Giardia duodenalis at the glutamate dehydrogenase locus using PCR-RFLP. *Infect. Genet. Evol.* **2004**, *4*, 125–130. [CrossRef]

58. Lalle, M.; Pozio, E.; Capelli, G.; Bruschi, F.; Crotti, D.; Cacciò, S.M. Genetic heterogeneity at the beta-giardin locus among human and animal isolates of Giardia duodenalis and identification of potentially zoonotic subgenotypes. *Int. J. Parasitol.* **2005**, *35*, 207–213. [CrossRef]

59. Sulaiman, I.M.; Fayer, R.; Bern, C.; Gilman, R.H.; Trout, J.M.; Schantz, P.M.; Das, P.; Al, I.M.S.E.; Xiao, L. Triosephosphate isomerase gene characterization and potential zoonotic transmission of Giardia duodenalis. *Emerg. Infect. Dis.* **2003**, *9*, 1444–1452. [CrossRef] [PubMed]

60. Tamura, K.; Stecher, G.; Peterson, D.; Filipski, A.; Kumar, S. MEGA6: Molecular evolutionary genetics analysis version 6.0. *Mol. Biol. Evol.* **2013**, *30*, 2725–2729. [CrossRef]

61. Thompson, J.D.; Gibson, T.J.; Plewniak, F.; Jeanmougin, F.; Higgins, D.G. The CLUSTAL_X windows interface: Flexible strategies for multiple sequence alignment aided by quality analysis tools. *Nucleic Acids Res.* **1997**, *25*, 4876–4882. [CrossRef]

62. Nicholas, K.B.; Nicholas, H.B., Jr. GeneDoc: A Tool for Editing and Annotating Multiple Sequence Alignments. (Distributed by the Author). 1997. Available online: https://www.scienceopen.com/document?vid=c8a87cd1-255f-4129-802b-d2382bb0fb37 (accessed on 11 February 2021).

63. Swofford, D.L. *PAUP* Version 4.0 b10. Phylogenetic Analysis Using Parsimony (* and Other Methods)*; Sinauer: Sunderland, UK, 2002.

64. Huelsenbeck, J.P.; Ronquist, F. MRBAYES: Bayesian inference of phylogenetic trees. *Bioinformatics* **2001**, *17*, 754–755. [CrossRef]

65. Pacheco, F.T.; Carvalho, S.S.; Teixeira, M.C.; Cardoso, L.S.; Andrade, L.S.; Das Chagas, G.M.; Gomes, D.C.; Mercês, C.F.; Rocha, F.C.; Silva, L.K.; et al. Immune response markers in sera of children infected with Giardia duodenalis AI and AII subassemblages. *Immunobiology* **2019**, *224*, 595–603. [CrossRef] [PubMed]

66. Santos, C.K.S.; Grama, D.F.; Limongi, J.E.; Costa, F.C.; Couto, T.R.; Soares, R.M.; Mundim, M.J.S.; Cury, M.C. Epidemiological, parasitological and molecular aspects of Giardia duodenalis infection in children attending public daycare centers in southeastern Brazil. *Trans. R. Soc. Trop. Med. Hyg.* **2012**, *106*, 473–479. [CrossRef]

67. Scalia, L.A.; Fava, N.M.; Soares, R.M.; Limongi, J.E.; da Cunha, M.J.; Pena, I.F.; Kalapothakis, E.; Cury, M.C. Multilocus genotyping of Giardia duodenalis in Brazilian children. *Trans. R. Soc. Trop. Med. Hyg.* **2016**, *110*, 343–349. [CrossRef] [PubMed]

68. Gomes, K.B.; Fernandes, A.P.; Menezes, A.; Júnior, R.A.; Silva, E.F.; Rocha, M.O. Giardia duodenalis: Genotypic comparison between a human and a canine isolates. *Rev. Soc. Bras. Med. Trop.* **2011**, *44*, 508–510. [CrossRef]

69. Uda-Shimoda, C.F.; Colli, C.M.; Pavanelli, M.F.; Falavigna-Guilherme, A.L.; Gomes, M.L. Simplified protocol for DNA extraction and amplification of 2 molecular markers to detect and type Giardia duodenalis. *Diagn. Microbiol. Infect. Dis.* **2014**, *78*, 53–58. [CrossRef] [PubMed]

70. Colli, C.M.; Bezagio, R.C.; Nishi, L.; Bignotto, T.S.; Ferreira, É.C.; Falavigna-Guilherme, A.L.; Gomes, M.L. Identical assemblage of Giardia duodenalis in humans, animals and vegetables in an urban area in Southern Brazil indicates a relationship among them. *PLoS ONE* **2015**, *10*, e0118065. [CrossRef]

71. Colli, C.M.; Bezagio, R.C.; Nishi, L.; Ferreira, É.C.; Falavigna-Guilherme, A.L.; Gomes, M.L. Food handlers as a link in the chain of transmission of Giardia duodenalis and other protozoa in public schools in southern Brazil: Table 1. *Trans. R. Soc. Trop. Med. Hyg.* **2015**, *109*, 601–603. [CrossRef] [PubMed]

72. Volotão, A.C.; Costa-Macedo, L.M.; Haddad, F.S.; Brandão, A.; Peralta, J.M.; Fernandes, O. Genotyping of Giardia duodenalis from human and animal samples from Brazil using beta-giardin gene: A phylogenetic analysis. *Acta Trop.* **2007**, *102*, 10–19. [CrossRef]

73. Faria, C.P.; Zanini, G.M.; Dias, G.S.; Da Silva, S.; Sousa, M.D.C. Molecular Characterization of Giardia lamblia: First Report of Assemblage B in Human Isolates from Rio de Janeiro (Brazil). *PLoS ONE* **2016**, *11*, e0160762. [CrossRef]

74. Faria, C.P.; Zanini, G.M.; Dias, G.S.; Sousa, M.D.C. Associations of Giardia lamblia assemblages with HIV infections and symptomatology: HIV virus and assemblage B were they born to each other? *Acta Trop.* **2017**, *172*, 80–85. [CrossRef]

75. Faria, C.P.; Zanini, G.M.; Dias, G.S.; Da Silva, S.; Sousa, M.D.C. New multilocus genotypes of Giardia lamblia human isolates. *Infect. Genet. Evol.* **2017**, *54*, 128–137. [CrossRef] [PubMed]

76. Cascais-Figueiredo, T.; Austriaco-Teixeira, P.; Fantinatti, M.; Silva-Freitas, M.L.; Santos-Oliveira, J.R.; Coelho, C.H.; Singer, S.M.; Da-Cruz, A.M. Giardiasis alters intestinal fatty acid binding protein (I-FABP) And plasma cytokines levels in children in Brazil. *Pathogens* **2019**, *9*, 7. [CrossRef] [PubMed]

77. Fantinatti, M.; Bello, A.R.; Fernandes, O.; Da-Cruz, A.M. Identification of Giardia lamblia assemblage E in humans points to a new anthropozoonotic cycle. *J. Infect. Dis.* **2016**, *214*, 1256–1259. [CrossRef]

78. De Quadros, R.M.; Weiss, P.H.E.; Marques, S.M.T.; Miletti, L.C. Potential cross-contamination of similar giardia duodenalis assemblage in children and pet dogs in Southern Brazil, as determined by PCR-RFLP. *Rev. Inst. Med. Trop. São Paulo* **2016**, *58*. [CrossRef]

79. De Godoy, E.A.M.; Junior, J.E.S.; Belloto, M.V.T.; De Moraes, M.V.P.; Cassiano, G.C.; Volotao, A.C.C.; Luvizotto, M.C.R.; Carareto, C.M.A.; Silva, M.C.D.M.; Machado, R.L.D. Molecular investigation of zoonotic genotypes of Giardia intestinalis isolates in humans, dogs and cats, sheep, goats and cattle in Araçatuba (São Paulo State, Brazil) by the analysis of ß-giardin gene fragments. *Microbiol. Res.* **2013**, *4*, e6. [CrossRef]

80. Coradi, S.; David, E.; Oliveira-Sequeira, T.; Ribolla, P.; Carvalho, T.; Guimarães, S. Genotyping of Brazilian Giardia duodenalis human axenic isolates. *J. Venom. Anim. Toxins Incl. Trop. Dis.* **2011**, *17*, 353–357. [CrossRef]

81. Durigan, M.; Abreu, A.G.; Zucchi, M.I.; Franco, R.M.B.; De Souza, A.P. Genetic diversity of Giardia duodenalis: Multilocus genotyping reveals zoonotic potential between clinical and environmental sources in a metropolitan region of Brazil. *PLoS ONE* **2014**, *9*, e115489. [CrossRef] [PubMed]

82. Corrêa, C.R.T.; Oliveira-Arbex, A.P.; David, É.B.; Guimarães, S. Genetic analysis of Giardia duodenalis isolates from children of low-income families living in an economically successful region in Southeastern Brazil. *Rev. Inst. Med. Trop. Sao Paulo* **2020**, *62*, e20. [CrossRef]

83. Volotão, A.C.; Ramos, N.M.; Fantinatti, M.; Moraes, M.V.; Netto, H.A.; Storti-Melo, L.M.; Godoy, E.A.; Rossit, A.R.; Fernandes, O.; Machado, R.L. Giardiasis as zoonosis: Between proof of principle and paradigm in the Northwestern region of São Paulo State, Brazil. *Braz. J. Infect. Dis.* **2011**, *15*, 382–383. [CrossRef] [PubMed]

84. David, É.B.; Guimarães, S.; de Oliveira, A.P.; de Oliveira-Sequeira, T.C.; Nogueira Bittencourt, G.; Moraes Nardi, A.R.; Martins Ribolla, P.E.; Bueno Franco, R.M.; Branco, N.; Tosini, F.; et al. Molecular characterization of intestinal protozoa in two poor communities in the State of São Paulo, Brazil. *Parasit Vectors* **2015**, *8*, 103. [CrossRef]

85. Souza, S.L.; Gennari, S.M.; Richtzenhain, L.J.; Pena, H.F.; Funada, M.R.; Cortez, A.; Gregori, F.; Soares, R.M. Molecular identification of Giardia duodenalis isolates from humans, dogs, cats and cattle from the state of São Paulo, Brazil, by sequence analysis of fragments of glutamate dehydrogenase (gdh) coding gene. *Veter. Parasitol.* **2007**, *149*, 258–264. [CrossRef] [PubMed]

86. Da Cunha, M.J.R.; Cury, M.C.; Santín, M. Molecular identification of Enterocytozoon bieneusi, Cryptosporidium, and Giardia in Brazilian captive birds. *Parasitol. Res.* **2016**, *116*, 487–493. [CrossRef]

87. Fava, N.M.N.; Soares, R.M.; Scalia, L.A.M.; Kalapothakis, E.; Pena, I.F.; Vieira, C.U.; Faria, E.S.M.; Cunha, M.J.; Couto, T.R.; Cury, M.C. Performance of glutamate dehydrogenase and triose phosphate isomerase genes in the analysis of genotypic variability of isolates of Giardia duodenalis from livestocks. *BioMed Res. Int.* **2013**, *2013*, 1–9. [CrossRef]

88. Fava, N.M.; Soares, R.M.; Scalia, L.A.; Da Cunha, M.J.R.; Faria, E.S.; Cury, M.C. Molecular typing of canine Giardia duodenalis isolates from Minas Gerais, Brazil. *Exp. Parasitol.* **2016**, *161*, 1–5. [CrossRef] [PubMed]

89. Meireles, P.; Montiani-Ferreira, F.; Thomaz-Soccol, V. Survey of giardiosis in household and shelter dogs from metropolitan areas of Curitiba, Paraná state, Southern Brazil. *Veter. Parasitol.* **2008**, *152*, 242–248. [CrossRef]

90. Uchôa, F.F.D.M.; Sudré, A.P.; Campos, S.D.E.; Almosny, N.R.P. Assessment of the diagnostic performance of four methods for the detection of Giardia duodenalis in fecal samples from human, canine and feline carriers. *J. Microbiol. Methods* **2018**, *145*, 73–78. [CrossRef]

91. Fantinatti, M.; Caseca, A.C.; Bello, A.R.; Fernandes, O.; Da-Cruz, A.M.; Filho, O.F.D.S. The presence of Giardia lamblia assemblage A in dogs suggests an anthropozoonotic cycle of the parasite in Rio de Janeiro, Brazil. *Infect. Genet. Evol.* **2018**, *65*, 265–269. [CrossRef]

92. Volotão, A.; Júnior, J.S.; Grassini, C.; Peralta, J.; Fernandes, O. Genotyping of Giardia duodenalis from Southern Brown Howler Monkeys (Alouatta clamitans) from Brazil. *Veter. Parasitol.* **2008**, *158*, 133–137. [CrossRef] [PubMed]

93. Soares, R.M.; De Souza, S.L.P.; Silveira, L.H.; Funada, M.R.; Richtzenhain, L.J.; Gennari, S.M. Genotyping of potentially zoonotic Giardia duodenalis from exotic and wild animals kept in captivity in Brazil. *Veter. Parasitol.* **2011**, *180*, 344–348. [CrossRef] [PubMed]

94. De Aquino, M.C.C.; Harvey, T.V.; Inácio, S.V.; Nagata, W.B.; Ferrari, E.D.; Oliveira, B.C.M.; Albuquerque, G.R.; Widmer, G.; Meireles, M.V.; Bresciani, K.D.S. First description of Giardia duodenalis in buffalo calves (Bubalus bubalis) in southwest region of São Paulo State, Brazil. *Food Waterborne Parasitol.* **2019**, *16*, e00062. [CrossRef] [PubMed]

95. David, E.B.; Patti, M.; Coradi, S.T.; Oliveira-Sequeira, T.C.G.; Ribolla, P.E.M.; Guimaraes, S. Molecular typing of Giardia duodenalis isolates from nonhuman primates housed in a Brazilian zoo. *Rev. Inst. Med. Trop. São Paulo* **2014**, *56*, 49–54. [CrossRef] [PubMed]

96. e Silva, F.M.P.; Monobe, M.M.; Lopes, R.S.; Araujo, J.P. Molecular characterization of Giardia duodenalis in dogs from Brazil. *Parasitol. Res.* **2012**, *110*, 325–334. [CrossRef] [PubMed]

97. Silva, F.M.P.; Lopes, R.S.; Araújo, J.P.; Paze, F.S.M. Genetic characterisation of Giardia duodenalis in dairy cattle in Brazil. *Folia Parasitol.* **2012**, *59*, 15–20. [CrossRef]

98. Silva, F.P.E.; Lopes, R.; Bresciani, K.D.S.; Amarante, A.; Araujo, J. High occurrence of Cryptosporidium ubiquitum and Giardia duodenalis genotype E in sheep from Brazil. *Acta Parasitol.* **2014**, *59*, 193–196. [CrossRef]

99. Ferreira, F.P.; Caldart, E.T.; Freire, R.L.; Mitsuka-Breganó, R.; De Freitas, F.M.; Miura, A.C.; Mareze, M.; Martins, F.D.C.; Urbano, M.R.; Seifert, A.L.; et al. The effect of water source and soil supplementation on parasite contamination in organic vegetable gardens. *Rev. Bras. Parasitol. Vet.* **2018**, *27*, 327–337. [CrossRef]

100. Almeida, J.C.; Martins, F.D.C.; Neto, J.M.F.; Dos Santos, M.M.; Garcia, J.L.; Navarro, I.T.; Kuroda, E.K.; Freire, R.L. Occurrence of Cryptosporidium spp. and Giardia spp. in a public water-treatment system, Paraná, Southern Brazil. *Rev. Bras. Parasitol. Vet.* **2015**, *24*, 303–308. [CrossRef]

101. Ladeia, W.A.; Martins, F.D.C.; e Silva, C.F.R.; Freire, R.L. Molecular surveillance of Cryptosporidium and Giardia duodenalis in sludge and spent filter backwash water of a water treatment plant. *J. Water Health* **2018**, *16*, 857–860. [CrossRef] [PubMed]

102. Souza, D.S.M.; Ramos, A.P.D.; Nunes, F.F.; Moresco, V.; Taniguchi, S.; Leal, D.A.G.; Sasaki, S.T.; Bícego, M.C.; Montone, R.C.; Durigan, M.; et al. Evaluation of tropical water sources and mollusks in southern Brazil using microbiological, biochemical, and chemical parameters. *Ecotoxicol. Environ. Saf.* **2012**, *76*, 153–161. [CrossRef]

103. De Araújo, R.S.; Aguiar, B.; Dropa, M.; Razzolini, M.T.P.; Sato, M.I.Z.; Lauretto, M.D.S.; Galvani, A.T.; Padula, J.A.; Matté, G.R.; Matté, M.H. Detection and molecular characterization of Cryptosporidium species and Giardia assemblages in two watersheds in the metropolitan region of São Paulo, Brazil. *Environ. Sci. Pollut. Res.* **2018**, *25*, 15191–15203. [CrossRef]
104. Ulloa-Stanojlović, F.M.; Aguiar, B.; Jara, L.M.; Sato, M.I.Z.; Guerrero, J.A.; Hachich, E.; Matté, G.R.; Dropa, M.; Matté, M.H.; De Araújo, R.S. Occurrence of Giardia intestinalis and Cryptosporidium sp. in wastewater samples from São Paulo State, Brazil, and Lima, Peru. *Environ. Sci. Pollut. Res.* **2016**, *23*, 22197–22205. [CrossRef] [PubMed]
105. Yamashiro, S.; Foco, M.L.R.; Pineda, C.O.; José, J.; Nour, E.A.A.; Siqueira-Castro, I.C.V.; Franco, R.M.B. Giardia spp. and Cryptosporidium spp. removal efficiency of a combined fixed-film system treating domestic wastewater receiving hospital effluent. *Environ. Sci. Pollut. Res.* **2019**, *26*, 22756–22771. [CrossRef]
106. Leal, D.A.G.; Souza, D.S.M.; Caumo, K.S.; Fongaro, G.; Panatieri, L.F.; Durigan, M.; Rott, M.B.; Barardi, C.R.M.; Franco, R.M.B. Genotypic characterization and assessment of infectivity of human waterborne pathogens recovered from oysters and estuarine waters in Brazil. *Water Res.* **2018**, *137*, 273–280. [CrossRef]
107. Fernandes, L.N.; De Souza, P.P.; De Araújo, R.S.; Razzolini, M.T.P.; Soares, R.M.; Sato, M.I.Z.; Hachich, E.M.; Cutolo, S.A.; Matté, G.R.; Matté, M.H. Detection of assemblages A and B of Giardia duodenalis in water and sewage from São Paulo state, Brazil. *J. Water Health* **2011**, *9*, 361–367. [CrossRef] [PubMed]
108. Tiyo, R.; De Souza, C.Z.; Piovesani, A.F.A.; Tiyo, B.T.; Colli, C.M.; Marchioro, A.A.; Gomes, M.L.; Falavigna-Guilherme, A.L. Predominance of Giardia duodenalis assemblage aii in fresh leafy vegetables from a market in Southern Brazil. *J. Food Prot.* **2016**, *79*, 1036–1039. [CrossRef] [PubMed]
109. Rafael, K.; Marchioro, A.A.; Colli, C.M.; Tiyo, B.T.; Evangelista, F.F.; Bezagio, R.C.; Falavigna-Guilherme, A.L. Genotyping of Giardia duodenalis in vegetables cultivated with organic and chemical fertilizer from street markets and community vegetable gardens in a region of Southern Brazil. *Trans. R. Soc. Trop. Med. Hyg.* **2017**, *111*, 540–545. [CrossRef] [PubMed]

pathogens

Article

Symptomatic and Asymptomatic Protist Infections in Hospital Inpatients in Southwestern China

Shun-Xian Zhang [1,2,†], David Carmena [3,†], Cristina Ballesteros [4], Chun-Li Yang [5], Jia-Xu Chen [2,6,7], Yan-Hong Chu [2,6,7], Ying-Fang Yu [2,6,7], Xiu-Ping Wu [2,6,7], Li-Guang Tian [2,6,7,*] and Emmanuel Serrano [3,4,*]

1   Clinical Research Center, LongHua Hospital Shanghai University of Traditional Chinese Medicine, Shanghai 200032, China; zhangshunxian110@163.com
2   Chinese Center for Tropical Diseases Research, National Institute of Parasitic Diseases, Chinese Center for Disease Control and Prevention, Shanghai 200025, China; chenjx@nipd.chinacdc.cn (J.-X.C.); chuyh@nipd.chinacdc.cn (Y.-H.C.); yuyf@nipd.chinacdc.cn (Y.-F.Y.); wuxp@nipd.chinacdc.cn (X.-P.W.)
3   Parasitology Reference and Research Laboratory, National Centre for Microbiology, Health Institute Carlos III, Majadahonda, 28220 Madrid, Spain; dacarmena@isciii.es
4   Wildlife Ecology & Health Group (WE&H), Servei d'Ecopatologia de Fauna Salvatge (SEFaS), Departament de Medicina i Cirurgia Animals, Universitat Autònoma de Barcelona (UAB), E-08193 Bellaterra, Spain; cballesteros6@rvc.ac.uk
5   Department of Clinical Research, the 903rd Hospital of People's Liberation Army of China, Hangzhou 310013, China; chunliyang2014@163.com
6   NHC Key Laboratory of Parasite and Vector Biology, National Institute of Parasitic Diseases, Chinese Center for Disease Control and Prevention, Shanghai 200025, China
7   School of Global Health, Chinese Center for Tropical Diseases Research-Shanghai Jiao Tong University School of Medicine, Shanghai 200025, China
*   Correspondence: tianlg@nipd.chinacdc.cn (L.-G.T.); emmanuel.serrano@uab.cat (E.S.)
†   These two authors contributed equally to this work.

Citation: Zhang, S.-X.; Carmena, D.; Ballesteros, C.; Yang, C.-L.; Chen, J.-X.; Chu, Y.-H.; Yu, Y.-F.; Wu, X.-P.; Tian, L.-G.; Serrano, E. Symptomatic and Asymptomatic Protist Infections in Hospital Inpatients in Southwestern China. *Pathogens* **2021**, *10*, 684. https://doi.org/10.3390/pathogens10060684

Academic Editors: Lawrence S. Young and Siddhartha Das

Received: 28 March 2021
Accepted: 19 May 2021
Published: 31 May 2021

Publisher's Note: MDPI stays neutral with regard to jurisdictional claims in published maps and institutional affiliations.

Abstract: *Cryptosporidium* spp., *Entamoeba histolytica*, *Giardia duodenalis*, and *Blastocystis* sp. infections have been frequently reported as etiological agents for gastroenteritis, but also as common gut inhabitants in apparently healthy individuals. Between July 2016 and March 2017, stool samples ($n = 507$) were collected from randomly selected individuals (male/female ratio: 1.1, age range: 38–63 years) from two sentinel hospitals in Tengchong City Yunnan Province, China. Molecular (PCR and Sanger sequencing) methods were used to detect and genotype the investigated protist species. Carriage/infection rates were: *Blastocystis* sp. 9.5% (95% CI: 7.1–12.4%), *G. duodenalis* 2.2% (95% CI: 1.1–3.8%); and *E. histolytica* 2.0% (95% CI: 0.9–3.6%). *Cryptosporidium* spp. was not detected at all. Overall, 12.4% (95% CI: 9.7–15.6) of the participants harbored at least one enteric protist species. The most common coinfection was *E. histolytica* and *Blastocystis* sp. (1.0%; 95% CI: 0.3–2.2). Sequence analyses revealed that 90.9% (10/11) of the genotyped *G. duodenalis* isolates corresponded to the sub-assemblage AI. The remaining sequence (9.1%, 1/11) was identified as sub-assemblage BIV. Five different *Blastocystis* subtypes, including ST3 (43.7%, 21/48), ST1 (27.1%, 13/48), ST7 (18.8%, 9/48), ST4 (8.3%, 4/48), and ST2 (2.1%, 1/48) were identified. Statistical analyses confirmed that (i) the co-occurrence of protist infections was purely random, (ii) no associations were observed among the four protist species found, and (iii) neither their presence, individually or jointly, nor the patient's age was predictors for developing clinical symptoms associated with these infections. Overall, these protist mono- or coinfections are asymptomatic and do not follow any pattern.

Keywords: coinfection; enteric protists; China; *Giardia duodenalis*; *Entamoeba histolytica*; *Cryptosporidium*; *Blastocystis* sp.; genotyping; molecular diversity

1. Introduction

Parasitic infections have been frequently reported as significant causes of gastrointestinal disorders and major contributors to the global burden of diarrheal disease globally [1–5].

Several diarrhea-causing enteric parasite species have been described in humans. Among them, the most relevant are the protozoa *Cryptosporidium* spp., *Entamoeba histolytica*, and *Giardia duodenalis*, and the stramenopile *Blastocystis* sp. [5,6]. However, the impact that these agents exert on public health has not been fully characterized yet as a large amount of the epidemiological information currently available is mainly based on single protist infections, and the role of coinfections is often understated. It is estimated that around 20% of child diarrheal episodes are reported in low and middle-income countries, and 9% of those documented in high-income settings are caused by *Cryptosporidium* spp. [1]. Amebiasis, the acute disease caused by *E. histolytica*, affects near 50 million people and causes 100,000 deaths each year [7]. Both *Cryptosporidium* spp. and *E. histolytica* have been recognized as significant causes of morbidity and mortality associated with diarrhea by the 2017 Global Burden Disease Study [4]. On the other hand, *G. duodenalis* infection is estimated to result in 280 million cases annually worldwide [8] and was included, together with *Cryptosporidium* spp., in the "Neglected Disease Initiative" launched by the World Health Organization in 2004 [9]. *Blastocystis* sp. is regarded as the most prevalent enteric protist isolated from diarrheal patients in high-income countries [1]. However, its pathogenicity remains controversial partly because it is also the most frequent non-fungal eukaryotic organism detected in fecal samples from apparently healthy individuals [10–12]. It has been argued that this protist could be used as an indicator of potential exposure to other pathogenic enteric protozoa [13]. Besides *Blastocystis* sp., *Cryptosporidium* spp., *E. histolytica*, and *G. duodenalis* infections are common in apparently healthy individuals [1,14,15]. There is a lack of reliable data on the true burden of asymptomatic infections due to the absence of monitoring programs, underreporting and the fact that carriage of subclinical stages is often underdiagnosed [1,16,17].

Even though there have been notable advances in this field, the factors that determine the course of enteric protozoan infections and the development (or not) of gastrointestinal symptoms remain poorly understood [18]. Protist species/genotypes have been suggested as predictors of pathogenicity/virulence. It is expected that the growing development of new molecular diagnostic tools contributes to clarifying the distinction between pathogenic and nonpathogenic lineages, as well as pathophysiological interactions and other epidemiological features of interest [19].

Many previous studies have explored and reported the presence of more than one pathogen in cases with diarrheal disease [6,19,20] as well as in healthy individuals [21,22]. Comparatively, less effort has been devoted to exploring the impact of concomitant enteric infections involving viral, bacterial, and/or parasitic agents and characterizing the enteric communities and the existence of specific interactions among them [15,23,24]. These interactions, at least between 2 specific enteric pathogens, are well documented in the veterinary field, such as the association of enterotoxigenic *Escherichia coli* and rotavirus that lead to severe diarrhea in piglets as an example [25]. However, it has not been addressed in immunocompetent humans until recently [15,26].

To improve our current understanding of the existence and impact of intestinal protozoan coinfections in humans, we conducted a hospital-based cross-sectional study. The aims of the survey were: (i) to investigate the occurrence of the four protist species most commonly associated with gastrointestinal disorders and to determine their molecular profile, (ii) to assess whether these protists co-occur by chance or are community-structured in hospital-based patients, and (iii) to explore the potential impact of coinfections on the clinical symptomatology among immunocompetent patients attending the People's Hospital of Tengchong City and the Chinese Medicine Hospital in Tengchong City.

2. Results

2.1. Characteristics of the Study Population

A total of 507 subjects participated in the study. The male:female ratio was 1.1 (260/247). Han nationality was predominant (94.9%, 481/507), and the median age of recruited participants was 52 years (interquartile range (IQR): 38–63 years). Most of the subjects lived in rural areas (78.1%, 396/507). Around two-thirds of them had completed primary education (65.6%, 332/507), and 63.1% of the participants were farmers (320/507).

The most common clinical symptom was decreased appetite (17.9%, 91/507), followed by abdominal pain (16.6%, 84/507) and nausea (16.6%, 84/507); in addition, 10.4% (59/507) of recruited patients presenting with acute diarrhea. Other less frequently reported symptoms included abdominal distension (12.4%, 63/507), itchy skin 12.2%, 62/507), constipation (10.8%, 55/507) and perianal pruritus (5.5%, 28/507). Overall, 50.0% (233/507) of the patients did not have any gastrointestinal symptoms.

2.2. Single Enteric Pathogen Infections and Coinfections

Blastocystis sp. was the most prevalent protist species found (9.5%, 48/507; 95% CI: 7.1–12.4), followed by *G. duodenalis* (2.2%, 11/507; 95% CI: 1.1–3.8) and *E. histolytica* (2.0%, 10/507; 95% CI: 0.9–3.6), whereas *Cryptosporidium* spp. was not detected at all. *Entamoeba histolytica* was more frequently found in diarrheal (11.9%, 7/59) than in non-diarrheal (0.7%, 3/448) cases (OR = 19.9, 95% CI: 5.0–79.5). The same was true for *G. duodenalis* (6.8%, 4/59 versus 1.6%, 7/448; OR = 4.6, 95% CI: 1.3–16.2). In contrast, no significant differences were observed on the distribution of *Blastocystis* sp. between diarrheal and non-diarrheal cases (13.6%, 8/59 versus 8.9%, 40/448; OR = 1.6, 95% CI: 0.7–3.6).

Overall, 12.4% of patients (63/507, 95% CI: 9.7–15.6) were infected by at least one enteric protist. Infections caused by two different enteric protist species were detected in 1.2% of patients (6/507, 95% CI: 0.4–2.6). The most common coinfection found was *E. histolytica* and *Blastocystis* sp. (1.0%, 5/507; 95% CI: 0.3–2.2), and the coinfection of *E. histolytica* and *Blastocystis* sp. in diarrheal cases was more common than in non-diarrheal patients (5.1%, 3/59; 0.4%, 2/448, chi-squared = 10.8, *p*-value = 0.001). Coinfection by *G. duodenalis* and *Blastocystis* sp. was identified in a single case (0.2%, 1/59; 95% CI: 0.01–1.1), and no significant difference of the coinfection by *G. duodenalis* and *Blastocystis* sp. was found in subjects with and without diarrhea (0.2%, 1/59; 0.0%, 0/449, chi-squared = 2.81, *p*-value = 0.116). No other coinfection of these four enteric protozoa was found in diarrhea individuals and healthy controls.

2.3. Genetic Characterization of Isolates

A total of 11 *G. duodenalis* isolates were successfully characterized at the triosephosphate isomerase (*tpi*) locus (Table 1). Sequence analysis allowed identifying assemblages A (90.9%, 10/11) and B (9.1%, 1/11). Further, all 10 assemblage A isolates were assigned to the sub-assemblage AI of the parasite, and the only one assemblage B isolate was identified as sub-assemblage BIV. *Giardia duodenalis* assemblage A was significantly more prevalent in individuals presenting with diarrhea (4/59) than in individuals without clinical manifestations (6/448, chi-squared = 7.37, *p*-value = 0.006; OR = 5.4, 95% CI: 1.5–19.6).

Table 1. Diversity, frequency, and molecular features of *Giardia duodenalis* sequences at the *tpi* locus generated in the present study. GenBank accession numbers are provided.

Locus	Assemblage	Sub-Assemblage	Isolates	Reference Sequence	Stretch	Single-Nucleotide Polymorphisms	GenBank ID
tpi	A	AI	1	L02120	594–1036	None	MW810321
			1	L02120	586–1060	G746A, A834G	MW810322
			1	L02120	595–1060	G915A, G996A, G1004	MW810323
			7 [1]	L02120	-	-	-
	B	BIV	1 [1]	AF069560	-	-	-

[1] isolates with associated sequences of insufficient quality to clearly determine the presence of single nucleotide polymorphisms.

Blastocystis sp. was detected by PCR amplification of the small subunit ribosomal RNA (*ssu* rRNA) gene in 48 samples (Table 2). Sequence analyses revealed the presence of five subtypes (ST) of this protist, with ST3 being the most prevalently found (43.7%, 21/48), followed by ST1 (27.1%, 13/48), ST7 (18.8%, 9/48), ST4 (8.3%, 4/48), and ST2 (2.1%, 1/48). Concerning intra-subtype genetic diversity, alleles 2, 4, and 88 were identified within ST1, allele 9 within ST2, allele 34 within ST3, alleles 42, 92, and 94 within ST4, and alleles 100 and 101 within ST7. No statistical differences in the distribution frequencies of *Blastocystis* STs were observed between diarrheal and non-diarrheal individuals (chi-squared = 3.73, *p*-value = 0.2913).

Table 2. Diversity, frequency, and molecular features of *Blastocystis* sp. sequences at the *ssu* rRNA locus generated in the present study. GenBank accession numbers are provided.

Subtype	Allele	Isolates	GenBank ID
ST1	2	3	MW798733
	4	7	MW798734
	88	3	MW798735
ST2	9	1	MW798736
ST3	34	21	MW798737
ST4	42	2	MW798738
	92	1	MW798739
	94	1	MW798740
ST7	100	3	MW798741
	101	6	MW798742

2.4. Impact of Coinfection on Diarrheal Symptomatology

According to our partial least square (PLS) analyses, no statistically significant associations were identified between the presence of *G. duodenalis*, *Blastocystis* sp. and *E. histolytica*, alone or in combination, and the occurrence of clinical manifestations (Stone-Geisser's Q2 test value <0.0975). Concerning the other predictor, age was not statistically associated with symptomatology in the study population either.

Furthermore, the presence of any of the three protist species investigated, jointly or separately, covaried negatively with symptomatology. The analysis revealed that most of the X's component variance was due to coinfection with enteric protists (43.1%), followed by *Blastocystis* sp. (27.6%) and *E. histolytica* (24.0%) (Table 3).

Table 3. Predictor weights of the PLS model explaining their association with clinical symptomatology in hospital-based patients in Tengchong City.

Predictor Variables	Loads	Weights	Percent	Cross-Correlation
Protozoa richness	−0.67	−0.66	43.1	−0.11
Giardia duodenalis	−0.2	−0.17	2.9	−0.03
Blastocystis sp.	−0.56	−0.53	27.6	−0.08
Entamoeba histolytica	−0.44	−0.5	24	−0.08

2.5. Co-Occurrence of Enteric Protozoa

The null model analysis showed that the observed C-score (265) was lower than expected by chance (267.45), indicating the existence of a random, noncompetitively structured protist community. No evidence of a statistically significant protist combination could be demonstrated in the surveyed population (SES = −0.33, *p*-value = 0.414).

3. Discussion

Our study confirmed the single prevalence rates of enteric protist species, including *G. duodenalis*, *Blastocystis* sp., and *E. histolytica* commonly reported in previous studies in similar settings characterized by adequate access to water sanitation and personal hygiene practices (WASH) [1,17,27]. In the present PCR-based report, *Blastocystis* sp. was the predominant intestinal protist species identified, followed by *G. duodenalis* and *E. histolytica*, whereas *Cryptosporidium* spp. was not detected at all. These data were also consistent with those documented in previous studies conducted in human populations in China [28].

Regarding the genetic diversity of the protist species investigated, out of the 11 *G. duodenalis* isolates successfully genotyped at the *tpi* locus, 10 (90.9%) were classified as assemblage A and the remaining one (9.1%) as assemblage B. All assemblage A sequences were identified as sub-assemblage AI. This finding is interesting as sub-assemblage AI is typically reported at much lower frequencies than sub-assemblage AII in European [29–31] and Asian [32] countries. This discrepancy may be due to differences in infection sources or transmission pathways (e.g., a variable proportion of infections of zoonotic nature). No mixed coinfections involving assemblages A and B were found. Similarly, no infections caused by animal-specific assemblages C–D (dogs), E (domestic and wild ungulates) and F (cats) were detected, suggesting that these host species play a limited role as a source of human giardiasis in this geographical area.

It has been suggested that the presence and proportion of *G. duodenalis* assemblages may present spatiotemporal variations [33,34]. Additionally, socioeconomic factors have also been suggested as potential drivers for *G. duodenalis* assemblage distribution [35]. Interestingly, some molecular epidemiological investigations have reported that assemblage's segregation may be involved in infection outcome and clinical presentation, with assemblage B resulting in more frequently symptomatic infection in endemic settings [21,36]. Other surveys have indicated that sub-assemblage AII may represent a more virulent genetic variant of *G. duodenalis* in humans [17,37,38]. The data mentioned above may explain, at least partially, the lack of evidence in support of a potential association between the presence of *G. duodenalis* and the occurrence of clinical symptomatology in the present study. It should be noted, however, that a recent case–control study assessing the frequency and genetic diversity of *G. duodenalis* infections in children younger than five years of age with and without diarrhea in southern Mozambique has demonstrated that the occurrence of gastrointestinal illness was not associated with a given genotype of the parasite [39].

Five distinct *Blastocystis* STs were identified in the hospital inpatient population under study, including ST1–ST4 and ST7. This is well in agreement with the available molecular data from China, where ST1–ST7 and ST12 have been described at variable frequency rates in different human populations [40]. Interestingly, clinical studies on patients presenting with diarrhea have shown that ST1 was related to clinical manifestations, including diarrhea and may have, therefore, potential pathogenicity. In contrast, ST3 was the *Blastocystis* genetic variant more frequently identified in asymptomatic infections [41–43]. This potential link between a given *Blastocystis* ST and the presence/absence of clinical manifestations could not be demonstrated in the present survey. Of note, a relatively high proportion of individuals carried *Blastocystis* ST7, a subtype mostly identified in birds and rarely found in mammals, including humans [44]. This finding clearly indicates that contact with poultry or captive avian species (or with their fecal material) was the most likely source of infection of these inpatients.

Remarkably, *Cryptosporidium* spp. was not detected in any of the recruited patients in this survey. In China, an overall *Cryptosporidium* infection rate of 3% has been estimated for 1987-2018 [45]. In hospital settings, documented infection rates vary greatly depending on the clinical population under study. Cryptosporidiosis cases have been reported with low frequencies in children hospitalized primarily for non-gastrointestinal illnesses (1.6%, 102/6284) in Shanghai [46] and in diarrheic children (10/500) in Wuhan, Hubei province [47]. In contrast, comparatively much higher (6-40%) infection rates have been identified in gastrointestinal (including esophageal, small intestine, colorectal, and liver) cancer patients [48]. These discrepancies may be associated with differences in the age and immunological status of the patients, the diagnostic methods used or even the geographical area considered.

As we have shown, it seems that, at the enteric protist level, there is not any specific community assemblage in the study population considering the four protist species considered. Our work showed that the occurrence of those pathogens was purely random and asymptomatic in the study population. Furthermore, not specific interactions were detected between them, and there was no evidence suggesting that their presence, jointly or separately, was associated with developing clinical symptomatology classically associated with infections by these species. Although all of them have been reported as causative agents of gastrointestinal disorders, especially in children in LMIC [20], asymptomatic carriage has also been commonly reported elsewhere. In the case of *Blastocystis* sp., its pathogenic relevance is still controversial [10,11] since this species is one of the most common enteric protists detected in humans [12,49]. Therefore, it is not surprising that their presence was not associated with clinical symptomatology in immunocompetent individuals.

Given the results obtained, different factors might provide an explanation that helps to understand the pathogenesis of enteric protozoan infections and what are the conditions that lead to disease. Some of them are related to the characteristics of the pathogen, such as different pathogenicity due to virulence variability of strains or the need of, at least, a second infection with another pathogen to cause clinical symptomatology. Moreover, host factors can play a key role, too, such as individual susceptibility or the presence of healthy functional barriers that protect the human intestine: the mucus layer, the intestinal epithelial layer and the intestinal microbiota [5,15,18,26]. However, further research needs to be performed to establish the ecology of enteric communities in healthy and unhealthy individuals.

We are aware that our research may have some limitations. First, this was a hospital-based cross-sectional study, so the surveyed population may not represent the general population in Tengchong City, and the generalizability to other populations may be rather limited. Although the occurrence rates of protist enteroparasites were consistent with those published in previous studies, low prevalences might have negatively influenced the reliability of the estimates in the PLS analyses. Second, the molecular methods (direct and nested PCRs) used here did not allow the quantification of parasites' DNA, so the direct association between parasites' burdens and occurrence of clinical manifestations could not be determined. In addition, acute bacterial or viral infections of the gastrointestinal tract were not found in this study. Additionally, molecular data on the diversity and frequency of species/genotypes were based on single-locus sequence analysis. Adopting multi-locus genotyping schemes (particularly for *G. duodenalis*) would likely improve the genetic data provided here. This is a task to be conducted soon.

Therefore, further research with a larger sample size, which considers a broader enteropathogen community, is needed to explore community assemblages present and their relationships with pathogenicity and clinical manifestations in rural and urban regions in China.

Undoubtedly, with the increasing awareness of the microbiota role in developing host immunity [5] and their potential relationship with many communicable and non-communicable diseases, such as irritable bowel syndrome [50], there is a growing need to understand the complex relationships between different pathogens and as well as between the intestinal microbiota and pathogens.

4. Materials and Methods

4.1. Study Design and Study Area

A cross-sectional hospital-based study was conducted from July 2016 to March 2017 in Tengchong (25°01′15″ N, 98°29′50″ E, 1596 m above sea level), a county-level city located in Yunnan Province, Southwest China. Tengchong has a tropical monsoon climate. The annual average temperature is 15 °C, and the average annual rainfall is 1535 mm with a year-round mean relative humidity of 77%. The total resident population is 659,000 (Census 2014), of which 60.5% live in rural areas. Two hospitals from this city (People's Hospital of Tengchong City and Chinese Medicine Hospital of Tengchong City) agreed to participate in the study.

The target sample size was $n = 423$ for an expected prevalence of 50%, 95% confidence interval and 5% precision. Finally, 507 patients were recruited.

4.2. Study Participants

Voluntary hospital inpatients (diarrhea cases and healthy controls) were recruited after a clear explanation of the study objectives provided by a member of the researcher team through personal interviews. Informed consent was obtained from the participants or their parents/legal guardians. Individuals presenting with diarrhea were recruited from the Gastroenterology Department. Non-diarrheal individuals were recruited from other hospital departments, including Respiratory Department, Trauma Department, and Emergency Department, among others. Enrolled patients were immunocompetent individuals with CD4 lymphocyte counts $\geq$500 cells/mm^3.

According to the World Health Organization, an acute diarrhea case was defined as an individual who had more than three episodes of abnormal stool within 24 h (e.g., loose, watery, bloody, or mucous stool) for any period lasting for <14 days [15]. A healthy subject was defined as a person who did not have any diarrhea symptoms in the 14 days before recruitment into the study. Subjects who met the following inclusion criteria were eligible for inclusion in this study: (i) patients presenting with acute diarrhea (cases), (ii) patients without gastrointestinal manifestations, including diarrhea (controls), (iii) patients with a negative result to the main bacterial (*Campylobacter* spp., *Clostridium* spp., *Escherichia coli*, *Salmonella* spp., *Shigella* spp., and *Yersinia enterocolitica*) and viral (rotavirus, adenovirus, sapovirus, norovirus, and astrovirus) enteric pathogens, and (iv) patients, from which written informed consents were available. Exclusion criteria included (i) subjects providing insufficient amount of fecal material, (ii) subjects having severe disorders of the cardiovascular or nervous systems, serious mental diseases, tumors, hepatitis A–C, and E virus infection, and/or human immunodeficiency virus (HIV) infection/acquired immune deficiency syndrome (AIDS), (iii) subjects, who did not provide written informed consent, (iv) lactating women, and (v) subjects taking antiparasitic drugs just before or during the recruitment period.

4.3. Specimen and Data Collection

Single stool samples (>3 g or >3 mL) were obtained from each participating subject using a sterile sampling cup during the study period at the two hospital settings and stored at 4 °C. Collected samples were shipped to the laboratory of the National Institute of Parasitic Diseases, Chinese Center for Disease Control and Prevention (Shanghai, China) without interrupting the cold chain and immediately stored at −70 °C until further processing.

A structured questionnaire was used to gather sociodemographic (including age, gender, educational level, occupation, nationality, and place of residence) and clinical (clinical manifestations associated with enteric protist infections, including abdominal distension, diarrhea, lack of appetite, itchy skin, perianal pruritus, constipation, nausea, abdominal pain, number of stools per day and type of stools) data from each recruited patient.

4.4. DNA Extraction and Purification

Genomic DNA was extracted from each stool sample (0.2 g or 0.2 mL) using a QIAamp DNA stool mini kit (Qiagen, Hilden, Germany) according to the manufacturer's protocol. Purified genomic DNA was stored at −70 °C until downstream analysis by polymerase chain reaction (PCR) amplification.

4.5. Molecular Detection of Intestinal Protozoan Species

A direct PCR protocol was used to detect *Blastocystis* sp. [51]. In contrast, nested PCR protocols were used for identifying *Cryptosporidium* spp., *G. duodenalis*, and *E. histolytica* [52–54]. The description of the primer pairs used, the expected size of the obtained amplicons and the cycling conditions of these protocols are provided in Table S1.

4.6. Sanger Sequencing Analysis

Positive-PCR products of the expected sizes were directly sequenced in both directions using appropriate internal primer sets (Table S1). DNA sequencing was conducted at

Sangon Biotech Company (Shanghai, China). Raw sequencing data were viewed using the Chromas Lite version 2.1 sequence analysis program (https://technelysium.com.au/wp/chromas/. Accessed on 19 May 2021). The BLAST tool (http://blast.ncbi.nlm.nih.gov/Blast.cgi. Accessed on 19 May 2021) was used to compare nucleotide sequences with sequences retrieved from the NCBI GenBank database. Generated DNA consensus sequences were aligned to appropriate reference sequences using the MEGA 6 software to identify *G. duodenalis* species and assemblages/sub-assemblages. *Blastocystis* sequences were submitted to the *Blastocystis* 18S database (http://pubmlst.org/blastocystis/. Accessed on 19 May 2021) for sub-type confirmation and allele identification. The sequences obtained in this study were deposited in GenBank under accession numbers MW810321-MW810323 (*G. duodenalis*) and MW798733-MW798742 (*Blastocystis* sp.).

4.7. Statistical Modeling

Data were analyzed using the software R version 4.0.5 [55]. The chi-squared or Fisher's exact test, Odd ratios (OR) and 95% confidence intervals (95% CIs) were used to compare and describe the qualitative variables.

Only participants with complete data records were included in the final analysis. A new variable was created according to the World Health Organization's definition of diarrhea and the information gathered about the type of stool and the number of depositions per day. A case of acute diarrhea was defined as a person with more than three episodes of liquid stools per day, lasting less than 2 weeks [56]. Prevalence rates at 95% CIs for single infections and coinfections in the study population were calculated using epiR library version 0.5-10 [57].

4.7.1. Co-Occurrence of Enteric Pathogens

Null model analysis was used to explore whether enteric protozoa coinfections were positive, negative, or randomly associated. Data were organized as a presence-absence 4×507 (row $\times$ columns) matrix, in which each row represented a protozoa species. Each column represented a study participant, "1" indicated that a species was present at a particular host and "0" indicated that a species was absent.

The C-score was the co-occurrence index used for co-occurrence patterns characterization. The algorithm chosen was the fixed row-equiprobable column [58]. The calculated C-score was compared with the expected C-score calculated for 5000 randomly assembled null matrices by Monte Carlo simulations. Furthermore, to compare the degree of co-occurrence across data, a standardized effect size (SES) was calculated, an index that measures the number of standard deviations that the observed index (C-score) is above or below the mean index of the simulated communities. The package "EcoSimR" version 0.1.0 was used to carry out the analysis [59].

4.7.2. Assessing the Impact of Coinfection with Enteric Pathogens on Diarrhea Severity

The partial least square (PLS) regression method was used to assess the impact of coinfection with enteric protozoa on developing clinical symptomatology. This technique was selected as it offers multiple advantages over other regression methods: it is the least restrictive of the multivariate techniques for exploring complex ecological patterns [60], including the impact of coinfections on the host's health [23], and its distribution is free and well suited to deal with multicollinearity [61]. In our analysis, we defined explanatory and response components or blocks. The explanatory block (PLS X's component) was defined by a presence-absence matrix representing the enteric protozoa community (*Blastocystis* sp., *G. duodenalis*, *E. histolytica*, and *Cryptosporidium* spp.). In addition, due to the previously mentioned age variability in the clinical presentation of diarrheal diseases, age in years was also included as a covariate in the explanatory block. Our response block (PLS's Y component) included the main symptoms described associated with the infections of those protist species (abdominal distension, lack of appetite, itchy skin, perianal pruritus, constipation, nausea, abdominal pain and acute diarrhea).

The significance of PLS models was assessed using Stone–Geisser's Q2 test, a cross-validation redundancy measure created to evaluate the predictive significance of exogenous variables. Values greater than 0.0975 indicate that predictors are statistically significant, whereas values below this threshold reveal no significance. Finally, the percentage of observed MNLS variability explained by the enteric pathogen block was also estimated. The "plspm" version 0.4.9 was used to perform the analysis [62].

5. Conclusions

The present study was the first to analyze the community assemblage of the four protist species commonly associated with human gastrointestinal disorders in immunocompetent individuals. Our results showed the absence of any structured community between them. Their occurrence was purely random. Moreover, there was no evidence of an association between their presence and developing clinical symptomatology.

Further research, including a broad range of enteric pathogens, is needed to disentangle the complex relationships and interactions of the intestinal ecosystem. This could eventually lead to a better understanding about what are the drivers behind gastrointestinal disorders.

Supplementary Materials: The following are available online at https://www.mdpi.com/article/10.3390/pathogens10060684/s1, Table S1: Oligonucleotides and PCR conditions used for the molecular identification and/or characterization of the intestinal protist parasites investigated in Tengchong City, southwest China.

Author Contributions: Conceptualization, L.-G.T., D.C., C.B., and E.S.; methodology, J.-X.C.; software, C.B.; validation, J.-X.C. and L.-G.T.; formal analysis, C.B. and D.C.; investigation, Y.-H.C., Y.-F.Y. and X.-P.W.; resources, C.-L.Y.; data curation, C.-L.Y. and D.C.; writing—original draft preparation, D.C.; writing—review and editing, C.B. and S.-X.Z.; visualization, D.C.; supervision, J.-X.C.; project administration, L.-G.T.; funding acquisition, E.S. and S.-X.Z. All authors have read and agreed to the published version of the manuscript.

Funding: This research was supported by the fund of the 13th Five-Year National Science and Technology Major Project for Infectious Diseases (No. 2017ZX10305501-002, No. 2018ZX10725-509), the fund of Chinese traditional medicine for treating the novel Coronavirus pneumonia patients in convalescence (No. JJ202002), the Emergency Project of Shanghai for the prevention and treatment of COVID-19 in traditional Chinese medicine (Grant No. 2020NCP001), the fund of the China Postdoctoral Science Foundation (No.2020T130022ZX), the fund of National Natural Science Foundation of China (No. 81473022). In addition, E.S. was a recipient of a Ramon y Cajal agreement (RYC-2016-21120) funded by the Spanish Ministry of Economy and Competitiveness (MINECO).

Institutional Review Board Statement: The study was conducted according to the guidelines of the Declaration of Helsinki, and approved by the Ethical Review Committee of the National Institute of Parasitic Diseases, Chinese Center for Disease Control and Prevention (No. 2014004; date of approval: 12 March 2014).

Informed Consent Statement: Informed consent was obtained from all subjects involved in the study.

Data Availability Statement: All relevant data are within the article and its additional files. The sequences obtained in this study were deposited in GenBank under accession numbers MW810321-MW810323 (*Giardia duodenalis* at the *tpi* locus) and MW798733-MW798742 (*Blastocystis* sp.).

Acknowledgments: We sincerely thank the efforts of the medical staff in collecting stool specimens and investigating and reporting on the subjects participating in this survey.

Conflicts of Interest: The authors declare no conflict of interest. The funders had no role in the design of the study; in the collection, analyses, or interpretation of data; in the writing of the manuscript, or in the decision to publish the results.

References

1. Fletcher, S.M.; Stark, D.; Harkness, J.; Ellis, J. Enteric protozoa in the developed world: A public health perspective. *Clin. Microbiol. Rev.* **2012**, *25*, 420–449. [CrossRef]
2. Shrestha, A.; Schindler, C.; Odermatt, P.; Gerold, J.; Erismann, S.; Sharma, S.; Koju, R.; Utzinger, J.; Cissé, G. Intestinal parasite infections and associated risk factors among schoolchildren in Dolakha and Ramechhap districts, Nepal: A cross-sectional study. *Parasites Vectors* **2018**, *11*, 532. [CrossRef]
3. Hotez, P.J.; Alvarado, M.; Basáñez, M.-G.; Bolliger, I.; Bourne, R.; Boussinesq, M.; Brooker, S.J.; Brown, A.S.; Buckle, G.; Budke, C.M.; et al. The Global Burden of Disease Study 2010: Interpretation and implications for the neglected tropical diseases. *PLoS Negl. Trop. Dis.* **2014**, *8*, e2865. [CrossRef]
4. GBD Causes of Death Collaborators. Global, regional, and national age-sex-specific mortality for 282 causes of death in 195 countries and territories, 1980–2017: A systematic analysis for the Global Burden of Disease Study 2017. *Lancet* **2018**, *392*, 1736–1788. [CrossRef]
5. Partida-Rodríguez, O.; Serrano-Vázquez, A.; Nieves-Ramírez, M.E.; Moran, P.; Rojas, L.; Portillo, T.; González, E.; Hernández, E.; Finlay, B.B.; Ximenez, C. Human intestinal microbiota: Interaction between parasites and the host immune response. *Arch. Med. Res.* **2017**, *48*, 690–700. [CrossRef]
6. Maas, L.; Dorigo-Zetsma, J.W.; de Groot, C.; Bouter, S.; Plötz, F.B.; van Ewijk, B. Detection of intestinal protozoa in paediatric patients with gastrointestinal symptoms by multiplex real-time PCR. *Clin. Microbiol. Infect.* **2014**, *20*, 545–550. [CrossRef]
7. Leung, P.-O.; Chen, K.-H.; Chen, K.-L.; Tsai, Y.-T.; Liu, S.-Y.; Chen, K.-T. Epidemiological features of intestinal infection with Entamoeba histolytica in Taiwan, 2002–2010. *Travel Med. Infect. Dis.* **2014**, *12*, 673–679. [CrossRef]
8. Esch, K.J.; Petersen, C.A. Transmission and epidemiology of zoonotic protozoal diseases of companion animals. *Clin. Microbiol. Rev.* **2013**, *26*, 58–85. [CrossRef]
9. Savioli, L.; Smith, H.; Thompson, A. *Giardia* and *Cryptosporidium* join the 'Neglected Diseases Initiative'. *Trends Parasitol.* **2006**, *22*, 203–208. [CrossRef]
10. Audebert, C.; Even, G.; Cian, A.; Loywick, A.; Merlin, S.; Viscogliosi, E.; Chabe, M. *Blastocystis* Investigation Group. Colonization with the enteric protozoa *Blastocystis* is associated with increased diversity of human gut bacterial microbiota. *Sci. Rep.* **2016**, *6*, 25255. [CrossRef]
11. Deng, Y.; Zhang, S.; Ning, C.; Zhou, Y.; Teng, X.; Wu, X.; Chu, Y.; Yu., Y.; Chen, J.; Tian, L.; Wang, W. Molecular Epidemiology and Risk Factors of Blastocystis sp. Infections Among General Populations in Yunnan Province, Southwestern China. *Risk Manag. Health Policy* **2020**, *13*, 1791–1801. [CrossRef]
12. Fouad, S.A.; Basyoni, M.M.; Fahmy, R.A.; Kobaisi, M.H. The pathogenic role of different *Blastocystis hominis* genotypes isolated from patients with irritable bowel syndrome. *Arab. J. Gastroenterol.* **2011**, *12*, 194–200. [CrossRef]
13. Schaechter, M. Intestinal protozoan. In *Encyclopedia of Microbiology*, 3th ed.; Schaechter, M., Ed.; Academis Press: Oxford, UK, 2009; pp. 696–705.
14. Olesen, B.; Neimann, J.; Böttiger, B.; Ethelberg, S.; Schiellerup, P.; Jensen, C.; Helms, M.; Scheutz, F.; Olsen, K.E.P.; Krogfelt, K.A.; et al. Etiology of diarrhea in young children in Denmark: A case-control study. *J. Clin. Microbiol.* **2005**, *43*, 3636–3641. [CrossRef]
15. Zhang, S.-X.; Zhou, Y.-M.; Xu, W.; Tian, L.-G.; Chen, J.-X.; Chen, S.-H.; Dang, Z.-S.; Gu, W.-P.; Yin, J.-W.; Serrano, E.; et al. Impact of co-infections with enteric pathogens on children suffering from acute diarrhea in southwest China. *Infect. Dis. Poverty* **2016**, *5*, 1–13. [CrossRef]
16. Newell, D.G.; Koopmans, M.; Verhoef, L.; Duizer, E.; Aidara-Kane, A.; Sprong, H.; Opsteegh, M.; Langelaar, M.; Threfall, J.; Scheutz, F.; et al. Food-borne diseases—The challenges of 20 years ago still persist while new ones continue to emerge. *Int. J. Food Microbiol.* **2010**, *139*, 3–15. [CrossRef]
17. Horton, B.; Bridle, H.; Alexander, C.L.; Katzer, F. *Giardia duodenalis* in the UK: Current knowledge of risk factors and public health implications. *Parasitology* **2018**, *146*, 413–424. [CrossRef]
18. Levine, M.M.; Robins-Browne, R.M. Factors that explain excretion of enteric pathogens by persons without diarrhea. *Clin. Infect. Dis.* **2012**, *55*, 303–311. [CrossRef] [PubMed]
19. Boughattas, S.; Behnke, J.M.; Al-Ansari, K.; Sharma, A.; Abu-Alainin, W.; Al-Thani, A.; Abu-Madi, M.A. Molecular analysis of the enteric protozoa associated with acute diarrhea in hospitalized children. *Front. Cell. Infect. Microbiol.* **2017**, *7*, 343. [CrossRef]
20. Kotloff, K.L.; Nataro, J.P.; Blackwelder, W.C.; Nasrin, D.; Farag, T.H.; Panchalingam, S.; Wu, Y.; Sow, S.O.; Sur, D.; Breiman, R.F.; et al. Burden and aetiology of diarrhoeal disease in infants and young children in developing countries (the Global Enteric Multicenter Study, GEMS): A prospective, case-control study. *Lancet* **2013**, *382*, 209–222. [CrossRef]
21. Puebla, L.E.J.; Núñez, F.A.; Santos, L.P.; Rivero, L.R.; Silva, I.M.; Valdés, L.A.; Millán, I.A.; Müller, N. Molecular analysis of *Giardia duodenalis* isolates from symptomatic and asymptomatic children from La Habana, Cuba. *Parasite Epidemiol. Control* **2017**, *2*, 105–113. [CrossRef]
22. Reh, L.; Muadica, A.S.; Köster, P.C.; Balasegaram, S.; Verlander, N.Q.; Chércoles, E.R.; Carmena, D. Substantial prevalence of enteroparasites *Cryptosporidium* spp., *Giardia duodenalis* and *Blastocystis* sp. in asymptomatic schoolchildren in Madrid, Spain, November 2017 to June 2018. *Eurosurveillance* **2019**, *24*, 1900241. [CrossRef]
23. Risco, D.; Serrano, E.; Fernández-Llario, P.; Cuesta, J.M.; Gonçalves, P.; García-Jiménez, W.L.; Martínez, R.; Cerrato, R.; Velarde, R.; Gómez, L.; et al. Severity of bovine tuberculosis is associated with co-infection with common pathogens in wild boar. *PLoS ONE* **2014**, *9*, e110123. [CrossRef]

24. Rivero-Juarez, A.; Dashti, A.; López-López, P.; Muadica, A.S.; Risalde, M.D.L.A.; Köster, P.C.; Machuca, I.; Bailo, B.; De Mingo, M.H.; Dacal, E.; et al. Protist enteroparasites in wild boar (*Sus scrofa ferus*) and black Iberian pig (*Sus scrofa domesticus*) in southern Spain: A protective effect on hepatitis E acquisition? *Parasites Vectors* **2020**, *13*, 1–9. [CrossRef]
25. Tzipori, S.; Chandler, D.; Makin, T.; Smith, M. *Escherichia coli* and rotavirus infections in four-week-old gnotobiotic piglets fed milk or dry food. *Aust. Veter. J.* **1980**, *56*, 279–284. [CrossRef] [PubMed]
26. Serrano, E.; Millán, J. What is the price of neglecting parasite groups when assessing the cost of co-infection? *Epidemiol. Infect.* **2013**, *142*, 1533–1540. [CrossRef] [PubMed]
27. Li, J.; Wang, H.; Wang, R.; Zhang, L. *Giardia duodenalis* infections in humans and other animals in China. *Front. Microbiol.* **2017**, *8*, 2004. [CrossRef]
28. Zhang, S.X.; Yang, C.L.; Gu, W.P.; Ai, L.; Serrano, E.; Yang, P.; Zhou, X.; Li, S.Z.; Lv, S.; Dang, Z.S.; et al. Case-control study of diarrheal disease etiology in individuals over 5 years in southwest China. *Gut Pathog.* **2016**, *8*, 1–11. [CrossRef]
29. Geurden, T.; Levecke, B.; Cacció, S.M.; Visser, A.; De Groote, G.; Casaert, S.; Vercruysse, J.; Claerebout, E. Multilocus genotyping of *Cryptosporidium* and *Giardia* in non-outbreak related cases of diarrhoea in human patients in Belgium. *Parasitology* **2009**, *136*, 1161–1168. [CrossRef]
30. De Lucio, A.; Martínez-Ruiz, R.; Merino, F.J.; Bailo, B.; Aguilera, M.; Fuentes, I.; Carmena, D. Molecular genotyping of *Giardia duodenalis* isolates from symptomatic individuals attending two major public hospitals in Madrid, Spain. *PLoS ONE* **2015**, *10*, e0143981. [CrossRef]
31. Minetti, C.; Lamden, K.; Durband, C.; Cheesbrough, J.; Fox, A.; Wastling, J.M. Determination of *Giardia duodenalis* assemblages and multi-locus genotypes in patients with sporadic giardiasis from England. *Parasites Vectors* **2015**, *8*, 1–10. [CrossRef]
32. Skhal, D.; Aboualchamat, G.; Al Mariri, A.; Al Nahhas, S. Prevalence of *Giardia duodenalis* assemblages and sub-assemblages in symptomatic patients from Damascus city and its suburbs. *Infect. Genet. Evol.* **2017**, *47*, 155–160. [CrossRef]
33. Bartelt, L.A.; Sartor, R.B. Advances in understanding *Giardia*: Determinants and mechanisms of chronic sequelae. *F1000Prime Rep.* **2015**, *7*, 62. [CrossRef]
34. Heyworth, M.F.F. Giardia duodenalis genetic assemblages and hosts. *Parasite* **2016**, *23*, 13. [CrossRef]
35. Sánchez, A.; Munoz, M.; Gómez, N.; Tabares, J.; Segura, L.; Salazar, Á.; Restrepo, C.; Ruíz, M.; Reyes, P.; Qian, Y.; Xiao, L.; et al. Molecular epidemiology of *Giardia*, *Blastocystis* and *Cryptosporidium* among indigenous children from the Colombian Amazon basin. *Front. Microbiol.* **2017**, *8*, 248. [CrossRef]
36. Hussein, E.M.; Ismail, O.A.; Mokhtar, A.B.; Mohamed, S.E.; Saad, R.M. Nested PCR targeting intergenic spacer (IGS) in genotyping of *Giardia duodenalis* isolated from symptomatic and asymptomatic infected Egyptian school children. *Parasitol. Res.* **2016**, *116*, 763–771. [CrossRef]
37. Xiao, L.; Fayer, R. Molecular characterisation of species and genotypes of *Cryptosporidium* and *Giardia* and assessment of zoonotic transmission. *Int. J. Parasitol.* **2008**, *38*, 1239–1255. [CrossRef] [PubMed]
38. Štrkolcová, G.; Goldová, M.; Mad'ar, M.; Čechová, L.; Halánová, M.; Mojzisova, J. *Giardia duodenalis* and *Giardia enterica* in children: First evidence of assemblages A and B in Eastern Slovakia. *Parasitol. Res.* **2016**, *115*, 1939–1944. [CrossRef] [PubMed]
39. Messa, A.; Köster, P.C.; Garrine, M.; Gilchrist, C.; Bartelt, L.A.; Nhampossa, T.; Massora, S.; Kotloff, K.; Levine, M.M.; Alonso, P.L.; et al. Molecular diversity of *Giardia duodenalis* in children under 5 years from the Manhiça district, Southern Mozambique enrolled in a matched case-control study on the aetiology of diarrhoea. *PLoS Negl. Trop. Dis.* **2021**, *15*, e0008987. [CrossRef] [PubMed]
40. Ning, C.-Q.; Hu, Z.-H.; Chen, J.-H.; Ai, L.; Tian, L.-G. Epidemiology of *Blastocystis* infection from 1990 to 2019 in China. *Infect. Dis. Poverty* **2020**, *9*, 1–14. [CrossRef] [PubMed]
41. Zhang, S.X.; Tian, L.G.; Lu, Y.; Li, L.H.; Chen, J.X.; Zhou, X.N. Epidemiological characteristics of *Blastocystis hominis* in urban region, southwestern China. *Zhongguo Ren Shou Gong Huan Bing Za Zhi* **2016**, *32*, 424–428.
42. Zhang, W.; Ren, G.; Zhao, W.; Yang, Z.; Shen, Y.; Sun, Y.; Liu, A.; Cao, J. Genotyping of *Enterocytozoon bieneusi* and subtyping of *Blastocystis* in cancer patients: Relationship to diarrhea and assessment of zoonotic transmission. *Front. Microbiol.* **2017**, *8*, 1835. [CrossRef] [PubMed]
43. Zhang, S.X.; Kang, F.Y.; Chen, J.X.; Tian, L.G.; Geng, L.L. Risk factors for Blastocystis infection in HIV/AIDS patients with highly active antiretroviral therapy in Southwest China. *Infect. Dis. Poverty* **2019**, *8*, 1–8. [CrossRef] [PubMed]
44. Hublin, J.S.; Maloney, J.G.; Santin, M. *Blastocystis* in domesticated and wild mammals and birds. *Res. Veter.-Sci.* **2021**, *135*, 260–282. [CrossRef]
45. Liu, A.; Gong, B.; Liu, X.; Shen, Y.; Wu, Y.; Zhang, W.; Cao, J. A retrospective epidemiological analysis of human *Cryptosporidium* infection in China during the past three decades (1987–2018). *PLoS Negl. Trop. Dis.* **2020**, *14*, e0008146. [CrossRef]
46. Feng, Y.; Wang, L.; Duan, L.; Gomez-Puerta, L.A.; Zhang, L.; Zhao, X.; Hu, J.; Zhang, N.; Xiao, L. Extended outbreak of cryptosporidiosis in a pediatric hospital, China. *Emerg. Infect. Dis.* **2012**, *18*, 312–314. [CrossRef] [PubMed]
47. Wang, T.; Fan, Y.; Koehler, A.V.; Ma, G.; Li, T.; Hu, M.; Gasser, R.B. First survey of *Cryptosporidium*, *Giardia* and *Enterocytozoon* in diarrhoeic children from Wuhan, China. *Infect. Genet. Evol.* **2017**, *51*, 127–131. [CrossRef] [PubMed]
48. Zhang, N.; Yu, X.; Zhang, H.; Cui, L.; Li, X.; Zhang, X.; Gong, P.; Li, J.; Li, Z.; Wang, X.; et al. Prevalence and genotyping of *Cryptosporidium parvum* in gastrointestinal cancer patients. *J. Cancer* **2020**, *11*, 3334–3339. [CrossRef] [PubMed]

49. Iseki, M.; Ali, I.K.M.D.; Hossain, M.B.; Zaman, V.; Haque, R.; Takahashi, Y.; Yoshikawa, H.; Wu, Z.; Kimata, I. Polymerase chain reaction-based genotype classification among human *Blastocystis hominis* populations isolated from different countries. *Parasitol. Res.* **2004**, *92*, 22–29. [CrossRef]

50. Cifre, S.; Gozalbo, M.; Ortiz, V.; Soriano, J.M.; Merino, J.F.; Trelis, M. *Blastocystis* subtypes and their association with Irritable Bowel Syndrome. *Med. Hypotheses* **2018**, *116*, 4–9. [CrossRef]

51. Menounos, P.G.; Spanakos, G.; Tegos, N.; Vassalos, C.M.; Papadopoulou, C.; Vakalis, N.C. Direct detection of *Blastocystis* sp. in human faecal samples and subtype assignment using single strand conformational polymorphism and sequencing. *Mol. Cell. Probes* **2008**, *22*, 24–29. [CrossRef]

52. Liu, H.; Shen, Y.; Yin, J.; Yuan, Z.; Jiang, Y.; Xu, Y.; Pan, W.; Hu, Y.; Cao, J. Prevalence and genetic characterization of *Cryptosporidium*, *Enterocytozoon*, *Giardia* and *Cyclospora* in diarrheal outpatients in china. *BMC Infect. Dis.* **2014**, *14*, 25. [CrossRef] [PubMed]

53. Sulaiman, I.M.; Fayer, R.; Bern, C.; Gilman, R.H.; Trout, J.M.; Schantz, P.M.; Das, P.; Al, I.M.S.E.; Xiao, L. Triosephosphate isomerase gene characterization and potential zoonotic transmission of *Giardia duodenalis*. *Emerg. Infect. Dis.* **2003**, *9*, 1444–1452. [CrossRef] [PubMed]

54. Evangelopoulos, G.S.A. A nested, multiplex, PCR assay for the simultaneous detection and differentiation of *Entamoeba histolytica* and *Entamoeba dispar* in faeces. *Ann. Trop. Med. Parasitol.* **2000**, *94*, 233–240. [CrossRef]

55. R Core Team. *R: A Language and Environment for Statistical Computing*; R Foundation for Statistical Computing: Vienna, Austria, 2021; Available online: https://www.R-project.org/ (accessed on 5 May 2021).

56. World Health Organization. WHO Diarrhea. 2018. Available online: https://www.who.int/topics/diarrhoea/en/ (accessed on 19 May 2021).

57. Aragon, T.J.E. Epidemiology Tools. 2017. Available online: https://cran.r-project.org/package=epitools (accessed on 19 May 2021).

58. Gotelli, N. Null model analysis of species co-occurrence patterns. *Ecology* **2000**, *9*, 2606–2621. [CrossRef]

59. Gotelli, N.; Graves, G.R. *Null Models in Ecology*; Smithsonian Institution Press: Washington, DC, USA, 1996; Available online: https://github.com/gotellilab/EcoSimR (accessed on 1 January 2015).

60. Gordo, L.M.C.I. Partial least squares regression as an alternative to current regression methods used in ecology. *OIKOS* **2009**, *118*, 681–690.

61. Salkind, N.J. *Encyclopedia of Measurement and Statistics*; Sage Publications: Thousand Oaks, CA, USA; London, UK, 2007.

62. Gaston Sanchez, L.T.G.R. Tools for Partial Least Squares Path Modeling (PLS-PM). 2017. Available online: https://rdrr.io/cran/plspm/ (accessed on 30 September 2019).

Article

High Diversity of *Cryptosporidium* Species and Subtypes Identified in Cryptosporidiosis Acquired in Sweden and Abroad

Marianne Lebbad [1], Jadwiga Winiecka-Krusnell [1], Christen Rune Stensvold [2] and Jessica Beser [1,*]

[1] Department of Microbiology, Public Health Agency of Sweden, 171 82 Solna, Sweden; marianne.lebbad@outlook.com (M.L.); ilia.krusnell@gmail.com (J.W.-K.)

[2] Department of Bacteria, Parasites and Fungi, Statens Serum Institut, DK-2300 Copenhagen S, Denmark; run@ssi.dk

* Correspondence: jessica.beser@fohm.se

Abstract: The intestinal protozoan parasite *Cryptosporidium* is an important cause of diarrheal disease worldwide. The aim of this study was to expand the knowledge on the molecular epidemiology of human cryptosporidiosis in Sweden to better understand transmission patterns and potential zoonotic sources. *Cryptosporidium*-positive fecal samples were collected between January 2013 and December 2014 from 12 regional clinical microbiology laboratories in Sweden. Species and subtype determination was achieved using small subunit ribosomal RNA and 60 kDa glycoprotein gene analysis. Samples were available for 398 patients, of whom 250 (63%) and 138 (35%) had acquired the infection in Sweden and abroad, respectively. Species identification was successful for 95% (379/398) of the samples, revealing 12 species/genotypes: *Cryptosporidium parvum* ($n = 299$), *C. hominis* ($n = 49$), *C. meleagridis* ($n = 8$), *C. cuniculus* ($n = 5$), *Cryptosporidium* chipmunk genotype I ($n = 5$), *C. felis* ($n = 4$), *C. erinacei* ($n = 2$), *C. ubiquitum* ($n = 2$), and one each of *C. suis*, *C. viatorum*, *C. ditrichi*, and *Cryptosporidium* horse genotype. One patient was co-infected with *C. parvum* and *C. hominis*. Subtyping was successful for all species/genotypes, except for *C. ditrichi*, and revealed large diversity, with 29 subtype families (including 4 novel ones: *C. parvum* IIr, IIs, IIt, and *Cryptosporidium* horse genotype VIc) and 81 different subtypes. The most common subtype families were IIa ($n = 164$) and IId ($n = 118$) for *C. parvum* and Ib ($n = 26$) and Ia ($n = 12$) for *C. hominis*. Infections caused by the zoonotic *C. parvum* subtype families IIa and IId dominated both in patients infected in Sweden and abroad, while most *C. hominis* cases were travel-related. Infections caused by non-*hominis* and non-*parvum* species were quite common (8%) and equally represented in cases infected in Sweden and abroad.

Keywords: molecular epidemiology; parasite; parasitology; epidemiology; genetic diversity; host specificity; Europe; Scandinavia; protist; sporozoa; zoonosis; zoonotic transmission

Citation: Lebbad, M.; Winiecka-Krusnell, J.; Stensvold, C.R.; Beser, J. High Diversity of *Cryptosporidium* Species and Subtypes Identified in Cryptosporidiosis Acquired in Sweden and Abroad. *Pathogens* **2021**, *10*, 523. https://doi.org/10.3390/pathogens10050523

Academic Editors: David Carmena, David González-Barrio and Pamela Carolina Köster

Received: 9 April 2021
Accepted: 23 April 2021
Published: 26 April 2021

Publisher's Note: MDPI stays neutral with regard to jurisdictional claims in published maps and institutional affiliations.

1. Introduction

Cryptosporidiosis is a global parasitic disease, which usually presents with self-limiting diarrhea. However, the disease can be severe, especially in immunocompromised and malnourished individuals [1]. The causative agent is the intestinal protozoan parasite *Cryptosporidium*, which infects a wide range of animals, including humans [2]. There are more than 40 recognized species described, but 2 in particular (*Cryptosporidium hominis* and *Cryptosporidium parvum*) account for most cases of human cryptosporidiosis [3]. In addition, around 20 other species/genotypes have also been observed in humans, including *Cryptosporidium meleagridis*, *Cryptosporidium cuniculus*, *Cryptosporidium ubiquitum*, *Cryptosporidium canis*, *Cryptosporidium felis*, *Cryptosporidium viatorum*, and *Cryptosporidium* chipmunk genotype I [3,4].

Cryptosporidiosis has been a notifiable disease in Sweden since 2004, with an increasing incidence from 69 (0.8 cases/100,000 inhabitants) cases annually in 2005 to 1088

127

(10.3/100,000 inhabitants) cases in 2019 [5]. There are several reasons for this increase, including better general knowledge of cryptosporidiosis, and the introduction of more sensitive diagnostic methods and, perhaps more importantly, multiple-agent diagnostic approaches, whereby clinicians do not need to specifically request testing for *Cryptosporidium*. Where this approach is used, the number of reported cases tends to increase [6,7]. Nevertheless, there is substantial local variation in the number of reported cases, which suggests that cryptosporidiosis is still underdiagnosed in Sweden [5].

Knowledge about the distribution of *Cryptosporidium* species and subtypes within a country is crucial for the management of cryptosporidiosis and to identify and understand transmission patterns and potential zoonotic sources [8]. The first larger molecular study of sporadic cryptosporidiosis cases in Sweden was conducted between April 2006 and November 2008. Patients from the Stockholm metropolitan area diagnosed with cryptosporidiosis were investigated, and 194 samples were successfully genotyped. A high occurrence of *C. parvum* was found; 111 cases vs. 65 of *C. hominis*. Only 17 (26%) of the *C. hominis* infections were acquired in Sweden, compared with 57 (51%) of the *C. parvum* infections. In addition, less common species such as *C. meleagridis* ($n = 11$), *C. felis* ($n = 2$), *C. viatorum* ($n = 2$), and *Cryptosporidium* chipmunk genotype I ($n = 2$) were found [9].

The largest known *Cryptosporidium* outbreaks in Sweden (and Europe) occurred in Östersund (Jämtland County) and Skellefteå (Västerbotten County) in 2010 and 2011. Both these outbreaks were caused by contamination of the municipal drinking water with *C. hominis* subtype IbA10G2, together affecting an estimated 45,500 persons [10,11]. *Cryptosporidium parvum* has been documented in water- and foodborne outbreaks as well as in outbreaks caused by animal contacts [12–14].

The primary aim of the present study was to identify if observations from the Stockholm metropolitan area (i.e., predominance of *C. parvum* but with relatively frequent detection of unusual species) reflected the situation in the rest of Sweden, and thereby to geographically expand the molecular investigation of cryptosporidiosis cases in Sweden. The second aim was to investigate if the *C. hominis* subtype IbA10G2 was established in the population following the large waterborne outbreaks in 2010 and 2011.

2. Materials and Methods

2.1. Invitation of Participating Laboratories

All clinical microbiology laboratories performing parasite diagnostic tests on human samples in Sweden were invited to participate in the study. The laboratories were asked to forward stool or fecal DNA from *Cryptosporidium*-positive cases to the Public Health Agency of Sweden (PHAS) for molecular species determination and subtyping, free of charge. Sample collection lasted for two years, from January 2013 to December 2014. All local departments of communicable disease control and prevention were also informed about the study, and the typing results were continuously submitted to the national mandatory notifications system (SmiNet).

2.2. Collection of Patient Data

Each submitted *Cryptosporidium*-positive sample was accompanied by information on the age, sex, and geographical location of the patient. Information concerning travel abroad within two weeks prior to onset of disease and, in some instances, assumed routes of transmission was retrieved from the referral and/or SmiNet and/or the local department of communicable disease control.

2.3. Laboratory Investigations

The original *Cryptosporidium* diagnoses were made at local clinical laboratories using modified Ziehl–Neelsen staining or real-time PCR. In total, samples from 398 patients from 12 different microbiological laboratories were forwarded to PHAS for molecular analysis (Table 1). In total, 70 of the stool samples had been fixed in sodium acetate–acetic acid–formalin (SAF), while 328 samples consisted of stool without preservative and/or

DNA extracted from unpreserved stool. Most of the SAF-fixed samples had been washed with phosphate buffered saline at the local laboratory before shipment.

Table 1. *Cryptosporidium* species in samples provided by 12 participating laboratories across 11 Swedish counties and locally identified as *Cryptosporidium*-positive.

County	Number of Laboratories	Number of Samples	Species (Number of Samples)
Halland	1	138	*C. parvum* (112), *C. hominis* (15), *C. cuniculus* (5), *C. erinacei* (2), *C. meleagridis* (1), non-typeable (3)
Jämtland	1	4	*C. parvum* (4)
Jönköping	1	76	*C. parvum* (58), *C. hominis* (5), *Cryptosporidium* chipmunk genotype I (3), non-typeable (10)
Kronoberg	1	1	*C. felis* (1)
Skåne	1	5	*C. parvum* (3), *C. felis* (1), *Cryptosporidium* horse genotype (1)
Stockholm	2	92	*C. parvum* (62), *C. hominis* (23), *C. ubiquitum* (2), *C. ditrichi* (1), *C. felis* (1), *C. meleagridis* (1), *C. viatorum* (1), non-typeable (1)
Uppsala	1	70	*C. parvum* (52), *C. hominis* (6), *C. meleagridis* (6), *Cryptosporidium* chipmunk genotype I (2), *C. felis* (1), *C. hominis* + *C. parvum* (1), *C. suis* (1), non-typeable (1)
Västerbotten	1	4	*C. parvum* (4)
Västernorrland	1	2	*C. parvum* (2)
Västra Götaland	1	5	*C. parvum* (2), non-typeable (3)
Örebro	1	1	non-typeable (1)
Total	12	398	*C. parvum* (299), *C. hominis* (49), *C. parvum* + *C. hominis* (1), *C. meleagridis* (8), *C. cuniculus* (5), *Cryptosporidium* chipmunk genotype I (5), *C. felis* (4), *C. erinacei* (2), *C. ubiquitum* (2), *C. ditrichi* (1), *C. suis* (1), *C. viatorum* (1), *Cryptosporidium* horse genotype (1), non-typeable (19)

DNA was extracted directly from the stool specimens using a QIAamp DNA mini kit (Qiagen, Germany) according to the manufacturer's recommendations. Prior to extraction, oocysts were disrupted using a bead-beater (Techtum, Sweden). DNA from external laboratories was extracted with local methods prior to submission to PHAS. To characterize *Cryptosporidium* species and subtypes, all samples were initially subjected to nested PCR of the small subunit rRNA (SSU rRNA) gene with subsequent restriction fragment length polymorphism (RFLP) and nested PCR of the 60 kDa glycoprotein (*gp60*) gene followed by sequencing [15–17]. In addition to RFLP analysis, bi-directional Sanger sequencing of the SSU rDNA amplicons was performed on (i) non-*hominis* and non-*parvum* isolates originally identified by RFLP, (ii) new or uncommon *gp60* subtype families, (iii) isolates where no amplification product was obtained at the *gp60* locus, and (iv) isolates where mixed species were suspected by RFLP. Species determination using a 70-kilo Dalton heat shock (*hsp70*) gene segment was performed on a limited number of samples where SSU rRNA PCR failed [18]. For the subtype determination of species not amplified with the Alves *gp60* primers [17], partial sequences of the *gp60* gene were amplified using different PCR protocols, depending on the investigated species, and products obtained by nested PCR were sequenced [19–23].

Cryptosporidium-specific actin and heat shock protein (*hsp70*) genes were amplified and sequenced in order to further characterize some uncommon species/genotypes [24,25].

Sequences were edited and analyzed using the BioEdit Sequence Alignment (version 7.0.9.0.) and compared with isolates in the GenBank database using the basic local alignment search tool (BLAST). All chromatograms were manually inspected for the presence of double peaks indicating mixed species/subtypes. In addition, *gp60* chromatograms and fasta files were examined using our in-house *gp60* sequence analyzer software CryptoTyper (unpublished).

Phylogenetic analysis was performed on SSU rDNA and *gp60* DNA sequences generated in the present study as well as sequences from known *Cryptosporidium* species and subtypes. Phylogenetic trees were generated using the neighbor-joining method based on Kimura's 2-parameter model [26]. To estimate robustness, bootstrap proportions were computed after 1000 replications. Evolutionary analyses were conducted in MEGA X (https://www.megasoftware.net/ accessed on 24 April 2021).

Unique and uncommon nucleotide sequences were deposited in GenBank under the following accession numbers: *gp60*—KU727289, KU852701-KU852740; SSU rRNA—KU892559-KU892561, KU892564-KU892566; actin—KU892568, KU892571 and KU892572; *hsp70*—KU892574 and KU892577.

3. Results

3.1. Participating Laboratories and Patient Demographics

Samples were provided from 12 of the 21 clinical laboratories performing *Cryptosporidium* diagnostics in Sweden at the time of the study (representing 11 of the 21 counties) (Table 1). Altogether, samples from 398 patients (124 collected in 2013 and 274 in 2014) were received, representing 63% (398/628) of all cases reported in Sweden during these two years [5] (Figure 1). Among the participants, 217 (55%) were women and 181 (45%) were men. The age range was 1 to 88 years (mean (standard deviation), 34.7 (18.09) years); see Figure 1. The majority of the 398 isolates came from sporadic cases (*n* = 369), while 22 were related to outbreaks and seven to family clusters.

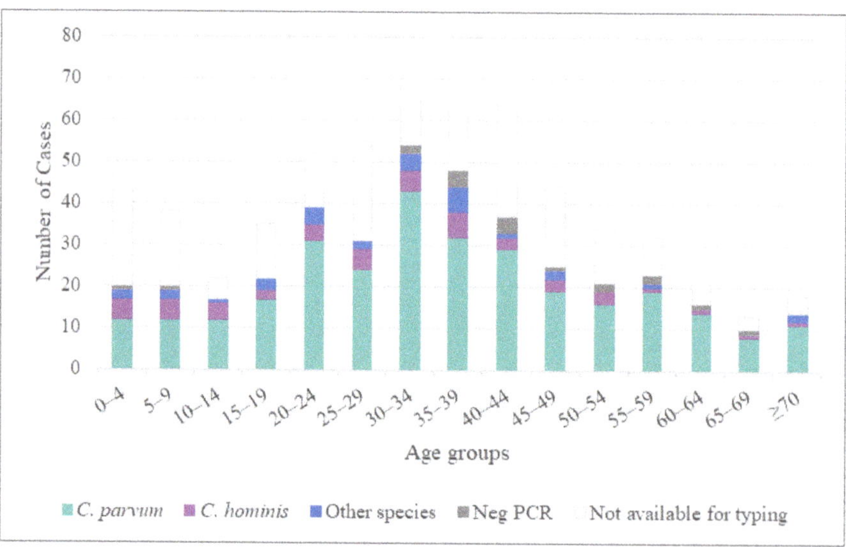

Figure 1. Number of reported cases of cryptosporidiosis detected in Sweden in 2013 and 2014 according to age group. One case with mixed *C. hominis* and *C. parvum* infection is not included in the figure.

3.2. Cryptosporidium Species Identified

Species identification was successful for 95% (379/398) of the samples, and 12 different species/genotypes were identified (Table 2). In total, 370 of the isolates were identified by RFLP and/or sequencing of the SSU rRNA gene, and 4 by sequencing of the *hsp70* gene (3 *C. parvum* and 1 *C. meleagridis*). For five samples with negative SSU rRNA PCR, species determination was based on the positive *gp60* result (4 *C. parvum* and 1 *C. hominis*). Of 19 samples that were PCR-negative and therefore not typeable (Tables 1 and 2), 14 had been preserved in SAF. *Cryptosporidium parvum* was the species most commonly observed (79%; 299/379), followed in frequency by *C. hominis* (13%; 49/379), *C. meleagridis* (*n* = 8), *C. cuniculus* (*n* = 5), *Cryptosporidium* chipmunk genotype I (*n* = 5), *C. felis* (*n* = 4), *Cryptosporidium erinacei* (*n* = 2), *C. ubiquitum* (*n* = 2), and 1 each of *Cryptosporidium suis*, *C. viatorum*, *Cryptosporidium ditrichi*, and *Cryptosporidium* horse genotype. One patient was co-infected with *C. parvum* and *C. hominis*, as evidenced by RFLP analysis of the SSU RNA gene followed by Sanger sequencing of the same gene.

Table 2. Distribution of *Cryptosporidium* spp. according to area of origin of infection among 398 patients with cryptosporidiosis diagnosed in Sweden from 2013 to 2014.

Species/Genotypes	Number of Samples (%)							
	Total	Sweden	Other European Countries	Africa	Asia	North America	South America	Unknown
C. parvum	299	211 (71)	60 (20)	9 (3)	5 (2)	3 (1)	1	10 (3)
C. hominis	49	8 (16)	10 (20)	14 (29)	13 (27)	3 (6)	1 (2)	0
C. parvum + *C. hominis*	1	0	0	0	1	0	0	0
C. meleagridis	8	0	0	0	8 (100)	0	0	0
C. cuniculus	5	3 (60)	2 (40)	0	0	0	0	0
Cryptosporidium chipmunk genotype I	5	5 (100)	0	0	0	0	0	0
C. felis	4	3 (75)	0	0	1 (25)	0	0	0
C. erinacei	2	1	1	0	0	0	0	0
C. ubiquitum	2	2	0	0	0	0	0	0
C. suis	1	0	1	0	0	0	0	0
C. viatorum	1	0	0	1	0	0	0	0
Cryptosporidium horse genotype	1	0	0	1	0	0	0	0
C. ditrichi	1	1	0	0	0	0	0	0
None-typable [1]	19	16 (84)	1 (5)	2 (11)	0	0	0	0
All cases	398	250 (63)	75 (19)	27 (7)	28 (7)	6 (3)	2 (1)	10 (3)

[1] Negative in all PCRs.

3.3. Origin of Infection

Out of the 398 patients, 250 (63%) were infected in Sweden, while 138 (35%) had acquired infection while traveling abroad to 55 different countries, representing all continents except Oceania and Antarctica, over the two weeks preceding disease onset. Ten cases had missing or uncertain data concerning origin of infection (Table 2). *Cryptosporidium parvum* was the most common species in cases of contracting infection in Sweden (84%; 211/250) and abroad (57%; 78/138); meanwhile, *C. hominis* was identified in 3% (8/250) of domestic cases and in 30% (41/138) of the cases infected abroad.

Cryptosporidium chipmunk genotype I (*n* = 5), *C. ubiquitum* (*n* = 2), and *C. ditrichi* (*n* = 1) were found only in cases infected in Sweden, while *C. meleagridis* (*n* = 8), as well as *C. suis*, *C. viatorum*, and *Cryptosporidium* horse genotype (*n* = 1 each), were exclusively found in travel-related cases. *Cryptosporidium felis*, *C. erinacei*, and *C. cuniculus* were diagnosed both in domestic and travel-related cases (Table 2).

3.4. Molecular Characterization of Cryptosporidium parvum

Subtyping using the *gp60* protocol was successful for 99% (296/300) of the *C. parvum*-positive samples, including the sample with mixed *C. hominis* and *C. parvum* (Table 3). In total, nine subtype families (IIa, IIc, IId, IIe, IIl, IIn, IIr, IIs, and IIt) and 42 subtypes were identified (Table 3). The most common subtype family was IIa, which was observed in 164 patients. It was represented by 22 different subtypes, of which IIaA16G1R1 (*n* = 42), IIaA15G2R1 (*n* = 31), and IIaA17G1R1 (*n* = 20) were the most common. The remaining 19 IIa subtypes were found in 1–13 patients each. Subtype family IId (*n* = 118 patients) was the second-most common subtype family, represented by 11 different subtypes, of which IIdA22G1 (*n* = 37) and IIdA20G1 (*n* = 24) were the most frequent. The remaining nine IId subtypes were found in 1–16 patients each. Three patients were co-infected with two *C. parvum* subtypes: one with IIaA14G2R1 and IIaA15G2R1, and two with IIaA15G2R1 and IIdA19G1 (Table 3). Subtypes from subtype families IIc, IIe, IIl, IIn, IIr, IIs, and IIt were found in one or two patients each (Table 3).

Table 3. *Cryptosporidium parvum gp60* sequences generated in this study.

Species (No. of Patients)	Subtype Family (No. of Patients)	Subtype (No. of Patients)	GenBank Acc. No. [1]	Origin of Infection (No. of Patients)
C. parvum (299)	IIa (164)	IIaA13G2R1 (2)	**KU852706**	Turkey (1)
			KU852701	Morocco (1)
		IIaA13R1 (1)	**KU852702**	Sweden (1)
		IIaA14G1R1 (13)	JQ030882	Cyprus (1) Georgia (2) Sweden (3)
			JX183798 (IIaA14G1R1b)	Sweden (7)
		IIaA14G1R1r1 (6)	**KU852703**	Sweden (4) Spain (1) Uzbekistan (1)
		IIaA14G2R1 (1)	KF128738	Greece (1)
		IIaA14R1 (2)	JX183797	Estonia (1) Sweden (1)
		IIaA15G1R1 (6)	AM937012	Sweden (4) unknown (1)
			KU852704 (IIaA15G1R1_variant)	Sweden (1)
		IIaA15G2R1 (31)	AF164490	Dominican Republic (1) France (1) Mexico (1) Portugal (9) Portugal/Spain (1) Sweden (16) Venezuela (1) unknown (1)
		IIaA16G1R1 (42)	EU647727 (IIaA16G1R1b)	Georgia (1) Italy (1) Serbia (1) Spain (6) Sweden (30 [2])
			KU852707	unknown (1)
			KT895368 (IIaA16G1R1b_variant)	Austria (1) Sweden (1)
		IIaA16G2R1 (3)	DQ192505	Sweden (3)
		IIaA16G3R1 (2)	DQ192506	Portugal (1) Sweden (1)

Table 3. *Cont.*

Species (No. of Patients)	Subtype Family (No. of Patients)	Subtype (No. of Patients)	GenBank Acc. No. [1]	Origin of Infection (No. of Patients)
		IIaA16R1 (1)	AM937010	Malta (1)
		IIaA17G1R1 (20)	GQ983359	Poland (1) Sweden (1)
			JX183801 (IIaA17G1R1c)	Italy (1) Sweden (11) South Africa (1)
			AF403168 (IIaA17G1R1c_variant)	Sweden (5)
		IIaA17R1 (1)	JX183800	Sweden (1)
		IIaA18G1R1 (12)	HQ005742	Finland (1)
			KF289038 (IIaA18G1R1b)	Sweden (2)
			KT895369 (IIaA18G1R1b_variant)	Portugal (1) Sweden (3)
			JX183803 (IIaA18G1R1d)	United Kingdom (1) Sweden (4)
		IIaA18R1 (1)	**KU852705**	Sweden (1)
		IIaA19G1R1 (3)	KC679056	Spain (1) Sweden (2)
		IIaA19G2R1 (1)	DQ630514	Mexico (1)
		IIaA20G1R1 (4)	KC995127	Croatia (1) Italy (1) Sweden (1) unknown (1)
		IIaA21G1R1 (5)	FJ917373	Sweden (5)
		IIaA22G1R1 (5)	JX183806	Sweden (4) unknown (1)
		IIaA23G1R1 (2)	KC995126	Sweden (2)
	IId (118)	IIdA16G1 (5)	FJ917372	Sweden (1)
			JX183808 (IIdA16G1b)	Spain (1) Sweden (3)
		IIdA17G1 (4)	**KU852708**	Italy (1) Norway (1) Spain (1) Sweden (1)
		IIdA18G1 (3)	AY738194	France (1)
			KU852709	Africa (1) Sweden (1)
		IIdA19G1 (16)	DQ280496	China (1) Morocco (2) Oman (1) Portugal (5) Sweden (2) Sweden/Portugal (1)
			KU852711	Sweden (3)
			KU852713	Sweden (1)
		IIdA20G1 (24)	AY738185 (IIdA20G1b)	Israel (1)
			JQ028866 (IIdA20G1e)	Sweden (20)
			KU852711	Croatia (1) Sweden (1)
			KU852713	Croatia (1)
		IIdA21G1 (3)	DQ280497	Portugal (1) Sweden (2)
		IIdA22G1 (37)	AY166806	Spain (1) Sweden (15)
			FJ917374 (IIdA22G1c)	Estonia (1) Greece (1) Sweden (15) Sweden/US (1) unknown (1)
			KR349103	France (1) Sweden (1)
		IIdA23G1 (5)	FJ917376	Ivory Coast (1) Sweden (4)
		IIdA24G1 (14)	JQ028865	Denmark (1) Germany (1) Sweden (11 [3])
			JX183810 (IIdA24G1c)	Sweden (1)

Table 3. *Cont.*

Species (No. of Patients)	Subtype Family (No. of Patients)	Subtype (No. of Patients)	GenBank Acc. No. [1]	Origin of Infection (No. of Patients)
		IIdA25G1 (6)	JX043492	Sweden (6)
		IIdA29G1 (1)	GU458803	Sweden (1)
	IIc (2)	IIcA5G3a (1)	AF164491	Germany (1)
		IIcA5G3j (1)	HQ005749	Sweden (1)
	IIe (2)	IIeA10G1 (1)	KM539058	Guinea (1)
		IIeA13G1 (1)	**KU852716**	Sweden (1)
	IIl (1)	IIlA16R2 (1)	AM937007	Europe/Asia (1)
	IIn (2)	IInA10 (2)	**KU852717**	Tanzania/Sweden (1) Tanzania (1)
	IIr (1)	IIrA5G1 (1)	**KU852719**	Sweden (1)
	IIs (1)	IIsA10G1 (1)	**KU852720**	Sweden (1)
	IIt (1)	IItA13R1 (1)	**KU852718**	Tanzania (1)
Mixed subtypes	IIa + IIa (1)	IIaA14G2R1 + IIaA15G2R1 (1)	KF128738, KF128738	Italy (1)
Mixed subtypes	IIa + IId (2)	IIaA15G2R1 + IIdA19G1 (2)	AF164490, JF691561	Portugal (2)
C. parvum + *C. hominis* (1)	Ia + IIa (1)	IaA18R3 + IIaA16R1 (1)	KM538987, AM937010	Syria (1)
	neg *gp60* PCR (4)			Sweden (4)

[1] Sequences submitted to GenBank during this study are indicated in boldface type. Accession numbers in non-boldface type refer to reference sequences from GenBank. [2] Four samples from an outbreak [27]; [3] two samples from an outbreak [27].

Among the IIa and IId subtypes identified in the present study, several sequence variations were observed, either in the conserved non-repetitive part or in the repetitive area. In Tables 3 and 4, all subtypes and subtype variants are referred to via a corresponding GenBank acc. no. with 100% identity. For instance, in subtype IIaA16G1R1, the most common variant is IIaA16G1R1b (EU647727), which was found in 39 patients, and was the most common subtype in patients infected in Sweden ($n = 30$). Subtype IIaA16G1R1b_variant (KT895368), wherein the TCG repeat is located at a different position in the repeat region compared to IIaA16G1R1b, was found in two patients. This type of sequence variation has been described by Alsmark et al. [14], and it was found in four IIa subtypes, but not in any other subtype family (Table 3). Subtype variations were also seen in IId subtypes, exemplified by IIdA22G1, where three variants were identified, differing by one single nucleotide polymorphism (SNP) in the conserved part, AY166806 ($n = 16$), FJ917374 (IIdA22G1c; $n = 19$), and KR349103 ($n = 2$). Another type of sequence variation was found in IIaA14G1R1r1, which has an interruption of the TCA repeats by an ACA. This subtype, which was found in six patients, was named according to the proposed nomenclature by Amer et al. [28], who described a similar sequence variation, IIaA14G1R1r1b, in a sample from a calf.

Three new *C. parvum gp60* subtype families, IIr, IIs and IIt, were identified in one patient each. The sequences showed 100% identity with *C. parvum* at the SSU rRNA gene: IIr and IIt to AF164102 (gene copy A), and IIs to LC270281 (gene copy B). At the *gp60* locus, subtype IItA13R1 (KU852718) was identified in a patient (Swec402) who had visited Tanzania prior to infection. The closest matches in GenBank were 93% to *C. parvum* subtype family IIb (AY166805) and 90% to IIp (MK956000). The other two patients (Swec447 and Swec434) were both infected in Sweden, the first with *gp60* subtype IIsA14G1 (KU852720), where the closest match was 95% identity to subtype family IIe (KU852716), and the second with subtype IIrA5G1 (KU852719), which showed 92% similarity to the *C. hominis* subtype family Ie (AY738184). This observation prompted us to investigate two additional genes, actin and *hsp70*. As regards the actin locus, the isolate was 100% identical to *C. parvum* sequence MH542350, and at the *hsp70* locus, where 1911 bp were successfully sequenced, it differed by eight SNPs compared with the IOWA *C. parvum* strain (XM625373) in the conserved part. These SNPs, however, did not display any changes in the amino acid sequence.

Table 4. *Cryptosporidium hominis gp60* sequences generated in this study.

Species (No. of Patients)	Subtype Family (No. of Patients)	Subtype (No. of Patients)	GenBank Acc. No. [1]	Origin of Infection (No. of Patients)
C. hominis (49)	Ia (12)	IaA17R3 (1)	**KU852723**	India (1)
		IaA18R3 (4)	JF927190	Sweden (2) Thailand (2)
		IaA18R4 (1)	FJ153246	Thailand (1)
		IaA20R3 (2)	**KU727289**	Tanzania (2)
		IaA23R3 (1)	JQ798143	India (1)
		IaA26R3 (1)	**KU852724**	Somalia (1)
		IaA28R4 (2)	KF682373	US (2)
	Ib (26)	IbA6G3 (1)	**KU852722**	Egypt (1)
		IbA9G3 (9)	DQ665688	Afghanistan (1) Ethiopia (1) Malawi (1) Mozambique (1) Somalia (1) Uzbekistan (1) Zambia (1)
			KF974523	Congo Republic (1) Uganda (1)
		IbA10G2 (15)	AY262031	Estonia (1) Greece (1) United Arab Emirates (2) Guatemala (1) Peru (1) Spain (7) Sweden (2)
		IbA13G3 (1)	KM539004	Burkina Faso (1)
	Id (3)	IdA16 (2)	HQ149034	China (1) Sri Lanka (1)
		IdA17 (1)	**KU852721**	Tanzania (1)
	Ie (1)	IeA11G3T3 (1)	GU214354	South Africa (1)
	If (2)	IfA12G1R5 (2)	HQ149036	Germany (1) Sweden (1)
	Ik (2)	IkA18G1 (2 [2])	KU727290	Sweden (2)
	Ii (2)	IiA17 (2 [2])	KF679724	Thailand (2)
	neg *gp60* PCR (1)			Sweden (1)

[1] Sequences submitted to GenBank during this study are indicated in boldface type. Accession numbers in non-boldface type refer to reference sequences from GenBank. [2] Cases described in Lebbad et al., 2018 [29].

3.5. Molecular Characterization of Cryptosporidium hominis

Gp60 subtyping was successful for 49 out of 50 *C. hominis* isolates (including the sample with mixed species) (Table 4). Seven subtype families, Ia, Ib, Id, Ie, If, Ii, and Ik, and 17 subtypes were identified. The most common subtype was IbA10G2 (*n* = 15), wherein 13 cases had contracted infection while abroad; the remaining 2 reflected domestic infections. The second-most common subtype was IbA9G3 (*n* = 9), detected in patients who were infected in nine different countries, mainly in Africa (Table 4). In addition to the two cases with IbA10G2, another three subtypes were found in patients infected with *C. hominis* in Sweden: IaA18R3 (n = 2), IfA12G1R5 (n = 1), and IkA18G1 (n = 2). The cases with the uncommon subtype families Ii and Ik have been described in a separate article [29].

3.6. Molecular Characterization of C. hominis/C. parvum Mixed Infection

One patient, a 4-year-old girl from Syria, had a mixed *C. hominis* and *C. parvum* infection with subtypes IaA18R3 and IIaA16R1.

3.7. Outbreaks and Family Clusters

Four outbreaks, all caused by *C. parvum*, were identified during the study period (Table 5). The first outbreak occurred among veterinary students in 2013, and two different subtypes were found, IIaA16G1R1 and IIdA24G1 [27]. Outbreaks 2, 3 and 4 were all foodborne and involved subtypes IIaA16G2R1, IIaA17R1, and IIdA17R1, respectively. Three small family clusters were also detected; two related to traveling, and one where a contaminated water well was the suspected vehicle (Table 5).

Table 5. Outbreaks and family clusters involving *C. parvum* and *C. hominis* identified in the study period.

Outbreak/ Cluster	Month/ Year	Number of Suspected Cases	Number of Confirmed Cases	Number of Cases Typed	Information	Species	Subtype	GenBank Acc. No.
Outbreak 1 [1]	Feb. 2013	13	6	6	Sweden: veterinary students	*C. parvum*	IIaA16G1R1 (4 cases) [1] IIdA24G1 (2 cases) [1]	EU647727 (IIaA16G1R1b) JQ028865
Outbreak 2	Jan. 2013	10	2	2	Sweden: foodborne, private dinner, salad was the suspected source of infection	*C. parvum*	IIaA16G2R1	DQ192505
Outbreak 3	May 2014	8	2	1	Sweden: foodborne, private dinner, no identified source of infection	*C. parvum*	IIaA17R1	JX183800
Outbreak 4	March 2014	-	23	13	Sweden: foodborne, restaurant, parsley was the suspected source	*C. parvum*	IIdA22G1	AY166806
Cluster 1	July 2014	5	3	3	Sweden: a family, suspected contaminated water well	*C. parvum*	IIaA15G2R1	AF164490
Cluster 2	Nov. 2014	2	2	2	Portugal: a couple traveling together	*C. parvum*	IIaA15G2R1 + IIdA19G1	AF164490 + JF691561
Cluster 3 [2]	Feb. 2013	3	2	2	Thailand: a father and his son traveling together [2]	*C. hominis*	IiA17 [2]	KF679724

[1] Outbreak described in Kinross et al., 2015 [27]; [2] described in Lebbad et al., 2018 [29].

3.8. Molecular Characterization of Additional Species

Of 30 isolates from additional species, 28 were successfully sequenced at the SSU rRNA locus. Subtyping with the *gp60* protocol was successful for 29 isolates; the only exception was *C. ditrichi*, wherein no suitable primers were available (Table 6).

Table 6. *Cryptosporidium gp60* sequences from non-*hominis* and non-*parvum Cryptosporidium* species generated in this study.

Species (No. of Patients)	Subtype Family (No. of Patients)	Subtype (No. of Patients)	GenBank Acc. No. [1]	Origin of Infection (No of Patients)
C. meleagridis (8)	IIIb (3)	IIIbA23G1R1 (2)	KJ210606 (IIIbA23G1R1a) [2]	Indonesia (1)
			KU852727 (IIIbA23G1R1c)	Malaysia (1)
		IIIbA24G1R1 (1)	**KU852729**	China (1)
	IIIe (4)	IIIeA17G2R1 (2)	**KU852726**	China (2)
		IIIeA19G2R1 (1)	KJ210620 [2]	Uzbekistan (1)
		IIIeA21G2R1 (1)	**KU852728**	Indonesia (1)
	IIIg (1)	IIIgA22G3R1 (1)	**KU852730**	Nepal (1)
C. cuniculus (5)	Va (1)	VaA19 (1)	**KU852733**	Sweden (1)
	Vb (4)	VbA20R2 (1)	**KU852735** (VbA20R2b)	Sweden (1)
		VbA25R3 (1)	**KU852731**	Spain (1)
		VbA29R4 (1)	**KU852734**	Sweden (1)
		VbA31R4 (1)	**KU852732**	Greece (1)
C. erinacei (2)	XIIIa (2)	XIIIaA23R12 (1)	**KU852736**	Sweden (1)
		XIIIaA24R9 (1)	**KU852737**	Greece (1)
C. ubiquitum (2)	XIIa (2)	XIIa-1 (2)	**KU852740**	Sweden (2)
C. viatorum (1)	XVa (1)	XVaA3b (1)	KP115937 [3]	Kenya (1)
Cryptosporidium chipmunk genotype I (5)	XIVa (5)	XIVaA20G2T1 (5)	KP099089	Sweden (5)
Cryptosporidium horse genotype (1)	VIc (1)	VIcA16 (1)	**KU852738**	Kenya (1)
C. felis (4) [4]	XIXa (3)	XIXa-39 (1)	MH240852 [4]	Indonesia (1)
		XIXa-43 (1)	MH240856 [4]	Sweden (1)
		XIXa-68 (1)	MH240883 [4,5]	Sweden (1)
	XIXb (1)	XIXb-1 (1)	MH240901 [4]	Sweden (1)
C. suis (1)	XXVa (1)	XXVaR37 (1)	MH187875	Lithuania (1)
C. ditrichi (1) [6]	*gp60* PCR neg			Sweden (1)

[1] Sequences submitted to GenBank during this study are indicated in boldface type. Accession numbers in non-boldface type refer to reference sequences from GenBank. [2] isolates included in Stensvold et al., 2014 [23]; [3] isolate included in Stensvold et al., 2015 [22]; [4] isolates included in Rojas et al., 2020 [21]; [5] case described in Beser et al., 2015 [30]; [6] case described in Beser et al., 2020 [31].

A recently adopted child from Lithuania was infected with *C. suis*. A *gp60* sequence was achieved using the *Cryptosporidium* chipmunk primers [19], and by means of additional sequencing primers (data not shown) subtype XXVaR37 was obtained. This case and method will be described in a separate article.

Cases with *Cryptosporidium* chipmunk genotype I subtype XIVaA20G2T1 and *C. ubiquitum* subtype XIIa-1 were found in five and two patients, respectively, all infected in Sweden.

Two patients were infected with *C. erinacei*. The SSU rDNA sequences from their samples were not identical; the sequence from patient Swec627 (KU892565) infected in Sweden differed by two SNPs from the closest *C. erinacei* match in GenBank (KC3056047), while the sequence from patient Swec653 infected in Greece was 100% identical to KC3056047 (Figure 2). The actin DNA sequence obtained from isolate Swec627 was 100% identical to a *C. erinacei* isolate (MN237648) from a European badger in Poland (unpublished). Based on *gp60* analysis, a new *C. erinacei* subtype, XIIIaA23R12, was identified in the first patient (Swec627), while subtype XIIIaA24R9 was found in the second patient (Swec653).

Figure 2. Phylogenetic relationships between partial SSU rDNA *Cryptosporidium* sequences obtained in the present study and sequences retrieved from the NCBI database. Trees were constructed using the neighbor-joining method based on genetic distance calculated by the Kimura's 2-parameter model as implemented in MEGA X. The final dataset included 749 positions. Bootstrap values ≥50% from 1000 replicates are indicated at each node. Isolates from this study are indicated in boldface.

Four patients were diagnosed with *C. felis*; three in Sweden, and one in Indonesia. The recently described *gp60* assay for *C. felis* [21] yielded four different sequence types.

One patient infected in Kenya with *C. viatorum* subtype XVaA3b was identified during the study period. Material from this case was used to develop a *C. viatorum gp60* subtyping assay [22].

The eight patients diagnosed with *C. meleagridis* were all infected in Asia. Six of the samples were identified as genotype 1 at the SSU rRNA locus (AF112574), while sequencing of this gene failed for the remaining two isolates. Investigation of the *gp60* gene was successful for all eight isolates and revealed three subtype families, IIIb, IIIe and IIIg, and six different subtypes (Table 6). Subtype IIIbA23G1R1c (KU852727) differed by eight SNPs in the post-repetitive part of the sequence from IIIbA23G1R1b (KJ210609), and was named IIIbA23G1R1c according to the proposed nomenclature [23].

Two subtype families of *C. cuniculus*, Va and Vb, and five different *gp60* subtypes were observed. It was noted that the *C. cuniculus* Vb sequences published in GenBank carried 2–5 ACA repeats just after the TCA repeats (data not shown); thus, we followed the recommendation by Nolan et al. from 2010 [32] and included an R in the subtype designation, as in VbA29R4 for isolate Swe658, with 29 TCA repeats followed by 4 ACA repeats (KU852734). One patient infected in Greece carried a new subtype, VbA31R4. The sequence from patient Swec678 had the same numbers of TCA and ACA repeats as a VbA20R2 strain from the UK (GU971649), but differed by six SNPs and a 3 bp deletion in the post-repetitive part of the sequence, resulting in five amino acid changes. The subtype variant was designated VbA20R2b (KU852735), according to the proposed *gp60* nomenclature [33].

A third subtype family of *Cryptosporidium* horse genotype, VIc, was identified in a patient (Swe490) who had visited Kenya. This subtype, referred to as VIcA16 (KU852738), showed 88% and 87% identity to subtype families VIa and VIb, respectively. Four ACA repeats followed just after the TCA repeats, a feature also observed in *C. cuniculus* subtype family Vb. An extended molecular investigation including the SSU rRNA, actin, and *hsp70* genes was performed. The SSU rDNA sequence (KU892564) differed by one SNP from *Cryptosporidium* horse genotype sequences deposited in GenBank (FJ435962, MK775041). The closest match for the actin sequence (KU892571) was a sequence from *Cryptosporidium tyzzeri* (AF382343), from which it differed by eight SNPs. No actin sequences from the horse genotype were available in GenBank for comparison. The *hsp70* sequence KU892577 (1895 bp) showed 100% identity with the four *Cryptosporidium* horse genotype sequences available in GenBank. However, none of these sequences (298–403 bp) were long enough to cover the area with the 12 bp segment repeats towards the 3' end of the *hsp70* gene. The horse genotype sequence from our study exhibited 10 repeats of a 12 bp segment with SNPs at the third and sixth bases—GG(C/T)GG(A/T)ATGCCA.

3.9. Phylogenetic Analyses

A phylogenetic tree, which contained representative SSU rDNA sequences from all 12 species and genotypes detected in the present study and published sequences from most *Cryptosporidium* species/genotypes hitherto detected in humans, was constructed (Figure 2). In the *gp60* tree, one representative sequence from each subtype family detected in this study, except *C. felis* and *C. suis*, was included (*n* = 26). Sequences from established subtype families are clustered with sequences from this study. The new *C. parvum* subtype family IIt clustered with IIb, subtype family IIr with Ie, and subtype family IIs with IIe, while the new *Cryptosporidium* horse genotype subtype family VIc clustered with VIa and VIb (Figure 3).

Figure 3. Phylogenetic relationships between partial *gp60 Cryptosporidium* sequences obtained in the present study and sequences retrieved from the NCBI database. Trees were constructed using the neighbor-joining method based on genetic distance calculated by the Kimura's 2-parameter model as implemented in MEGA X. The final dataset included 840 positions. Bootstrap values ≥50% from 1000 replicates are indicated at each node. New subtype families observed in this study are indicated by filled circles. All isolates from this study are indicated in boldface.

4. Discussion

In the present study, performed between January 2013 and December 2014 and including samples from 398 patients with cryptosporidiosis, a high diversity of *Cryptosporidium* species and subtypes was identified. *Cryptosporidium parvum* was still the dominant species, and even fewer *C. hominis* cases were identified compared with a previous study performed in the Stockholm metropolitan area between April 2006 and November 2008. Meanwhile, the total number of patients infected with non-*hominis* and non-*parvum* species was quite high (8%), corroborating earlier observations [9].

Species determination was successful for 95% of the samples. *Cryptosporidium parvum* (79%) dominated both in patients infected in Sweden (84%) and abroad (57%), while *C. hominis* was much less common (13%) and identified in only 3% of the domestic cases and in 30% of the cases infected abroad. Mixed *C. hominis* and *C. parvum* infection was observed only in 1 patient, and for 10 of the patients infected with *C. parvum*, the origin of infection was unknown or uncertain. The high occurrence of *C. parvum* compared with *C. hominis* observed in Sweden is similar to the situation in other industrial countries, such as France, Ireland, and Canada [34–36], although a higher percentage of *C. hominis* has been observed in Spain and Australia [37,38]. Shifting trends have been seen over time in the Netherlands and New Zealand, showing the importance of longitudinal studies [39,40].

In terms of the analysis of cases according to age, the observed bimodal distribution (Figure 1) supports previous observations [9], with a relatively high number of cases observed among infants and toddlers and with a second peak—and the largest one—in the 30–44-year-olds. Quite similar age distributions have been observed in studies from Denmark, France, and Canada, but with the second peak in the 20–35-year-olds [34,36,41]. The bimodal age distribution may reflect transmission between parents and their children; however, no such family clusters were detected during the present study. There was a slight difference in the proportions of female and male cases: 55% and 45%, respectively. Interestingly, this difference in gender distribution has been seen in cryptosporidiosis cases in Sweden every year since 2005 (57% and 43% on average for women and men, respectively) [5]. A similar gender distribution was observed in neighboring Denmark in 2010–2014 [41]. Whether this difference in gender distribution reflects a higher awareness among females of the need to seek medical care, or something else, remains unknown.

A few *C. parvum* outbreaks were included in the present study (Table 5), but even if those patients (*n* = 22) are excluded from the dataset, *C. parvum* remains the dominant species in patients infected in Sweden, at least in the four areas that contributed most of the samples (Table 1). Differences in the geographical distribution of *C. parvum* and *C. hominis*, with more *C. parvum* in rural regions and more *C. hominis* in urban settings, have been described [42,43], and a similar tendency was seen in our study, where the rural region of Halland had relatively more cases of *C. parvum* and fewer cases of *C. hominis* compared with the metropolitan region of Stockholm (Table 1). These differences are often attributed to the closer contact with farm animals in rural areas and more traveling activity for people living in urban areas [44].

The subtyping of *C. parvum* showed a great variability; nine subtype families and 42 subtypes were observed. Two *C. parvum* subtype families, IIa (*n* = 164) and IId (*n* = 118), dominated both in patients infected in Sweden and abroad, while the other subtype families (IIc, IIe, IIl, IIn, IIr, IIs, IIt) were found only in one or two isolates each. The most common *C. parvum* subtype in the present study was IIaA16G1R1b (EU647727), observed in 39 patients. Detected in 24 sporadic cases and 4 cases related to an outbreak among veterinary students, it was also previously observed as the most common subtype amongst people infected in Sweden [9,27] (Table 5). This subtype is common in Swedish cattle and has been involved in earlier outbreaks and family clusters in Sweden [9,45], but was not reported amongst 48 subtyped *C. parvum* outbreaks between 2009 and 2017 in the UK [46].

Subtype IIaA15G2R1 is the dominating *C. parvum* subtype in many countries, both in cattle and in humans, and is responsible for numerous outbreaks of human cryptosporidiosis, probably due to its biological fitness and high virulence [41,46–49]. This subtype was found in 33 patients, of whom 16 were infected in Sweden; 13 were sporadic cases, and 3 cases were part of a family cluster (Table 5). This subtype has been detected in a few Swedish calves [50], but has not been associated with any known larger *Cryptosporidium* outbreak in Sweden.

Many of the IIa and IId subtype sequences in the present study exhibited examples of polymorphism in the non-repetitive part, which is sometimes indicated by a lower-case letter when sequences are reported to GenBank [33], but this is not always the case. To overcome this dilemma, all subtypes and subtype variants in Tables 3 and 4 are referred to via a specific GenBank acc. no. As an example, the most common IId subtype, IIdA22G1 ($n = 37$), occurred in three variants in this study, and one of them (AY166806) ($n = 15$) was only detected in domestic cases, most of them in connection to a foodborne outbreak, in which parsley was the suspected vehicle (Table 5). Another variant of this subtype, IIdA22G1c (FJ917374), was identified in 19 cases, of which 15 represented sporadic cases infected in Sweden. Both these variants of IIdA22G1 have been seen in earlier Swedish studies of cattle and humans [9,50]. The third variant of this subtype, KR349104, was seen in only two cases; one infected in France and one in Sweden.

Subtype IIdA24G1 gained attention in 2011 in connection to a foodborne outbreak linking two Swedish cities, and it was also involved in an outbreak in 2013 among veterinary students occurring during the present study [12,27]. The same subtype, which is considered to be quite rare, was linked to 1 out of 48 *C. parvum* outbreaks in the UK where *gp60* subtyping was performed [46].

The subtypes mentioned above, IIdA22G1c and IIdA24G1, were involved in two foodborne outbreaks, which occurred simultaneously in late 2019 in 10 and 12 Swedish counties, respectively. Subtype IIdA22G1c was identified in 122 cases and IIdA24G1 in 86 cases. Spinach juice was the suspected vehicle for subtype IIdA22G1c, while no specific food item could be identified as a source of infection for IIdA24G1 [51].

Two new *C. parvum* subtypes were identified within the subtype families IIe and IIn. Subtype IIeA13G1 (KU852716) was found in a patient infected in Sweden, and subtype IInA10 (KU852717) in two patients; one infected in Tanzania, and one with unclear origin of infection (either Tanzania or Sweden (secondary infection)). *Cryptosporidium parvum* IIe is a well-known anthroponotic subtype family, but since only two cases (both humans from India) have been reported from subtype family IIn [52], it is impossible to say whether this one might be an anthroponotic or zoonotic subtype family. The same is true for the 3 new *C. parvum* subtype families, IIr, IIs and IIt, that were added to the 16 families already described [48]. Subtype IIrA5G1 and IIsA14G1 were found in patients infected in Sweden. Interestingly, a patient with subtype IIsA10G1 was recently described in a publication from Zambia [53]. The patient with subtype IItA13R1 was infected in Tanzania.

One of the objectives of the present study was to investigate whether *C. hominis* subtype IbA10G2 had proliferated in Sweden after the large waterborne *C. hominis* outbreaks in Östersund and Skellefteå in 2010–2011, wherein an estimated 45,500 persons developed cryptosporidiosis. In the earlier study by Insulander et al. [9], performed from 2006 to 2008, 12 of the 17 *C. hominis* samples from patients infected in the Stockholm area carried subtype IbA10G2, while in the present study only two of the eight *C. hominis* patients infected in different parts of Sweden carried this subtype. The higher frequency of this subtype in the earlier study could be explained by the fact that several outbreaks at day care centers, with an index person infected abroad, appeared during that study period, while no such outbreaks were recorded in the current study. Few samples were submitted from the former outbreak areas; four from Jämtland County (Östersund) and four from Västerbotten County (Skellefteå), with all cases representing *C. parvum* (Table 1). The number of reported cases from these two counties during 2013 and 2014 was also low, at eight and five, respectively, and remained quite low during the following years up till now [5]. Therefore, even with these limitations in mind, we speculate that subtype IbA10G2 has failed to establish itself in Sweden following the large outbreaks.

In the US, where IbA10G2 used to be the most common outbreak-related subtype, a new *C. hominis* subtype, IaA28R4, emerged in 2007 as a major subtype in sporadic cases and waterborne outbreaks [54]. Since 2013, subtype IfA12G1R5 has emerged as the dominant *C. hominis* subtype in the US, as well as in Western Australia [37,55]. In the present study, IaA28R4 was observed in two patients, both of whom had recently visited the US, and subtype IfA12G1R5 in two patients, one of whom was infected in Sweden and one in Germany.

Cryptosporidium hominis is generally regarded as an anthroponotic species only occasionally infecting other animals. However, recent studies have shown that for a certain subtype family, Ik, equines are natural hosts [56]. During the present study, the first human cases infected with this subtype family, Ik, were detected [29] (Table 4). Two unrelated patients, both infected in Sweden, carried the same *gp60* subtype, IkA18G1. This subtype was recently detected in a horse from China (MK770627)—information reinforcing the suspicion that zoonotic transmission might have occurred. Another uncommon *C. hominis* subtype family, Ii, was found in a father and son after returning from Thailand, where they visited a monkey farm. This *C. hominis* variant was previously described as *C. hominis* monkey genotype, and has rarely been detected in humans [29].

In the present study, 10 *Cryptosporidium* species/genotypes were found in addition to *C. parvum* and *C. hominis*, corresponding to 8% (30/379) of the total number of genotyped samples. This agrees with an earlier Swedish investigation performed on *Cryptosporidium* patients from the Stockholm metropolitan area, where 9% (17/194) of the genotyped samples represented species other than *C. hominis* and *C. parvum*. However, in the Stockholm study, the majority of these infections were acquired abroad; only three patients, two with *Cryptosporidium* chipmunk genotype I and one with *C. felis*, were infected in Sweden [9]. In the present study, an equal number of patients with species other than *C. hominis* or *C. parvum* were infected in Sweden ($n = 15$) and abroad ($n = 15$) (Table 1). This difference might reflect differences in study populations; in the first study, most patients originated from an urban area, while in the present study, many of the patients were from rural areas and probably more exposed to zoonotic *Cryptosporidium* species endemic to Sweden.

The most common cause of non-*hominis* and non-*parvum* infections acquired in Sweden was *Cryptosporidium* chipmunk genotype I, which was diagnosed in five adults; four women and one man. Cryptosporidiosis caused by the chipmunk genotype is considered an emerging infection in the US [19,43]. Meanwhile, in Europe, only one human case (from France) has been reported outside Sweden [19]. The first Swedish cases were diagnosed in September 2007 and August 2008, at a time when no *gp60* subtype method was available for the chipmunk genotype I [57]. However, later analyses have shown that they carried the same subtype as the patients from the present study, XIVaA20G2T1, a subtype that recently was detected in red squirrels in Sweden (unpublished information). Red squirrels are consequently the most possible source of infection for our patients; the other described host animals—chipmunks, grey squirrels, and deer mice— are not native to the country [19,58]. Recent observations have shown a rising number of chipmunk genotype I infections acquired in Sweden, reflecting an emerging infection in this country (manuscript submitted).

Cryptosporidium erinacei was first described in European hedgehogs, but has also been identified in other animals, such as horses and rats [59–61]. Most human cases have been reported from New Zealand ($n = 13$), while reports from Europe are scarce, including one case from the Czech Republic and one from France [40,62,63]. The European hedgehog is considered an endangered species in Sweden and other parts of Europe, which might explain the low number of reports from Europe compared with New Zealand, where it has become an invasive species since its introduction about 150 years ago. The two cases found during the present study are the first human *C. erinacei* cases reported from Sweden. The patient infected in Sweden carried a unique *gp60* subtype, XIIIaA23R12, while the patient infected in Greece carried subtype XIIIaA24R9, which has recently been reported in New Zealand [40].

Cryptosporidium cuniculus is a common species among rabbits worldwide, while human infection with this species has gained special attention in the UK, where, in addition to a documented outbreak, sporadic cases are quite common [64]. Reports of human infections from other parts of the world are increasing, and sporadic cases have been described in Nigeria, France, Spain, Australia, and New Zealand [40,48]. Two *C. cuniculus* subtype families have been described, Va and Vb. Four of the study patients were infected with Vb, which is the most commonly reported subtype family both in rabbits and humans, as well as in a few other animals, such as kangaroo and alpaca [65,66]. Subtype family Va, which was observed in one of the patients infected in Sweden, appears to be less frequent globally, and previous human infections were all reported in the UK [64]. These are the first documented human cases of *C. cuniculus* infection in Sweden.

One patient was infected with *C. viatorum* while visiting Kenya [22]. This species was initially thought to be anthroponotic because only human cases had been detected until 2018, when the first non-human host, an Australian swamp rat, was identified [67,68]. This swamp rat isolate was genetically quite different from all known human isolates, but recent studies from China focusing on different rat species have found *C. viatorum* isolates genetically similar to human isolates, and subtyping with *gp60* has demonstrated the same subtype, XVaA3g, in wild rats from China and a human from Australia [37,69].

Another visitor to Kenya was diagnosed with *Cryptosporidium* horse genotype. All horse and donkey samples positive for this species when sequencing of the *gp60* locus was performed carried subtype family VIa [70,71], which has also been found in one human case from Poland (MK784560) (unpublished). Only four cases with subtype family VIb have been described: two human cases, one in the UK and one in the US [54,72], and two cases in four-toed hedgehogs from Japan [73,74]. No reports of the VIb subtype family in equine hosts are available. The patient from our study carried a new *gp60* subtype family (VIc) of *Cryptosporidium* horse genotype. Since no other reports of this subtype family have been published to date, we cannot speculate on how this patient contracted the infection.

Two children of 3 and 5 years of age, respectively, were infected in Sweden with *C. ubiquitum*, *gp60* subtype XIIa-1. These cases were unrelated, and no probable source of infection was identified. *Cryptosporidium ubiquitum* has a wide host range and has been found in wild and domestic ruminants, rodents, and primates, including humans. In the UK, it is believed that most human infections with *C. ubiquitum* are related to exposure to sheep, which are common carriers of subtype XIIa, while infected humans in the US mainly carry the rodent-associated subtypes XIIb and XIIc [20]. The only published *C. ubiquitum* case amongst Swedish animals is from a calf [45], and the occurrence in other animals in Sweden, including sheep, remains unknown and should be investigated.

The natural hosts of *C. suis* are domesticated pigs and wild boars. Human infections are not frequently diagnosed; only around 12 cases were reported prior to our case, which was a two 2-year-old adopted child diagnosed with cryptosporidiosis upon arrival in Sweden from Lithuania [48,75].

Human infection with the feline parasite *C. felis* is not uncommon; indeed, it is regarded as one of the five most common *Cryptosporidium* species infecting humans worldwide [48]. During the study period, four *C. felis* cases were detected, with three patients being infected in Sweden and one in Indonesia. Two of the Swedish patients had known contact with cats, and analyses of fecal samples from cats and their owners using a newly described *gp60* method showed that they shared *gp60* subtypes; thus, zoonotic infection was confirmed [21,30].

One study patient was infected with *C. ditrichi*, a recently described *Cryptosporidium* species in *Apodemus* spp. in Europe [76]. This was the first time *C. ditrichi* was diagnosed in a human, and later on, two more Swedish cases were identified [31].

Eight patients were infected with *C. meleagridis*, the third most common *Cryptosporidium* species infecting humans [77]. *Cryptosporidium meleagridis* has a wide host range, ranging from various birds to humans, and in some countries (e.g., Thailand and Peru), this species is reportedly more common in humans than *C. parvum* [77]. All patients infected with *C. meleagridis* in the present study had acquired infection while traveling in different Asian countries. The same *gp60* subtype was identified in two patients who had made the same journey to China, while the remaining six patients carried different subtypes (Table 6). Subtypes IIIeA17G2R1, IIIeA19G2R1 and IIIeA21G2R1 have been observed in chicken in China, and subtype IIIbA24G1R1 in poultry from Brazil, indicating zoonotic transmission as a possible route of infection [78,79]. Zoonotic transmission has been documented at a

Swedish organic farm, where the infected person and the chickens carried the same *hsp70* and *gp60* subtypes [23,80].

When this study was initiated, the intention was to cover the entire country of Sweden, and all 21 clinical laboratories carrying out parasitology diagnostic tests in Sweden on a routine basis at the time of the study were invited to participate. One limitation is that by the end of the study period, only 12 of the regional laboratories (representing 11 different counties) had provided samples for the study. Our results, however, could very well be considered to reflect the reality, as nine counties only reported zero to one patient each with cryptosporidiosis during the study period (2013 and 2014) [5]. However, the number of referred samples varied considerably between the laboratories, with only four of them accounting for 95% of the submitted samples (Table 1). With this skewed distribution, we cannot conclude that the results of this study might be generalizable to the whole country. Another reason for this biased referral of samples could be the limited access to unfixed stool samples at some laboratories recommending that stool samples for parasitology be fixed in SAF. The intention at the beginning of the study was to avoid receiving samples fixed in SAF or formalin, which has a known negative effect on the PCR success rate [81], but in the second year, one laboratory with a high detection rate of *Cryptosporidium* and limited access to native (i.e., non-preserved) samples was invited to provide fixed specimens. In total, we received 70 samples fixed in SAF, 63 from this laboratory and 7 from other participating laboratories. The success rate of PCR for SAF-fixed samples was 80% (56/70), compared with 99% (323/328) for native stool samples. One recent study pointed out the limited availability of native stool material for the molecular analyses of stool parasites [34], but with the increased use of DNA-based diagnostics, the availability of unfixed material has increased, and the problem with fixed samples is in decline.

Another potential bias is that the clinical laboratories performing the primary diagnostics used different testing strategies and detection methods. Some diagnostic real-time PCR methods are designed to primarily detect *C. parvum* and *C. hominis*, and do not target some of the less common and genetically distinct species, which might have led to an underrepresentation of these in some of the counties [82].

5. Conclusions

Cryptosporidium parvum was the dominant species both in cases infected abroad and in domestic cases. The observed occurrence of *C. hominis* was generally low, and no indication of an expansion of the subtype IbA10G2 previously causing the vast *Cryptosporidium* outbreaks in Sweden was found. There was a high diversity of species and subtypes, with 8% of the cases reflecting species other than *C. parvum* and *C. hominis*, some of which were found for the first time in humans in Sweden (e.g., *C. cuniculus*, *C. erinacei*, and *C. ubiquitum*). Our study also identified humans as a new host of *C. ditrichi*. Overall, zoonotic species and subtypes plays a major role in human cryptosporidiosis in Sweden.

Author Contributions: Conceptualization, J.B. and J.W.-K.; methodology, J.B. and M.L.; software J.B. and M.L.; validation, J.B. and M.L.; formal analysis, J.B., J.W.-K. and M.L.; investigation, J.B., J.W.-K. and M.L.; resources, J.B.; data curation, J.B., J.W.-K. and M.L.; writing—original draft preparation, J.B., M.L. and C.R.S.; writing—review and editing, J.B., M.L. and C.R.S.; visualization, J.B., M.L. and C.R.S.; supervision, J.B.; project administration, J.B.; funding acquisition, J.B. All authors have read and agreed to the published version of the manuscript.

Funding: The study was funded by a grant from the Swedish Civil Contingencies Agency (Grant number: 2012-172).

Institutional Review Board Statement: Approval (registration number 2013/201-31/4) was obtained from the Ethical Review Board at Karolinska Institutet, Stockholm, Sweden, including approval to publish the data (date of approval: 27 February 2013).

Informed Consent Statement: Patient consent was waived since all samples were from patients with cryptosporidiosis, which is a notifiable disease in Sweden according to the Swedish Disease Act

(2004:168) and included in the national surveillance of microbial pathogens. Species identification and genotyping of *Cryptosporidium* and communication of these data form part of the surveillance and do not require consent from the patient (https://www.folkhalsomyndigheten.se/the-public-health-agency-of-sweden/communicable-disease-control/surveillance-of-communicable-diseases/notifiable-diseases/ accessed on 24 April 2021).

Data Availability Statement: Data presented in this study are available on request from the corresponding author, Jessica Beser. Due to existing general data protection rules, the data are not publicly available.

Acknowledgments: The authors are grateful to all the clinical microbiological laboratories that provided samples to the study and the laboratory personnel at PHAS that performed analyses. The authors would also like to thank Margareta Löfdahl for providing epidemiological data and Erik Alm for the development of the software CryptoTyper.

Conflicts of Interest: The authors declare no conflict of interest.

References

1. Kotloff, K.L.; Nataro, J.P.; Blackwelder, W.C.; Nasrin, D.; Farag, T.H.; Panchalingam, S.; Wu, Y.; Sow, S.O.; Sur, D.; Breiman, R.F.; et al. Burden and aetiology of diarrhoeal disease in infants and young children in developing countries (the Global Enteric Multicenter Study, GEMS): A prospective, case-control study. *Lancet* **2013**, *382*, 209–222. [CrossRef]
2. Ryan, U.; Fayer, R.; Xiao, L. *Cryptosporidium* species in humans and animals: Current understanding and research needs. *Parasitology* **2014**, *141*, 1667–1685. [CrossRef]
3. Feng, Y.; Ryan, U.M.; Xiao, L. Genetic diversity and population structure of *Cryptosporidium*. *Trends Parasitol.* **2018**, *34*, 997–1011. [CrossRef] [PubMed]
4. Zahedi, A.; Ryan, U. *Cryptosporidium*—An update with an emphasis on foodborne and waterborne transmission. *Res. Vet. Sci.* **2020**, *132*, 500–512. [CrossRef] [PubMed]
5. Public Health Agency of Sweden. Surveillance of Communicable Diseases Sweden. Available online: https://www.folkhalsomyndigheten.se/folkhalsorapportering-statistik/statistik-a-o/sjukdomsstatistik/cryptosporidiuminfektion/ (accessed on 24 April 2021).
6. Harvala, H.; Ogren, J.; Boman, P.; Riedel, H.M.; Nilsson, P.; Winiecka-Krusnell, J.; Beser, J. *Cryptosporidium* infections in Sweden-understanding the regional differences in reported incidence. *Clin. Microbiol. Infect.* **2016**, *22*, 1012–1013. [CrossRef] [PubMed]
7. Ogren, J.; Dienus, O.; Beser, J.; Henningsson, A.J.; Matussek, A. Protozoan infections are under-recognized in Swedish patients with gastrointestinal symptoms. *Eur. J. Clin. Microbiol. Infect. Dis.* **2020**, *39*, 2153–2160. [CrossRef]
8. Waldron, L.S.; Dimeski, B.; Beggs, P.J.; Ferrari, B.C.; Power, M.L. Molecular epidemiology, spatiotemporal analysis, and ecology of sporadic human cryptosporidiosis in Australia. *Appl. Environ. Microbiol.* **2011**, *77*, 7757–7765. [CrossRef]
9. Insulander, M.; Silverlås, C.; Lebbad, M.; Karlsson, L.; Mattsson, J.; Svenungsson, B. Molecular epidemiology and clinical manifestations of human cryptosporidiosis in Sweden. *Epidemiol. Infect.* **2013**, *141*, 1009–1020. [CrossRef] [PubMed]
10. Bjelkmar, P.; Hansen, A.; Schönning, C.; Bergström, J.; Löfdahl, M.; Lebbad, M.; Wallensten, A.; Allestam, G.; Stenmark, S.; Lindh, J. Early outbreak detection by linking health advice line calls to water distribution areas retrospectively demonstrated in a large waterborne outbreak of cryptosporidiosis in Sweden. *BMC Public Health* **2017**, *17*, 328. [CrossRef] [PubMed]
11. Widerström, M.; Schönning, C.; Lilja, M.; Lebbad, M.; Ljung, T.; Allestam, G.; Ferm, M.; Bjorkholm, B.; Hansen, A.; Hiltula, J.; et al. Large outbreak of *Cryptosporidium hominis* infection transmitted through the public water supply, Sweden. *Emerg. Infect. Dis.* **2014**, *20*, 581. [CrossRef] [PubMed]
12. Gherasim, A.; Lebbad, M.; Insulander, M.; Decraene, V.; Kling, A.; Hjertqvist, M.; Wallensten, A. Two geographically separated food-borne outbreaks in Sweden linked by an unusual *Cryptosporidium parvum* subtype, October 2010. *Eurosurveillance* **2012**, *17*, 20318. [CrossRef] [PubMed]
13. Insulander, M.; Lebbad, M.; Stenström, T.A.; Svenungsson, B. An outbreak of cryptosporidiosis associated with exposure to swimming pool water. *Scand. J. Infect. Dis.* **2005**, *37*, 354–360. [CrossRef]
14. Alsmark, C.; Nolskog, P.; Angervall, A.L.; Toepfer, M.; Winiecka-Krusnell, J.; Bouwmeester, J.; Bjelkmar, P.; Troell, K.; Lahti, E.; Beser, J. Two outbreaks of cryptosporidiosis associated with cattle spring pasture events. *Vet. Parasitol. Reg. Stud. Rep.* **2018**, *14*, 71–74. [CrossRef]
15. Xiao, L.; Escalante, L.; Yang, C.; Sulaiman, I.; Escalante, A.A.; Montali, R.J.; Fayer, R.; Lal, A.A. Phylogenetic analysis of *Cryptosporidium* parasites based on the small-subunit rRNA gene locus. *Appl. Environ. Microbiol.* **1999**, *65*, 1578–1583. [CrossRef]
16. Xiao, L.; Bern, C.; Limor, J.; Sulaiman, I.; Roberts, J.; Checkley, W.; Cabrera, L.; Gilman, R.H.; Lal, A.A. Identification of 5 types of *Cryptosporidium* parasites in children in Lima, Peru. *J. Infect. Dis.* **2001**, *183*, 492–497. [CrossRef]
17. Alves, M.; Xiao, L.; Sulaiman, I.; Lal, A.A.; Matos, O.; Antunes, F. Subgenotype analysis of *Cryptosporidium* isolates from humans, cattle, and zoo ruminants in Portugal. *J. Clin. Microbiol.* **2003**, *41*, 2744–2747. [CrossRef] [PubMed]
18. Morgan, U.M.; Monis, P.T.; Xiao, L.; Limor, J.; Sulaiman, I.; Raidal, S.; O'Donoghue, P.; Gasser, R.; Murray, A.; Fayer, R.; et al. Molecular and phylogenetic characterisation of *Cryptosporidium* from birds. *Int. J. Parasitol.* **2001**, *31*, 289–296. [CrossRef]

19. Guo, Y.; Cebelinski, E.; Matusevich, C.; Alderisio, K.A.; Lebbad, M.; McEvoy, J.; Roellig, D.M.; Yang, C.; Feng, Y.; Xiao, L. Subtyping novel zoonotic pathogen *Cryptosporidium* chipmunk genotype I. *J. Clin. Microbiol.* **2015**, *53*, 1648–1654. [CrossRef] [PubMed]

20. Li, N.; Xiao, L.; Alderisio, K.; Elwin, K.; Cebelinski, E.; Chalmers, R.; Santin, M.; Fayer, R.; Kvac, M.; Ryan, U.; et al. Subtyping *Cryptosporidium ubiquitum*, a zoonotic pathogen emerging in humans. *Emerg. Infect. Dis.* **2014**, *20*, 217. [CrossRef] [PubMed]

21. Rojas-Lopez, L.; Elwin, K.; Chalmers, R.M.; Enemark, H.L.; Beser, J.; Troell, K. Development of a gp60-subtyping method for *Cryptosporidium felis*. *Parasites Vectors* **2020**, *13*, 1–8. [CrossRef] [PubMed]

22. Stensvold, C.; Elwin, K.; Winiecka-Krusnell, J.; Chalmers, R.; Xiao, L.; Lebbad, M. Development and application of a gp60-based typing assay for *Cryptosporidium viatorum*. *J. Clin. Microbiol.* **2015**, *53*, 1891–1897. [CrossRef] [PubMed]

23. Stensvold, C.R.; Beser, J.; Axén, C.; Lebbad, M. High applicability of a novel method for gp60-based subtyping of *Cryptosporidium meleagridis*. *J. Clin. Microbiol.* **2014**, *52*, 2311–2319. [CrossRef]

24. Sulaiman, I.M.; Lal, A.A.; Xiao, L. Molecular phylogeny and evolutionary relationships of *Cryptosporidium* parasites at the actin locus. *J. Parasitol.* **2002**, *88*, 388–394. [CrossRef]

25. Sulaiman, I.M.; Morgan, U.M.; Thompson, R.C.; Lal, A.A.; Xiao, L. Phylogenetic relationships of *Cryptosporidium* parasites based on the 70-kilodalton heat shock protein (HSP70) gene. *Appl. Environ. Microbiol.* **2000**, *66*, 2385–2391. [CrossRef] [PubMed]

26. Kumar, S.; Stecher, G.; Li, M.; Knyaz, C.; Tamura, K. MEGA X: Molecular evolutionary genetics analysis across computing platforms. *Molecular Biol. Evol.* **2018**, *35*, 1547–1549. [CrossRef] [PubMed]

27. Kinross, P.; Beser, J.; Troell, K.; Axen, C.; Björkman, C.; Lebbad, M.; Winiecka-Krusnell, J.; Lindh, J.; Lofdahl, M. *Cryptosporidium parvum* infections in a cohort of veterinary students in Sweden. *Epidemiol Infect.* **2015**, *143*, 2748–2756. [CrossRef] [PubMed]

28. Amer, S.; Zidan, S.; Adamu, H.; Ye, J.; Roellig, D.; Xiao, L.; Feng, Y. Prevalence and characterization of *Cryptosporidium* spp. in dairy cattle in Nile River delta provinces, Egypt. *Exp. Parasitol.* **2013**, *135*, 518–523. [CrossRef] [PubMed]

29. Lebbad, M.; Winiecka-Krusnell, J.; Insulander, M.; Beser, J. Molecular characterization and epidemiological investigation of *Cryptosporidium hominis* IkA18G1 and *C. hominis* monkey genotype IiA17, two unusual subtypes diagnosed in Swedish patients. *Exp. Parasitol.* **2018**, *188*, 7. [CrossRef] [PubMed]

30. Beser, J.; Toresson, L.; Eitrem, R.; Troell, K.; Winiecka-Krusnell, J.; Lebbad, M. Possible zoonotic transmission of *Cryptosporidium felis* in a household. *Infect. Ecol. Epidemiol.* **2015**, *5*, 28463. [CrossRef] [PubMed]

31. Beser, J.; Bujila, I.; Wittesjo, B.; Lebbad, M. From mice to men: Three cases of human infection with *Cryptosporidium ditrichi*. *Infect. Genet. Evol.* **2020**, *78*, 104120. [CrossRef] [PubMed]

32. Nolan, M.J.; Jex, A.R.; Haydon, S.R.; Stevens, M.A.; Gasser, R.B. Molecular detection of *Cryptosporidium cuniculus* in rabbits in Australia. *Infect. Genet. Evol.* **2010**, *10*, 1179–1187. [CrossRef]

33. Sulaiman, I.M.; Hira, P.R.; Zhou, L.; Al-Ali, F.M.; Al-Shelahi, F.A.; Shweiki, H.M.; Iqbal, J.; Khalid, N.; Xiao, L. Unique endemicity of cryptosporidiosis in children in Kuwait. *J Clin. Microbiol.* **2005**, *43*, 2805–2809. [CrossRef] [PubMed]

34. Guy, R.A.; Yanta, C.A.; Muchaal, P.K.; Rankin, M.A.; Thivierge, K.; Lau, R.; Boggild, A.K. Molecular characterization of *Cryptosporidium* isolates from humans in Ontario, Canada. *Parasites Vectors* **2021**, *14*, 69. [CrossRef] [PubMed]

35. O'Leary, J.K.; Blake, L.; Corcoran, G.D.; Sleator, R.D.; Lucey, B. Increased diversity and novel subtypes among clinical *Cryptosporidium parvum* and *Cryptosporidium hominis* isolates in Southern Ireland. *Exp. Parasitol.* **2020**, *218*, 107967. [CrossRef] [PubMed]

36. Costa, D.; Razakandrainibe, R.; Valot, S.; Vannier, M.; Sautour, M.; Basmaciyan, L.; Gargala, G.; Viller, V.; Lemeteil, D.; Ballet, J.-J. Epidemiology of Cryptosporidiosis in France from 2017 to 2019. *Microorganisms* **2020**, *8*, 1358. [CrossRef]

37. Braima, K.; Zahedi, A.; Oskam, C.; Reid, S.; Pingault, N.; Xiao, L.; Ryan, U. Retrospective analysis of *Cryptosporidium* species in Western Australian human populations (2015–2018), and emergence of the *C. hominis* IfA12G1R5 subtype. *Infect. Genet. Evol.* **2019**, *73*, 306–313. [CrossRef] [PubMed]

38. De Lucio, A.; Merino, F.J.; Martínez-Ruiz, R.; Bailo, B.; Aguilera, M.; Fuentes, I.; Carmena, D. Molecular genotyping and sub-genotyping of *Cryptosporidium* spp. isolates from symptomatic individuals attending two major public hospitals in Madrid, Spain. *Infect. Genet. Evol.* **2016**, *37*, 49–56. [CrossRef] [PubMed]

39. Nic Lochlainn, L.M.; Sane, J.; Schimmer, B.; Mooij, S.; Roelfsema, J.; Van Pelt, W.; Kortbeek, T. Risk Factors for Sporadic Cryptosporidiosis in the Netherlands: Analysis of a 3-Year Population Based Case-Control Study Coupled with Genotyping, 2013–2016. *J. Infect. Dis.* **2019**, *219*, 1121–1129. [CrossRef] [PubMed]

40. Garcia-R, J.C.; Pita, A.B.; Velathanthiri, N.; French, N.P.; Hayman, D.T. Species and genotypes causing human cryptosporidiosis in New Zealand. *Parasitol. Res.* **2020**, *119*, 2317–2326. [CrossRef] [PubMed]

41. Stensvold, C.R.; Ethelberg, S.; Hansen, L.; Sahar, S.; Voldstedlund, M.; Kemp, M.; Hartmeyer, G.T.; Otte, E.; Engsbro, A.L.; Nielsen, H.V.; et al. *Cryptosporidium* infections in Denmark, 2010–2014. *Dan. Med. J.* **2015**, *62*, A5086. [PubMed]

42. Chalmers, R.; Elwin, K.; Thomas, A.; Guy, E.; Mason, B. Long-term *Cryptosporidium* typing reveals the aetiology and species-specific epidemiology of human cryptosporidiosis in England and Wales, 2000 to 2003. *Euro Surveill.* **2009**, *14*, 19086. [CrossRef] [PubMed]

43. Loeck, B.K.; Pedati, C.; Iwen, P.C.; McCutchen, E.; Roellig, D.M.; Hlavsa, M.C.; Fullerton, K.; Safranek, T.; Carlson, A.V. Genotyping and Subtyping *Cryptosporidium* to Identify Risk Factors and Transmission Patterns—Nebraska, 2015–2017. *Morb. Mortal. Wkly. Rep.* **2020**, *69*, 335. [CrossRef]

44. Chalmers, R.; Smith, R.; Elwin, K.; Clifton-Hadley, F.; Giles, M. Epidemiology of anthroponotic and zoonotic human cryptosporidiosis in England and Wales, 2004–2006. *Epidemiol. Infect.* **2011**, *139*, 700–712. [CrossRef]

45. Björkman, C.; Lindström, L.; Oweson, C.; Ahola, H.; Troell, K.; Axén, C. *Cryptosporidium* infections in suckler herd beef calves. *Parasitology* **2015**, *142*, 1108–1114. [CrossRef]

46. Chalmers, R.M.; Robinson, G.; Elwin, K.; Elson, R. Analysis of the *Cryptosporidium* spp. and gp60 subtypes linked to human outbreaks of cryptosporidiosis in England and Wales, 2009 to 2017. *Parasites Vectors* **2019**, *12*, 95. [CrossRef] [PubMed]

47. Thomas-Lopez, D.; Müller, L.; Vestergaard, L.S.; Christoffersen, M.; Andersen, A.-M.; Jokelainen, P.; Agerholm, J.S.; Stensvold, C.R. Veterinary students have a higher risk of contracting cryptosporidiosis when calves with high fecal *Cryptosporidium* loads are used for fetotomy exercises. *Appl. Environ. Microbiol.* **2020**, *86*. [CrossRef]

48. Xiao, L.; Feng, Y. Molecular epidemiologic tools for waterborne pathogens *Cryptosporidium* spp. and *Giardia duodenalis*. *Food Waterborne Parasitol.* **2017**, *8*, 14–32. [CrossRef] [PubMed]

49. Feng, Y.; Torres, E.; Li, N.; Wang, L.; Bowman, D.; Xiao, L. Population genetic characterisation of dominant *Cryptosporidium parvum* subtype IIaA15G2R1. *Int. J. Parasitol.* **2013**, *43*, 1141–1147. [CrossRef] [PubMed]

50. Silverlås, C.; Bosaeus-Reineck, H.; Näslund, K.; Björkman, C. Is there a need for improved *Cryptosporidium* diagnostics in Swedish calves? *Int. J. Parasitol.* **2013**, *43*, 155–161. [CrossRef]

51. Whithworth, J. Five Foodborne Outbreaks Added to Cryptosporidium Rise in Sweden. Food Safety News, 13 March 2020. Available online: https://www.foodsafetynews.com/2020/03/five-foodborne-outbreaks-added-to-cryptosporidium-rise-in-sweden/ (accessed on 24 April 2021).

52. Ajjampur, S.S.; Liakath, F.B.; Kannan, A.; Rajendran, P.; Sarkar, R.; Moses, P.D.; Simon, A.; Agarwal, I.; Mathew, A.; O'Connor, R.; et al. Multisite study of cryptosporidiosis in children with diarrhea in India. *J Clin. Microbiol.* **2010**, *48*, 2075–2081. [CrossRef]

53. Mulunda, N.R.; Hayashida, K.; Yamagishi, J.; Sianongo, S.; Munsaka, G.; Sugimoto, C.; Mutengo, M.M. Molecular characterization of *Cryptosporidium* spp. from patients with diarrhoea in Lusaka, Zambia. *Parasite* **2020**, *27*, 53. [CrossRef]

54. Xiao, L.; Hlavsa, M.C.; Yoder, J.; Ewers, C.; Dearen, T.; Yang, W.; Nett, R.; Harris, S.; Brend, S.M.; Harris, M.; et al. Subtype analysis of *Cryptosporidium* specimens from sporadic cases in Colorado, Idaho, New Mexico, and Iowa in 2007: Widespread occurrence of one *Cryptosporidium hominis* subtype and case history of an infection with the *Cryptosporidium* horse genotype. *J. Clin. Microbiol.* **2009**, *47*, 3017–3020. [CrossRef] [PubMed]

55. Hlavsa, M.C.; Roellig, D.M.; Seabolt, M.H.; Kahler, A.M.; Murphy, J.L.; McKitt, T.K.; Geeter, E.F.; Dawsey, R.; Davidson, S.L.; Kim, T.N.; et al. Using Molecular Characterization to Support Investigations of Aquatic Facility-Associated Outbreaks of Cryptosporidiosis—Alabama, Arizona, and Ohio, 2016. *MMWR Morb. Mortal. Wkly. Rep.* **2017**, *66*, 493–497. [CrossRef] [PubMed]

56. Jian, F.; Liu, A.; Wang, R.; Zhang, S.; Qi, M.; Zhao, W.; Shi, Y.; Wang, J.; Wei, J.; Zhang, L.; et al. Common occurrence of *Cryptosporidium hominis* in horses and donkeys. *Infect. Genet. Evol.* **2016**, *43*, 261–266. [CrossRef] [PubMed]

57. Lebbad, M.; Beser, J.; Insulander, M.; Karlsson, L.; Mattsson, J.G.; Svenungsson, B.; Axén, C. Unusual cryptosporidiosis cases in Swedish patients: Extended molecular characterization of *Cryptosporidium viatorum* and *Cryptosporidium* chipmunk genotype I. *Parasitology* **2013**, *140*, 1735–1740. [CrossRef]

58. Kváč, M.; Hofmannová, L.; Bertolino, S.; Wauters, L.; Tosi, G.; Modrý, D. Natural infection with two genotypes of *Cryptosporidium* in red squirrels (*Sciurus vulgaris*) in Italy. *Folia Parasitol.* **2008**, *55*, 95. [CrossRef]

59. Zhao, W.; Zhou, H.; Huang, Y.; Xu, L.; Rao, L.; Wang, S.; Wang, W.; Yi, Y.; Zhou, X.; Wu, Y.; et al. *Cryptosporidium* spp. in wild rats (*Rattus* spp.) from the Hainan Province, China: Molecular detection, species/genotype identification and implications for public health. *Int. J. Parasitol. Parasites Wildl.* **2019**, *9*, 317–321. [CrossRef]

60. Laatamna, A.E.; Wagnerova, P.; Sak, B.; Kvetonova, D.; Aissi, M.; Rost, M.; Kvac, M. Equine cryptosporidial infection associated with *Cryptosporidium* hedgehog genotype in Algeria. *Vet. Parasitol.* **2013**, *197*, 350–353. [CrossRef]

61. Kváč, M.; Hofmannová, L.; Hlásková, L.; Květoňová, D.; Vítovec, J.; McEvoy, J.; Sak, B. *Cryptosporidium erinacei* n. sp. (Apicomplexa: Cryptosporidiidae) in hedgehogs. *Vet. Parasitol.* **2014**, *201*, 9–17. [CrossRef]

62. Costa, D.; Razakandrainibe, R.; Sautour, M.; Valot, S.; Basmaciyan, L.; Gargala, G.; Lemeteil, D.; Favennec, L.; Dalle, F. Human cryptosporidiosis in immunodeficient patients in France (2015–2017). *Exp. Parasitol.* **2018**, *192*, 108–112. [CrossRef] [PubMed]

63. Kvac, M.; Sakova, K.; Kvetonova, D.; Kicia, M.; Wesolowska, M.; McEvoy, J.; Sak, B. Gastroenteritis caused by the *Cryptosporidium* hedgehog genotype in an immunocompetent man. *J. Clin. Microbiol.* **2014**, *52*, 347–349. [CrossRef]

64. Puleston, R.L.; Mallaghan, C.M.; Modha, D.E.; Hunter, P.R.; Nguyen-Van-Tam, J.S.; Regan, C.M.; Nichols, G.L.; Chalmers, R.M. The first recorded outbreak of cryptosporidiosis due to *Cryptosporidium cuniculus* (formerly rabbit genotype), following a water quality incident. *J Water Health* **2014**, *12*, 41–50. [CrossRef] [PubMed]

65. Koehler, A.V.; Rashid, M.H.; Zhang, Y.; Vaughan, J.L.; Gasser, R.B.; Jabbar, A. First cross-sectional, molecular epidemiological survey of *Cryptosporidium*, *Giardia* and *Enterocytozoon* in alpaca (*Vicugna pacos*) in Australia. *Parasites Vectors* **2018**, *11*, 498. [CrossRef] [PubMed]

66. Koehler, A.V.; Whipp, M.J.; Haydon, S.R.; Gasser, R.B. *Cryptosporidium cuniculus*–new records in human and kangaroo in Australia. *Parasites Vectors* **2014**, *7*, 492. [CrossRef]

67. Elwin, K.; Hadfield, S.J.; Robinson, G.; Crouch, N.D.; Chalmers, R.M. *Cryptosporidium viatorum* n. sp. (Apicomplexa: Cryptosporidiidae) among travellers returning to Great Britain from the Indian subcontinent, 2007–2011. *Int. J. Parasitol.* **2012**, *42*, 675–682. [CrossRef] [PubMed]

68. Koehler, A.V.; Wang, T.; Haydon, S.R.; Gasser, R.B. *Cryptosporidium viatorum* from the native Australian swamp rat *Rattus lutreolus*-an emerging zoonotic pathogen? *Int. J. Parasitol. Parasites Wildl.* **2018**, *7*, 18–26. [CrossRef] [PubMed]
69. Chen, Y.-W.; Zheng, W.-B.; Zhang, N.-Z.; Gui, B.-Z.; Lv, Q.-Y.; Yan, J.-Q.; Zhao, Q.; Liu, G.H. Identification of *Cryptosporidium viatorum* XVa subtype family in two wild rat species in China. *Parasites Vectors.* **2019**, *12*, 1–5. [CrossRef] [PubMed]
70. Burton, A.; Nydam, D.; Dearen, T.; Mitchell, K.; Bowman, D.; Xiao, L. The prevalence of *Cryptosporidium*, and identification of the *Cryptosporidium* horse genotype in foals in New York State. *Vet. Parasitol.* **2010**, *174*, 139–144. [CrossRef]
71. Caffara, M.; Piva, S.; Pallaver, F.; Iacono, E.; Galuppi, R. Molecular characterization of *Cryptosporidium* spp. from foals in Italy. *Vet. J.* **2013**, *198*, 531–533. [CrossRef] [PubMed]
72. Robinson, G.; Elwin, K.; Chalmers, R.M. Unusual *Cryptosporidium* genotypes in human cases of diarrhea. *Emerg. Infect. Dis.* **2008**, *14*, 1800. [CrossRef]
73. Abe, N.; Matsubara, K. Molecular identification of *Cryptosporidium* isolates from exotic pet animals in Japan. *Vet. Parasitol.* **2015**, *209*, 254–257. [CrossRef]
74. Takaki, Y.; Takami, Y.; Watanabe, T.; Nakaya, T.; Murakoshi, F. Molecular identification of *Cryptosporidium* isolates from ill exotic pet animals in Japan including a new subtype in *Cryptosporidium fayeri*. *Vet. Parasitol. Reg. Stud. Rep.* **2020**, *21*, 100430. [CrossRef]
75. Sannella, A.R.; Suputtamongkol, Y.; Wongsawat, E.; Caccio, S.M. A retrospective molecular study of *Cryptosporidium* species and genotypes in HIV-infected patients from Thailand. *Parasites Vectors* **2019**, *12*, 91, Epub 2019/03/15. [CrossRef] [PubMed]
76. Condlova, S.; Horcickova, M.; Sak, B.; Kvetonova, D.; Hlaskova, L.; Konecny, R.; Stanko, M.; McEvoy, J.; Kvac, M. *Cryptosporidium apodemi* sp. n. and *Cryptosporidium ditrichi* sp. n. (Apicomplexa: Cryptosporidiidae) in *Apodemus* spp. *Eur. J. Protistol.* **2018**, *63*, 1–12. [CrossRef]
77. Xiao, L.; Feng, Y. Zoonotic cryptosporidiosis. *FEMS Immunol. Med Microbiol.* **2008**, *52*, 309–323. [CrossRef] [PubMed]
78. Da Cunha, M.J.R.; Cury, M.C.; Santín, M. Molecular characterization of *Cryptosporidium* spp. in poultry from Brazil. *Res. Vet. Sci.* **2018**, *118*, 331–335. [CrossRef] [PubMed]
79. Liao, C.; Wang, T.; Koehler, A.V.; Fan, Y.; Hu, M.; Gasser, R.B. Molecular investigation of *Cryptosporidium* in farmed chickens in Hubei Province, China, identifies 'zoonotic' subtypes of *C. meleagridis*. *Parasites Vectors* **2018**, *11*, 1–8. [CrossRef] [PubMed]
80. Silverlås, C.; Mattsson, J.G.; Insulander, M.; Lebbad, M. Zoonotic transmission of *Cryptosporidium meleagridis* on an organic Swedish farm. *Int. J. Parasitol.* **2012**, *42*, 963–967. [CrossRef] [PubMed]
81. Abdelsalam, I.M.; Sarhan, R.M.; Hanafy, M.A. The impact of different copro-preservation conditions on molecular detection of *Cryptosporidium* species. *Iran. J. Parasitol.* **2017**, *12*, 274.
82. Autier, B.; Belaz, S.; Razakandrainibe, R.; Gangneux, J.P.; Robert-Gangneux, F. Comparison of three commercial multiplex PCR assays for the diagnosis of intestinal protozoa. *Parasite* **2018**, *25*, 48. [CrossRef] [PubMed]

pathogens

Article

Molecular Investigation of Zoonotic Intestinal Protozoa in Pet Dogs and Cats in Yunnan Province, Southwestern China

Yu-Gui Wang [1,2,3], Yang Zou [2,*], Ze-Zhong Yu [4], Dan Chen [5], Bin-Ze Gui [3], Jian-Fa Yang [1], Xing-Quan Zhu [1,6], Guo-Hua Liu [3] and Feng-Cai Zou [1,*]

[1] Key Laboratory of Veterinary Public Health of Yunnan Province, College of Veterinary Medicine, Yunnan Agricultural University, Kunming 650201, China; wyg18894032073@outlook.com (Y.-G.W.); 2002012@ynau.edu.cn (J.-F.Y.); xingquanzhu1@hotmail.com (X.-Q.Z.)

[2] State Key Laboratory of Veterinary Etiological Biology, Key Laboratory of Veterinary Parasitology of Gansu Province, Lanzhou Veterinary Research Institute, Chinese Academy of Agricultural Sciences, Lanzhou 730046, China

[3] Hunan Provincial Key Laboratory of Protein Engineering in Animal Vaccines, College of Veterinary Medicine, Hunan Agricultural University, Changsha 410128, China; GBZ296936@163.com (B.-Z.G.); liuguohua@hunau.edu.cn (G.-H.L.)

[4] Department of Animal Science, Yuxi Agricultural Vocation Technical College, Yuxi 653106, China; yzznzy2021@163.com

[5] School of Life Science, Fudan University, Shanghai 200438, China; 20110700150@fudan.edu.cn

[6] College of Veterinary Medicine, Shanxi Agricultural University, Taigu 030801, China

[*] Correspondence: zouyangdr@163.com (Y.Z.); zoufengcai@ynau.edu.cn (F.-C.Z.)

Citation: Wang, Y.-G.; Zou, Y.; Yu, Z.-Z.; Chen, D.; Gui, B.-Z.; Yang, J.-F.; Zhu, X.-Q.; Liu, G.-H.; Zou, F.-C. Molecular Investigation of Zoonotic Intestinal Protozoa in Pet Dogs and Cats in Yunnan Province, Southwestern China. *Pathogens* **2021**, *10*, 1107. https://doi.org/10.3390/pathogens10091107

Academic Editors: David Carmena, David González-Barrio and Pamela Carolina Köster

Received: 4 August 2021
Accepted: 23 August 2021
Published: 30 August 2021

Publisher's Note: MDPI stays neutral with regard to jurisdictional claims in published maps and institutional affiliations.

Abstract: *Giardia duodenalis*, *Enterocytozoon bieneusi* and *Cryptosporidium* spp. are common enteric pathogens that reside in the intestines of humans and animals. These pathogens have a broad host range and worldwide distribution, but are mostly known for their ability to cause diarrhea. However, very limited information on prevalence and genotypes of *G. duodenalis*, *E. bieneusi* and *Cryptosporidium* spp. in pet dogs and cats are available in China. In the present study, a total of 433 fecal samples were collected from 262 pet dogs and 171 pet cats in Yunnan province, southwestern China, and the prevalence and the genotypes of *G. duodenalis*, *E. bieneusi* and *Cryptosporidium* spp. were investigated by nested PCR amplification and DNA sequencing. The prevalence of *G. duodenalis*, *E. bieneusi* and *Cryptosporidium* spp. was 13.7% (36/262), 8.0% (21/262), and 4.6% (12/262) in dogs, and 1.2% (2/171), 2.3% (4/171) and 0.6% (1/171) in cats, respectively. The different living conditions of dogs is a risk factor that is related with the prevalence of *G. duodenalis* and *E. bieneusi* ($p < 0.05$). However, there were no statistically significant difference in prevalence of three pathogens in cats. DNA sequencing and analyses showed that four *E. bieneusi* genotypes (PtEb IX, CD9, DgEb I and DgEb II), one *Cryptosporidium* spp. (*C. canis*) and two *G. duodenalis* assemblages (C and D) were identified in dogs; two *E. bieneusi* genotypes (Type IV and CtEb I), one *Cryptosporidium* spp. (*C. felis*) and one *G. duodenalis* assemblage (F) were identified in cats. Three novel *E. bieneusi* genotypes (DgEb I, DgEb II and CtEb I) were identified, and the human-pathogenic genotypes/species Type IV *C. canis* and *C. felis* were also observed in this study, indicating a potential zoonotic threat of pet dogs and cats. Our results revealed the prevalence and genetic diversity of *G. duodenalis*, *E. bieneusi* and *Cryptosporidium* spp. infection in pet dogs and cats in Yunnan province, southwestern China, and suggested the potential threat of pet dogs and cats to public health.

Keywords: *Giardia duodenalis*; *Enterocytozoon bieneusi*; *Cryptosporidium* spp.; zoonotic genotypes; pet dogs and cats; Yunnan province; China

1. Introduction

Giardia duodenalis, *Cryptosporidium* spp. and *Enterocytozoon bieneusi* are three eukaryotic unicellular protozoans, which are the causative pathogens of giardiasis, cryptosporidiosis, and microsporidiosis, respectively [1–4]. These pathogens can cause many gastrointestinal symptoms such as abdominal pain, nausea, vomiting, anorexia and weight loss especially acute and chronic diarrhea [5–10]. Humans and various animals can be infected by *G. duodenalis*, *Cryptosporidium* spp. and *E. bieneusi* through fecal-oral transmission of their cysts or spores [11,12].

At present, eight *G. duodenalis* assemblages (A–H) have been identified by the molecular biological detection method [13]. Among these genotypes, assemblages A and B are regarded as zoonotic assemblages which mainly infect humans and other mammals [14]. Other *G. duodenalis* assemblages (C–H) are commonly considered as host-specific, while assemblages C and D are usually canine-specific assemblages, and assemblage F is usually a feline-specific assemblage [15,16]. However, assemblages E and F have also been detected in humans [17,18]. In total, over 40 *Cryptosporidium* species have been reported, and over 21 species have been reported in humans, including *C. canis* and *C. felis*, which cause the vast majority of infections in dogs and cats, respectively [12,19]. Moreover, *Cryptosporidium muris*, *Cryptosporidium parvum* and *Cryptosporidium ubiquitum* have also been reported in dogs and cats [6,7,20–24]. *Enterocytozoon bieneusi* is the most common species causing human gut infections among nearly 1500 microsporidian species [23]. At least 500 *E. bieneusi* genotypes have been defined thus far, which can be divided into several genetically isolated groups, including zoonotic groups (Group 1 and Group 2) and host adapted groups (Groups 3 to 11) [23,24].

Due to the closer relationships between humans with pet dogs and cats, many pathogens can be transmitted to humans through pet dogs and cats, including *G. duodenalis*, *Cryptosporidium* spp. and *E. bieneusi*. Therefore, investigation of the prevalence and genotypes/species of *G. duodenalis*, *Cryptosporidium* spp. and *E. bieneusi* in pet dogs and cats will improve our understanding of the potential threat posed by these pathogens in companion animals in Yunnan province, China.

2. Results

2.1. Prevalence of G. duodenalis, E. Bieneusi and Cryptosporidium spp. in Pet Dogs and Cats

The prevalence of *G. duodenalis*, *Cryptosporidium* spp. and *E. bieneusi* was 13.7% (95%CI 9.6–17.9), 4.6% (95%CI 2.0–7.1), 8.0% (95%CI 4.7–11.3) in dogs; and it was 1.2% (95%CI 0–2.8), 0.6% (95%CI 0–1.7) and 2.3% (95%CI 0.1–4.6) in cats, respectively (Table 1). Among three regions, the prevalence of *G. duodenali* in dogs in Kunming city was significantly higher than that in Chuxiong city and Lijiang city ($p < 0.05$). Moreover, the prevalence of *G. duodenalis* in dogs in shelter dogs (27.8%, 20/72, 95%CI 17.4–38.1) was higher than that in pet markets (2.9%, 1/34, 95%CI 0–8.6) and pet hospitals (9.6%, 15/156, 95%CI 5.0–14.2), and the difference was statistically significant ($p < 0.001$). However, no statistically significant difference in prevalence of *G. duodenalis* in pet cats was observed (Table 1).

Among the different living conditions of dogs, the difference in *E. bieneusi* prevalence was statistically significant ($p < 0.001$). The prevalence of *E. bieneusi* in dogs aged more than 6 months was 10.3% (95%CI 6.0–14.6), which was significantly higher than that in dogs aged less than 6 months (1.5%, 95%CI 0–4.3) (Table 1). Also, the prevalence of *E. bieneusi* in female dogs was 10.3% (95%CI 5.5–15.1), which was higher than that in male dogs (4.7%, 95%CI 0.7–8.7), but the difference in prevalence was not statistically significant ($p = 0.098$). Similarly, the prevalence of *E. bieneusi* in female cats (3.3%, 95%CI 0–7.9) was slightly higher than that in male cats (1.8%, 95%CI 0–4.3) (Table 1).

Furthermore, the prevalence of *Cryptosporidium* spp. in dogs in shelter (15.3%, 95%CI 7.0–23.6) was higher than that in pet markets (no detection) and pet hospitals (0.6%, 95%CI 0–1.9). Between two gender groups, the prevalence of *Cryptosporidium* spp. in male and female dogs was not significantly different (Table 1).

Table 1. Factors associated with *Giardia duodenalis*, *Enterocytozoon bieneusi* and *Cryptosporidium* spp. prevalence in pet dogs and cats in Yunnan province, southwestern China.

Animals	Factors	Category	No. Sample	Giardia duodenalis			Enterocytozoon bieneusi			Cryptosporidium spp.		
				No. Positive	% (95% CI)	p-Value	No. Positive	% (95% CI)	p-Value	No. Positive	% (95% CI)	p-Value
Dogs	Age	<6 months	68	8	11.8 (4.1–19.4)	0.582	1	1.5 (0–4.3)	0.021	1	1.5 (0–4.3)	0.154
		>6 months	194	28	14.4 (9.5–19.4)		20	10.3 (6.0–14.6)		11	5.7 (2.4–8.9)	
	Region	Kunming	134	26	19.4 (12.7–26.1)	0.013	18	13.4 (7.7–19.2)	-	11	8.2 (3.6–12.9)	-
		Lijiang	90	9	10.0 (3.8–16.2)		0	0		1	1.1 (0–3.3)	
		Chuxiong	38	1	2.6 (0–7.7)		3	7.9 (0–16.5)		0	0	
	Gender	Male	107	16	15.0 (8.2–21.7)	0.636	5	4.7 (0.7–8.7)	0.098	5	4.7 (0.7–8.7)	0.95
		Female	155	20	12.9 (7.6–18.2)		16	10.3 (5.5–15.1)		7	4.5 (1.2–7.8)	
	Living condition	Pet hospital	156	15	9.6 (5.0–14.2)	<0.001	1	0.6 (0–1.9)	<0.001	1	0.6 (0–1.9)	-
		Pet market	34	1	2.9 (0–8.6)		3	8.8 (0–18.4)		0	0	
		Shelter	72	20	27.8 (17.4–38.1)		17	23.6 (13.8–33.4)		11	15.3 (7.0–23.6)	
		Subtotal	262	36	13.7 (9.6–17.9)		21	8.0 (4.7–11.3)		12	4.6 (2.0–7.1)	
Cats	Age	< 6 months	145	2	1.4 (0–3.3)	-	4	2.8 (0.1–5.4)	-	1	0.7 (0–2.0)	-
		> 6 months	26	0	0		0	0		0	0	
	Region	Kunming	36	1	2.8 (13.1–42.4)	-	0	0	-	0	0	-
		Lijiang	110	0	0		0	0		1	0.9 (0–2.7)	
		Chuxiong	25	1	4.0 (0–11.7)		4	16.0 (1.6–30.4)		0	0	
	Gender	Male	111	2	1.8 (0–4.3)	-	2	1.8 (0–4.3)	-	1	0.9 (0–2.7)	-
		Female	60	0	0		2	3.3 (0–7.9)		0	0	
	Living condition	Pet hospital	154	2	1.3 (0–3.1)	-	4	2.6 (0.1–5.1)	-	1	2.6 (0–1.9)	-
		Shelter	17	0	0		0	0		0	0	
		Subtotal	171	2	1.2 (0–2.8)		4	2.3 (0.1–4.6)		1	0.6 (0–1.7)	
Total			433	38	8.8 (6.1–11.4)		25	5.8 (3.6–8.0)		13	3.0 (1.4–4.6)	

2.2. Assemblages and Subtypes of G. duodenalis in Pet Dogs and Cats

PCR amplification and DNA sequencing showed that 38 positive samples (36 from dogs and 2 from cats) of *G. duodenalis* were detected at bg locus, resulting three assemblages, namely C (4 from dogs), D (32 from dogs) and F (2 from cats). In addition, at the gdh locus, the 19 gdh-positive samples were identified as assemblage C (4 from dogs), D (13 from dogs) and F (2 from cats). Only one tpi-positive sample (1 from dogs) was identified as assemblage C.

Sequence alignment analysis revealed some single nucleotide polymorphisms at bg-sequences, gdh-sequences and tpi-sequences, respectively. At bg locus, one subtype of assemblage C, 7 subtypes of assemblage D and one subtype of assemblage F were identified, including five novel (Da4 * ~ Da7 *, Fa1 *) and four known sub-assemblages (Table 2). Also, at gdh gene locus, three subtypes of assemblage C, seven subtypes of assemblage D and one subtype of assemblage F were identified, including four novel (Cb3 *, Db5 * ~ Db7 *) and six known subtypes (Table 2). Only one novel subtype (Cc1 *) of assemblage C was found at tpi gene locus (Table 2). Moreover, one sample were successfully amplified and sequenced at three gene loci (bg, gdh and tpi), forming one mixed infection (Table 3).

Table 2. Variations in nucleotide sequences of assemblages of *Giardia duodenalis* in pet dogs and cats in Yunnan province, southwestern China.

Locus	Host (Subtypes)	Nucleotide at Position						No. Positive	Accession Number
	(a) Variations in bg nucleotide sequences among assemblage D								
		31	61	103	109	203			
	Reference sequences	G	A	G	C	A			MG873354
	Dog (Da1)							20	MN734349
	Dog (Da2)			A				5	MN734350
	Dog (Da3)				T			3	MN734353
	Dog (Da4 *)			A	T			1	MN734351
	Dog (Da5 *)	A			A			1	MN734354
bg	Dog (Da6 *)		C	A				1	MN734352
	Dog (Da7 *)	A			A	G		1	MN734355
	(b) Variations in bg nucleotide sequences among assemblage F								
		55							
	Reference sequences	C							KX960131
	Cat (Fa1 *)	T						2	MN734356
	(c) Variations in bg nucleotide sequences among assemblage C								
	Reference sequences								KY979502
	Dog (Ca1)							4	MN734348
	(a) Variations in gdh nucleotide sequences among assemblage C								
		340	589	600	603	693			
	Reference sequences	A	G	C	A	G			MF990016
	Dog (Cb1)	G	A	T	G			2	MN734358
	Dog (Cb2)	G	A	T	G	T		1	MN734359
	Dog (Cb3 *)				G			1	MN734357
	(b) Variations in gdh nucleotide sequences among assemblage D								
		356	368	386	506	509	654		
	Reference sequences	C	A	T	A	C	A		MF990017
gdh	Dog (Db1)							1	MN734366
	Dog (Db2)	T	G					3	MN734362
	Dog (Db3)			A			G	5	MN734364
	Dog (Db4)	T						1	MN734363
	Dog (Db5*)		G					1	MN734361
	Dog (Db6*)			A				1	MN734360
	Dog (Db7*)	T	G		T	T		1	MN734365
	(c) Variations in gdh nucleotide sequences among assemblage F								
	Reference sequences								KM977649
	Cat (Fb1)							2	MN734367
	Variations in tpi nucleotide sequences among assemblage C								
		135	315						
tpi	Reference sequences	G	T						KY979494
	Dog (Cc1 *)	T	C					1	MN734368

* means novel subtypes of assemblage.

Table 3. Multilocus characterization of *Giardia duodenalis* isolates based on the bg, tpi and gdh genes.

Isolate	Assemblage			No. Sequences	MLG Type
	bg	tpi	gdh		
XSQG34	D	C	C	1	Mixed

2.3. Genotypes of Enterocytozoon bieneusi and Cryptosporidium spp. in Pet Dogs and Cats

Based on the ITS sequence, a total of four genotypes, including two known genotypes PtEb IX (n = 18), CD9 (n = 1) in dogs and two novel genotypes DgEb I (n = 1) and DgEb II (n = 1) were identified in pet dogs, and one known genotype Type IV (n = 3) and one novel genotype CtEb I (n = 1) were identified in pet cats (Table 4). The phylogenetic tree showed that genotypes DgEb I, DgEb II, PtEb IX and CD9 all belonged to the dog-specific group. However, genotypes Type IV and CtEb I belonged to the zoonotic Group 1 (Figure 1). Moreover, mixed infections with more than one genotype of *E. bieneusi* in dogs and cats were not detected.

Two *Cryptosporidium* species were identified among the 13 *Cryptosporidium*-positive samples, including 12 samples of *C. canis* in dogs and one sample of *C. canis* in cats (Table 4). Five nucleotide sequences of *C. canis* obtained in this study had 100% similarity to those deposited sequences in GenBank under accession numbers MN696800. Other sequences of *C. canis* had 99% similarity to those deposited sequences in GenBank under accession number KR999984 and KT749818, respectively (Table 4). Moreover, only one *C. canis* sequence had 97% similarity to those deposited sequences in GenBank under accession number KM977642 (Table 4).

Table 4. Species or genotypes of *Cryptosporidium* spp. and *Enterocytozoon bieneusi* in pet dogs and cats in Yunnan province, southwestern China.

Hosts	*Enterocytozoon bieneusi* Genotype (No.)	GenBank Accession Number in This Study	
Dog	DgEb I * (1)	MZ542370	
Dog	CD9 (1)	MZ542369	
Dog	DgEb II * (1)	MZ542373	
Dog	PtEb IX (1)	MZ542371	
Dog	PtEb IX (17)	MZ542372	
Cat	Type IV (3)	MZ542374	
Cat	CtEb I * (1)	MZ542375	
Hosts	***Cryptosporidium* spp. Genotype (No.)**	**Reference Sequences GenBank Accession Number**	**Similarity**
Dog	*C. canis* (5)	MN696800	100%
Dog	*C. canis* (4)	KR999984	99%
Dog	*C. canis* (3)	KT749818	99%
Cat	*C. felis* (1)	KM977642	97%

Note: * represent novel genotype.

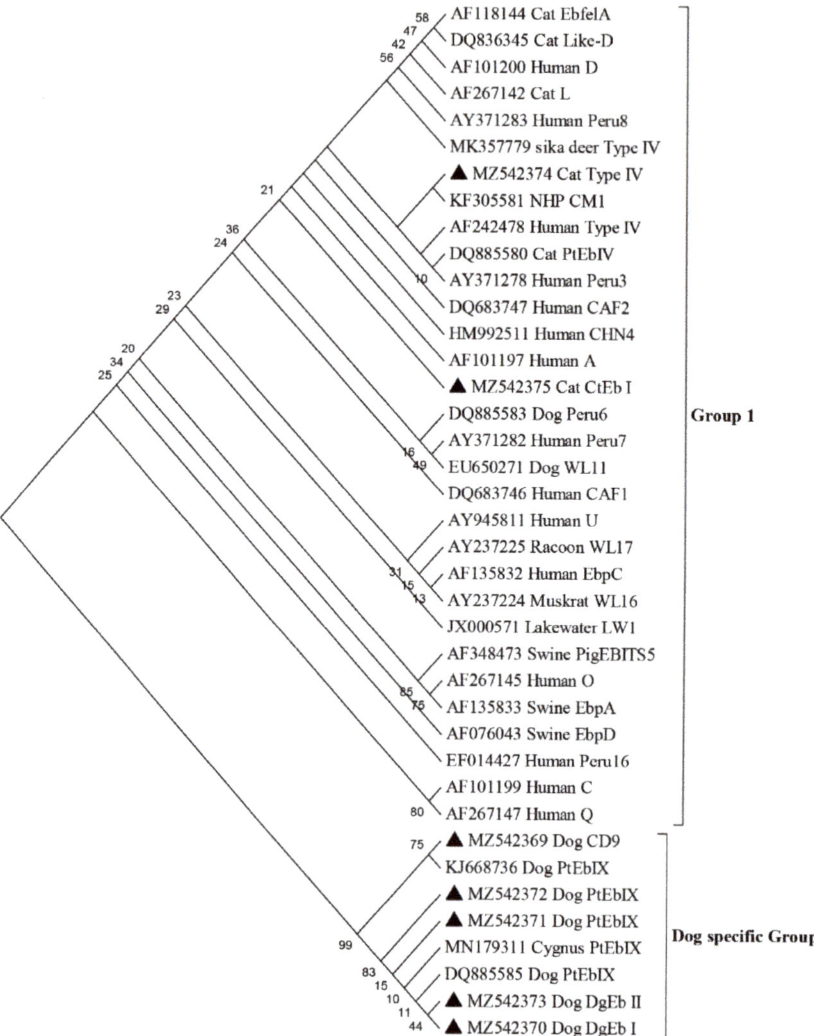

Figure 1. Phylogenetic relationship based on ITS sequences of *Enterocytozoon bieneusi* in pet dogs and cats in Yunnan province, southwestern China. (Note: The samples in this study are indicated by triangles).

3. Discussion

Dogs and cats, as domestic animals, share a common environment with humans and other animals, and can infect them with various unicellular zoonotic pathogens. Thus far, many studies about the infection of *G. duodenalis*, *Cryptosporidium* spp. and *E. bieneusi* in dogs and cats have been recorded worldwide, such as Asia, Europe and Latin America, although only a few have been reported in Africa (Table 5) [6,7,16,20,25–53]. According to the studies in China, the prevalence of *G. duodenalis* ranges from 4.5–26.2% in dogs and 1.9–13.1% in cats [6,7,25,26]; the prevalence of *Cryptosporidium* spp. ranges from 3.1–7.5% in dogs and 5.6–5.8% in cats [6,7,46,47]; and the prevalence of *E. bieneusi* ranges from 6.0–13.9% in dogs and 1.4–11.5% in cats [6,7,33–35], respectively (Table 5).

Table 5. Prevalence of *Giardia duodenalis, Enterocytozoon bieneusi* and *Cryptosporidium* spp. in dogs and cats in different regions of the world.

Regions	Hosts	Prevalence (%)	Hosts	Prevalence (%)	Reference
(a) Prevalence of *Giardia duodenalis* in dogs and cats in different regions of the world.					
China					
Shanghai	Dogs	26.2%	Cats	13.1%	[7]
Guangdong	Dogs	10.8%	Cats	5.8%	[25]
Heilongjiang	Dogs	4.5%	Cats	1.9%	[6]
Sichuan	Dogs	11.3%	-	-	[26]
Henan	Dogs	14.3%	-	-	[27]
Hangzhou	-	-	Cats	1.2%	[28]
Yunnan	Dogs	13.7%	Cats	1.2%	Present study
Other countries					
Australia	Dogs	6.3%	Cats	2.0%	[20]
Greece	Dogs	25.2%	Cats	20.5%	[29]
Spain	Dogs	33%	Cats	9.2%	[30]
Ontario	Dogs	64.0%	Cats	87.0%	[31]
Brazil	Dogs	19.6%	-	-	[32]
(b) Prevalence of *Enterocytozoon bieneusi* in dogs and cats in different regions of the world.					
China					
Shanghai	Dogs	6.0%	Cats	5.6%	[7]
Heilongjiang	Dogs	6.7%	Cats	5.8%	[6]
Henan	Dogs	13.9%	Cats	11.5%	[33]
Eastern China	Dogs	8.6%	Cats	1.4%	[34]
Changchun	Dogs	7.8%	-	-	[35]
Yunnan	Dogs	8.0%	Cats	2.3%	Present study
Other countries					
Colombia	Dogs	15.0%	Cats	17.4%	[36,37]
Egypt	Dogs	13.0%	Cats	12.5%	[38]
Germany	Dogs	0.0%	Cats	5.0%	[39]
Spain	Dogs	0.8%	Cats	3.0%	[40]
Japan	Dogs	2.5%	Cats	14.3%	[41]
Poland	Dogs	4.9%	Cats	9.1%	[42]
Thailand	Dogs	0.0%	Cats	31.3%	[43]
Portugal	Dogs	100.0%	Cats	100.0%	[44]
Iran	Dogs	25.8%	Cats	7.5%	[45]
(c) Prevalence of *Cryptosporidium* spp. in dogs and cats in different regions of the world.					
China					
Shanghai	Dogs	6.0%	Cats	5.6%	[7]
Heilongjiang	Dogs	6.7%	Cats	5.8%	[6]
Zhengzhou	Dogs	3.1%	-	-	[46]
Ya'an	Dogs	7.5%	-	-	[47]
Yunnan	Dogs	4.6%	Cats	0.6%	Present study
Other countries					
Japan	-	-	Cats	1.4%	[48]
Spain	Dogs	5.5%	Cats	8.8%	[16]
Germany	Dogs	1.2%	Cats	5.3%	[49]
Greece	Dogs	5.9%	Cats	6.8%	[29]
Thailand	Dogs	2.1%	Cats	2.5%	[50]
Brasil	Dogs	24.5%	Cats	11.1%	[51]
Italy	Dogs	1.7%	-	-	[52]
Netherland	Dogs	8.7%	Cats	4.6%	[53]

In the present study, the prevalence of *G. duodenalis* in dogs is higher than that in Heilongjiang (4.5%) [6], Guangdong (10.8%) [25] and Sichuan (11.3%) [26] provinces, China, and is also higher than other zoonotic pathogens in dogs, such as 10.3% for *Babesia canis*, 9.1% for *Anaplasma* spp., 4.5% for *Leishmania infantum*, 1.7% for *Borrelia burgdorferi*, 0.4% for *Ehrlichia* spp. and 1.7% for *Dirofifilaria immitis* in Italy [54], but is lower than that in Henan province (14.3%) [27], Shanghai city (26.2%) in China [7] and other countries (Table 5).

Similarly, the *G. duodenalis* prevalence in pet cats is consistent with that in Hangzhou city (1.2%) [28], China; but is lower than that in Heilongjiang (1.9%) [6] and Guangdong (5.8%) provinces [25] and Shanghai city (13.1%) [7] in China and other countries (Table 5), and is also lower than *L. infantum* (3.0%) in Greece and Italy; *Rickettsia felis* (10.8%), *Rickettisa typhi* (4.2%), *Anaplasma phagocytophilum* (2.4%) and *Ehrlichia canis* (2.4%) in cats in Italy [55,56]. The reason is complicated among different studies because many factors could affect the prevalences such as sample sizes, sample sources, environments, animal welfare, hygiene conditions, age and sex of samples, and the sensitivity of tested methods. Moreover, the living condition is a risk factor ($p < 0.05$) that is significantly related to the prevalence of *G. duodenalis* in pet dogs in this study. We suspect that the poor sanitation of shelters contributes significantly to nosocomial transmission, adding to the prevalence of *G. duodenalis* in pet dogs. Furthermore, the higher prevalence of *G. duodenalis* was detected in pet dogs in Kunming city ($p < 0.05$) (Table 1), which suggests that the region is also a risk factor significantly associated with *G. duodenalis* infection in this study. In addition, the prevalence of *G. duodenalis* in male dogs was higher than that in female dogs in the present study, which is consistent with observations in other previous studies [2,57], although the difference was not statistically significant ($p > 0.05$). Compared with dogs, cats seem to be less susceptible to infection with *G. duodenalis* (Table 1). This might be explained by the different living habits of these two animals.

Similar to *G. duodenalis*, the prevalences of *E. bieneusi* in pet dogs and cats in different regions are different (Table 5). This is probably because the route and source of infection for dogs or cats in each region may be different. In addition, other factors can also affect the prevalence of *E. bieneusi* in dogs and cats. Furthermore, statistical analysis showed that a significant difference was observed among pet dogs in shelters, pet markets and pet hospitals (Table 1), which indicates that dogs living in shelters are more easily infected with *E. bieneusi* than those dogs in pet hospitals and markets. The reason may be the poorer hygiene conditions in shelters compared with pet markets and pet hospitals. Pet dogs aged more than 6 mouths seemed to be more susceptible to infection with *E. bieneusi* ($p < 0.05$) (Table 1), suggesting that further relevant research should pay more attention to the adult dogs. Additionally, only cats in Chuxiong city were found to be infected by *E. bieneusi* (Table 1); thus, we speculate that the regional factors may have a significant effect on the prevalence of *E. bieneusi* in cats. But this hypothesis needs to be tested. Additionally, there was no significant difference in the prevalence of *Cryptosporidium* spp. rate in pet dogs or cats (Table 1).

Up to now, six assemblages (assemblage A, B, C, D, E and F) have been identified in dogs and cats in previous studies [6,7,25–27,31], and canine-specific and feline-specific assemblages C, D and F are also found in other animals [11]. These findings indicate that both dogs and cats are a reservoir of *G. duodenalis*, which has risk of transmission among different animals. In the present study, only two assemblages (C and D) were identified in pet dogs, which is similar to previous studies [26,27]. Furthermore, a previous work demonstrated that the assemblages C and D are more sensitive than assemblage A in pet dogs [58]. Moreover, we found nine subtypes of assemblage (at bg locus, $n = 4$, at gdh locus, $n = 4$ and at tpi locus, $n = 1$) in dogs and one subtype of assemblage (at bg locus, $n = 1$) in cats (Table 2). The assemblage of *G. duodenalis* in dogs in the current study seems to more likely to mutate, thus further studies need to examine the genetic structure of these subtypes. Also, one mixed genotype of *G. duodenalis* was found in dogs in this study (Table 3), revealing the diversity of *G. duodenalis* in our investigation area.

Early studies have reported that genotypes of *E. bieneusi* CD1 to CD8, D, O, PigEBITSS, EbpA, CMl, Peru8 and EbpC are identified in dogs, and genotypes D, BEB6, I, CC1, CC2, CC3, CC4 are identified in cats in other provinces of China [33,59]. In the present study, the dominant genotype of *E. bieneusi* PtEb IX (18/21) is a common dog-specific *E. bieneusi* genotype identified in dogs (Table 4). Additionally, two novel genotypes (DgEb I and DgEb II) were also identified in dogs in our study, which enrich the genotype variety of *E. bieneusi* in dogs. *E. bieneusi* genotype Type IV and novel genotype CtEb I in pet cats belonged to Group 1 of zoonotic potential (Figure 1), which imply that pet cats may be a potential source of human infection with *E. bieneusi* in Yunnan province, China.

According to previous studies, *C. ubiquitum* and *C. canis* are commonly found in dogs, and *C. parvum* and *C. felis* are commonly found in cats in Heilongjiang, Shanghai and other cities or provinces of China [6,7]. In the present study, we only identified *C. canis* and *C. felis* in pet dogs and cats, respectively (Table 4). By contrast with the current study, the *C. parvum* and *C. muris* have been found in dogs or cats in other countries [20–22,36,60]. Despite our results revealing the presence of host-specific *Cryptosporidium* spp. species (*C. canis* and *C. felis*) in pet dogs and cats, these two species have been reported in humans and mainly in developing countries [6]. This finding suggests that people still need to take further precautions when they are in close contact with their pets. In addition, some nucleotide sequences of *Cryptosporidium* spp. obtained in pet dogs and cats have mutations in this study (Table 4).

4. Materials and Methods

4.1. Study Sites

The fecal samples of pet dogs and cats were collected in Kunming city, Lijiang city and Chuxiong city in Yunnan province (Location: $21°8'$ N to $29°15'$ N and $97°31'$ E to $106°11'$ E), southwestern China, which covers more than 390,000 square kilometers and has a population of approximately 48 million.

4.2. Sampling

During August to September 2018, a total of 433 fresh fecal samples were collected from pet dogs and cats in three cities of Yunnan province, including Kunming city (134 dogs and 36 cats), Lijiang city (90 dogs and 110 cats) and Chuxiong city (38 dogs and 25 cats). The Kunming, Lijiang and Chuxiong cities have more numbers of pet dogs and cats than other cities of Yunnan province, and all the samples of the cats and dogs were randomly collected from the biggest pet hospital, pet market and shelter in each city (i.e., Kunming city, Lijiang city and Chuxiong city), respectively. Moreover, the information regarding regions, ages, genders and living conditions were recorded. All the fecal samples were saved into 15 mL centrifuge tube with 2.5% potassium dichromate, and then were stored at 4 °C until for DNA extraction.

4.3. Genomic DNA Extraction and PCR Amplification

Each fecal sample was washed three times with distilled water by centrifuging at 13,000 *g* for 5 min to remove potassium dichromate, and 300 mg of the precipitated samples were used for DNA extraction using the E.Z.N.A. Stool DNA kit (OMEGA, Biotek Inc. USA) according to the manufacturer's instructions. The genomic DNA was stored at –20 °C before PCR amplification. The *G. duodenalis* identification was performed by nested PCR amplification of bg, gdh and tpi gene loci according to previous reports [25,61], *Cryptosporidium* spp. identification was conducted by nested PCR amplification of the 18S ribosomal RNA [62], and *E. bieneusi* identification was carried out by nested PCR amplification of ITS rDNA sequences as previously described [63]. The positive and negative controls were included in each PCR reaction. All the secondary PCR products were checked by 2% (*w*/*v*) agarose gel electrophoresis after ethidium bromide staining and visualized under UV light.

4.4. Sequence Analysis

The PCR-positive products were sent to Tsingke Biological Technology Company (Xi'an, China) for two-directional sequencing. The obtained sequences were spliced together after initial collation with their DNA peak form graph by Chromas v.2.6. The genotypes/species of *G. duodenalis*, *Cryptosporidium* spp. and *E. bieneusi* were identified by aligning the obtained sequences with corresponding sequences in the GenBank database (http://www.ncbi.lm.nih.gov/GenBank/, accessed on 11 July 2021). The phylogenetic tree was established by neighbor-joining method (NJ) with Kimura 2-parameter model in MEGA 7.0 (http://www.megasoftware.net/, accessed on 11 July 2021). The novel genotypes of *E. bieneusi* were decided by the ~243-bp ITS region [64,65].

4.5. Statistical Analysis

Prevalence of *G. duodenalis*, *Cryptosporidium* spp. and *E. bieneusi* in age, regio, gender and living conditions groups were analyzed using Chi-square test in SPSS 24.0 (SPSS Inc., Chicago, IL, USA). The 95% confidence intervals (CIs) were estimated. The difference was considered statistically significant when *p*-value < 0.05.

5. Conclusions

The present investigation revealed the prevalence and assemblages/genotypes/species of *G. duodenalis*, *E. bieneusi* and *Cryptosporidium* spp. in pet dogs and cats in Yunnan province, China. The infection with *G. duodenalis*, *E. bieneusi* and *Cryptosporidium* spp. in dogs and cats suggests that we should take measures to prevent and control those pathogens from being transmitted to other animals and humans. Our data provided the valuable information for a better understanding of the epidemiology and public health threat of *Giardiasis*, *E. bieneusi* and *Cryptosporidium* spp. in pet dogs and cats in southwestern China.

Author Contributions: F.-C.Z., Y.Z. and X.-Q.Z. conceived and designed the study. Y.-G.W. performed the experiments, analyzed the data and drafted the manuscript. Y.Z., Z.-Z.Y., D.C., B.-Z.G., J.-F.Y. and G.-H.L. participated in the implementation of the study. X.-Q.Z., F.-C.Z., Y.Z. and Y.-G.W. critically revised the manuscript. All authors have read and agreed to the published version of the manuscript.

Funding: This work was supported by the Yunnan Expert Workstation (Grant No. 202005AF150041), the Veterinary Public Health Innovation Team of Yunnan Province (Grant No. 202105AE160014), the Fund for Shanxi "1331 Project" (Grant No. 20211331-13) and the Agricultural Science and Technology Innovation Program (ASTIP) (Grant No. CAAS-ASTIP-2016-LVRI-03). The funders had no role in the design of the study; in the collection, analyses, or interpretation of data; in the writing of the manuscript, or in the decision to publish the results.

Institutional Review Board Statement: The study protocol has been reviewed and approved by the institutional animal ethical committee of Lanzhou Veterinary Research Institute, Chinese Academy of Agricultural Sciences. The approval code: AECLVRI-2018-003.

Informed Consent Statement: Not applicable.

Data Availability Statement: The data that support the figures within this paper and other finding of this study are available from the corresponding authors upon reasonable request. All of the obtained representative *G. duodenalis* bg, gdh and tpi nucleotide sequences were deposited in GenBank (https://www.ncbi.nlm.nih.gov/ accessed on 26 November 2019) under the accession numbers MN734348- MN734356, MN734357-MN734367 and MN734368, respectively. The nucleotide sequences of *Cryptosporidium* spp. and *E. bieneusi* were deposited in GenBank under accession numbers MZ540366-MZ540371 and MZ542369-MZ542375, respectively.

Conflicts of Interest: The authors declare no conflict of interest.

References

1. Lagunas-Rangel, F.A.; Bermúdez-Cruz, R.M. Epigenetics in the early divergent eukaryotic *Giardia duodenalis*: An update. *Biochimie* **2019**, *156*, 123–128. [CrossRef]
2. Li, J.; Zhang, P.; Wang, P.; Alsarakibi, M.; Zhu, H.; Liu, Y.; Meng, X.; Li, J.; Guo, J.; Li, G. Genotype identification and prevalence of *Giardia duodenalis* in pet dogs of Guangzhou, Southern China. *Vet. Parasitol.* **2012**, *188*, 368–371. [CrossRef]
3. Ortega-Pierres, M.G.; Jex, A.R.; Ansell, B.R.E.; Svärd, S.G. Recent advances in the genomic and molecular biology of Giardia. *Acta Trop.* **2018**, *184*, 67–72. [CrossRef]
4. Li, W.; Li, Y.; Song, M.; Lu, Y.; Yang, J.; Tao, W.; Jiang, Y.; Wan, Q.; Zhang, S.; Xiao, L. Prevalence and genetic characteristics of *Cryptosporidium, Enterocytozoon bieneusi* and *Giardia duodenalis* in cats and dogs in Heilongjiang province, China. *Vet. Parasitol.* **2015**, *208*, 125–134. [CrossRef]

5.   Tungtrongchitr, A.; Sookrung, N.; Indrawattana, N.; Kwangsi, S.; Ongrotchanakun, J.; Chaicumpa, W. *Giardia intestinalis* in Thailand: Identification of genotypes. *J. Health Popul. Nutr.* **2010**, *28*, 42–52. [CrossRef] [PubMed]

6.   Vivancos, V.; González-Alvarez, I.; Bermejo, M.; Gonzalez-Alvarez, M. Giardiasis: Characteristics, Pathogenesis and New Insights about Treatment. *Curr. Top. Med. Chem.* **2018**, *18*, 1287–1303. [CrossRef] [PubMed]

7.   Xu, H.; Jin, Y.; Wu, W.; Li, P.; Wang, L.; Li, N.; Feng, Y.; Xiao, L. Genotypes of *Cryptosporidium* spp. Enterocytozoon bieneusi and Giardia duodenalis in dogs and cats in Shanghai, China. *Parasit. Vectors* **2016**, *9*, 121.

8.   Lucio-Forster, A.; Griffiths, J.K.; Cama, V.A.; Xiao, L.; Bowman, D.D. Minimal zoonotic risk of cryptosporidiosis from pet dogs and cats. *Trends Parasitol.* **2010**, *26*, 174–179. [CrossRef] [PubMed]

9.   Ballweber, L.R.; Xiao, L.; Bowman, D.D.; Kahn, G.; Cama, V.A. Giardiasis in dogs and cats: Update on epidemiology and public health significance. *Trends Parasitol.* **2010**, *26*, 180–189. [CrossRef]

10. Rossle, N.F.; Latif, B. Cryptosporidiosis as threatening health problem: A review. *Asian Pac. J. Trop. Biomed.* **2013**, *3*, 916–924. [CrossRef]

11. Heyworth, M.F. *Giardia duodenalis* genetic assemblages and hosts. *Parasite* **2016**, *23*, 13. [CrossRef]

12. Ryan, U.; Hijjawi, N.; Xiao, L. Foodborne cryptosporidiosis. *Int. J. Parasitol.* **2018**, *48*, 1–12. [CrossRef]

13. Monis, P.T.; Caccio, S.M.; Thompson, R.C. Variation in *Giardia*: Towards a taxonomic revision of the gene. *Trends Parasitol.* **2009**, *25*, 93–100. [CrossRef] [PubMed]

14. Vanni, I.; Cacciò, S.M.; van Lith, L.; Lebbad, M.; Svärd, S.G.; Pozio, E.; Tosini, F. Detection of *Giardia duodenalis* assemblages A and B in human feces by simple, assemblage-specific PCR assays. *PLoS Negl. Trop. Dis.* **2012**, *6*, e1776. [CrossRef] [PubMed]

15. Adell-Aledón, M.; Köster, P.C.; de Lucio, A.; Puente, P.; Hernández-de-Mingo, M.; Sánchez-Thevenet, P.; Dea-Ayuela, M.A.; Carmena, D. Occurrence and molecular epidemiology of *Giardia duodenalis* infection in dog populations in eastern Spain. *BMC Vet. Res.* **2018**, *14*, 26. [CrossRef] [PubMed]

16. de Lucio, A.; Bailo, B.; Aguilera, M.; Cardona, G.A.; Fernández-Crespo, J.C.; Carmena, D. No molecular epidemiological evidence supporting household transmission of zoonotic *Giardia duodenalis* and *Cryptosporidium* spp. from pet dogs and cats in the province of Álava, Northern Spain. *Acta Trop.* **2017**, *170*, 48–56. [CrossRef]

17. Abdel-Moein, K.A.; Saeed, H. The zoonotic potential of *Giardia intestinalis* assemblage E in rural settings. *Parasitol. Res.* **2016**, *115*, 3197–3202. [CrossRef]

18. Pipiková, J.; Papajová, I.; Majláthová, V.; Šoltys, J.; Bystrianska, J.; Schusterová, I.; Vargová, V. First report on *Giardia duodenalis* assemblage F in Slovakian children living in poor environmental conditions. *J. Microbiol. Immunol. Infect.* **2020**, *53*, 148–156. [CrossRef]

19. Feng, Y.; Ryan, U.M.; Xiao, L. Genetic diversity and population structure of Cryptosporidium. *Trends Parasitol.* **2018**, *34*, 997–1011. [CrossRef]

20. Palmer, C.S.; Traub, R.J.; Robertson, I.D.; Devlin, G.; Rees, R.; Thompson, R.C.A. Determining the zoonotic significance of *Giardia* and *Cryptosporidium* in Australian dogs and cats. *Vet. Parasitol.* **2008**, *154*, 142–147. [CrossRef]

21. Lupo, P.J.; Langer-Curry, R.C.; Robinson, M.; Okhuysen, P.C.; Chappell, C.L. *Cryptosporidium muris* in a Texas canine population. *Am. J. Trop. Med. Hyg.* **2008**, *78*, 917–921. [CrossRef]

22. Giangaspero, A.; Iorio, R.; Paoletti, B.; Traversa, D.; Capelli, G. Molecular evidence for *Cryptosporidium* infection in dogs in Central Italy. *Parasitol. Res.* **2006**, *99*, 297–299. [CrossRef] [PubMed]

23. Li, W.; Feng, Y.; Santin, M. Host Specificity of *Enterocytozoon bieneusi* and Public Health Implications. *Trends Parasitol.* **2019**, *35*, 436–451. [CrossRef] [PubMed]

24. Li, W.; Xiao, L. Ecological and public health significance of *Enterocytozoon bieneusi*. *One Health* **2020**, *12*, 100209. [CrossRef]

25. Pan, W.; Wang, M.; Abdullahi, A.Y.; Fu, Y.; Yan, X.; Yang, F.; Shi, X.; Zhang, P.; Hang, J.; Li, G. Prevalence and genotypes of *Giardia lamblia* from stray dogs and cats in Guangdong, China. *Vet. Parasitol. Reg. Stud. Rep.* **2018**, *13*, 30–34. [CrossRef] [PubMed]

26. Zhang, Y.; Zhong, Z.; Deng, L.; Wang, M.; Li, W.; Gong, C.; Fu, H.; Cao, S.; Shi, X.; Wu, K.; et al. Detection and multilocus genotyping of *Giardia duodenalis* in dogs in Sichuan province, China. *Parasite* **2017**, *24*, 31. [CrossRef]

27. Qi, M.; Dong, H.; Wang, R.; Li, J.; Zhao, J.; Zhang, L.; Luo, J. Infection rate and genetic diversity of Giardia duodenalis in pet and stray dogs in Henan Province, China. *Parasitol. Int.* **2016**, *65*, 159–162. [CrossRef]

28. Li, W.; Liu, X.; Gu, Y.; Liu, J.; Luo, J. Prevalence of *Cryptosporidium, Giardia, Blastocystis*, and *trichomonads* in domestic cats in East China. *J. Vet. Med. Sci.* **2019**, *81*, 890–896. [CrossRef]

29. Kostopoulou, D.; Claerebout, E.; Arvanitis, D.; Ligda, P.; Voutzourakis, N.; Casaert, S.; Sotiraki, S. Abundance, zoonotic potential and risk factors of intestinal parasitism amongst dog and cat populations: The scenario of Crete, Greece. *Parasit. Vectors* **2017**, *10*, 43. [CrossRef]

30. Gil, H.; Cano, L.; de Lucio, A.; Bailo, B.; de Mingo, M.H.; Cardona, G.A.; Fernández-Basterra, J.A.; Aramburu-Aguirre, J.; López-Molina, N.; Carmena, D. Detection and molecular diversity of *Giardia duodenalis* and *Cryptosporidium* spp. in sheltered dogs and cats in Northern Spain. *Infect. Genet. Evol.* **2017**, *50*, 62–69. [CrossRef]

31. McDowall, R.M.; Peregrine, A.S.; Leonard, E.K.; Lacombe, C.; Lake, M.; Rebelo, A.R.; Cai, H.Y. Evaluation of the zoonotic potential of *Giardia duodenalis* in fecal samples from dogs and cats in Ontario. *Can. Vet. J.* **2011**, *52*, 1329–1333.

32. Fava, N.M.; Soares, R.M.; Scalia, L.A.; Cunha, M.J.; Faria, E.S.; Cury, M.C. Molecular typing of canine *Giardia duodenalis* isolates from Minas Gerais, Brazil. *Exp. Parasitol.* **2016**, *161*, 1–5. [CrossRef] [PubMed]

33. Karim, M.R.; Don, H.; Yu, F.; Jian, F.; Zhang, L.; Wang, R.; Zhang, S.; Rume, F.I.; Ning, C.; Xiao, L. Genetic diversity in *Enterocytozoon bieneusi* isolates from dogs and cats in China: Host specificity and public health implications. *Clin. Microbiol.* **2014**, *52*, 3297–3302. [CrossRef]
34. Li, W.C.; Qin, J.; Wang, K.; Gu, Y.F. Genotypes of *Enterocytozoon bieneusi* in dogs and cats in eastern China. *Iran. J. Parasitol.* **2018**, *13*, 457–465. [PubMed]
35. Zhang, X.; Wang, Z.; Su, Y.; Liang, X.; Sun, X.; Peng, S.; Lu, H.; Jiang, N.; Yin, J.; Xiang, M.; et al. Identification and genotyping of *Enterocytozoon bieneusi* in China. *J. Clin. Microbiol.* **2011**, *49*, 2006–2008. [CrossRef] [PubMed]
36. Santín, M.; Trout, J.M.; Vecino, J.A.; Dubey, J.P.; Fayer, R. *Cryptosporidium, Giardia* and *Enterocytozoon bieneusi* in cats from Bogota (Colombia) and genotyping of isolates. *Vet. Parasitol.* **2006**, *141*, 334–339. [CrossRef] [PubMed]
37. Santín, M.; Cortés Vecino, J.A.; Fayer, R. *Enterocytozoon bieneusi* genotypes in dogs in Bogota, Colombia. *Am. J. Trop. Med. Hyg.* **2008**, *79*, 215–217. [CrossRef]
38. Al-Herrawy, A.Z.; Gad, M.A. Microsporidial spores in fecal samples of some domesticated animals living in Giza, Egypt. *Iran. J. Parasitol.* **2016**, *11*, 195–203.
39. Dengjel, B.; Zahler, M.; Hermanns, W.; Heinritzi, K.; Spillmann, T.; Thomschke, A.; Löscher, T.; Gothe, R.; Rinder, H. Zoonotic potential of *Enterocytozoon bieneusi*. *J. Clin. Microbiol.* **2001**, *39*, 4495–4499. [CrossRef]
40. Dashti, A.; Santín, M.; Cano, L.; de Lucio, A.; Bailo, B.; de Mingo, M.H.; Köster, P.C.; Fernández-Basterra, J.A.; Aramburu-Aguirre, J.; López-Molina, N.; et al. Occurrence and genetic diversity of *Enterocytozoon bieneusi* (Microsporidia) in owned and sheltered dogs and cats in Northern Spain. *Parasitol. Res.* **2019**, *118*, 2979–2987. [CrossRef]
41. Abe, N.; Kimata, I.; Iseki, M. Molecular evidence of *Enterocytozoon bieneusi* in Japan. *J. Vet. Med. Sci.* **2009**, *71*, 217–219. [CrossRef]
42. Piekarska, J.; Kicia, M.; Wesołowska, M.; Kopacz, Z.; Gorczykowski, M.; Szczepankiewicz, B.; Kváč, M.; Sak, B. Zoonotic Microsporidia in dogs and cats in Poland. *Vet. Parasitol.* **2019**, *246*, 108–111. [CrossRef] [PubMed]
43. Mori, H.; Mahittikorn, A.; Thammasonthijarern, N.; Chaisiri, K.; Rojekittikhun, W.; Sukthana, Y. Presence of zoonotic *Enterocytozoon bieneusi* in cats in a temple in Central Thailand. *Vet. Parasitol.* **2013**, *197*, 696–701. [CrossRef]
44. Lobo, M.L.; Xiao, L.; Cama, V.; Stevens, T.; Antunes, F.; Matos, O. Genotypes of *Enterocytozoon bieneusi* in mammals in Portugal. *J. Eukaryot. Microbiol.* **2006**, *53*, S61–S64. [CrossRef]
45. Jamshidi, S.H.; Tabrizi, A.S.; Bahrami, M.; Momtaz, H. Microsporidia in household dogs and cats in Iran; A zoonotic concern. *Vet. Parasitol.* **2012**, *185*, 121–123. [CrossRef] [PubMed]
46. Zhang, J.; Dong, H. Investigation of Parasite Infection in the Digestive Tract of Domestic Dogs in Zhengzhou. In Proceedings of the Academic Symposium of the Fourth Branch Association of Small Animal Medicine & The Sixteenth Branch Association of Veterinary Surgery, Hohhot, China, 18 August 2009; pp. 226–229. Available online: https://kns.cnki.net/kcms/detail/detail.aspx?FileName=ZGXJ200908003047&DbName=CPFD2010 (accessed on 11 July 2021). (In Chinese).
47. Hu, L.; Qi, M.; Peng, G.; Zhong, Z.; Dong, H.; Zhang, L.; Liu, Y. Investigation of intestinal parasites infection in Ya'an City. *Anim. Husb. Vet. Med.* **2011**, *10*, 80–81.
48. Ito, Y.; Itoh, N.; Kimura, Y.; Kanai, K. Molecular detection and characterization of *Cryptosporidium* spp. among breeding cattery cats in Japan. *Parasitol. Res.* **2016**, *115*, 2121–2123. [CrossRef] [PubMed]
49. Sotiriadou, I.; Pantchev, N.; Gassmann, D.; Karanis, P. Molecular identifification of *Giardia* and *Cryptosporidium* from dogs and cats. *Parasite* **2013**, *20*, 8. [CrossRef]
50. Koompapong, K.; Mori, H.; Thammasonthijarern, N.; Prasertbun, R.; Pintong, A.; Popruk, S.; Rojekittikhun, W.; Chaisiri, K.; Sukthana, Y.; Mahittikorn, A. Molecular identifification of *Cryptosporidium* spp. in seagulls, pigeons, dogs, and cats in Thailand. *Parasite* **2014**, *21*, 52. [CrossRef] [PubMed]
51. Moreira, A.D.S.; Baptista, C.T.; Brasil, C.L.; Valente, J.S.S.; Bruhn, F.P.P.; Pereira, D.I.B. Risk factors and infection due to *Cryptosporidium* spp. in dogs and cats in southern Rio Grande do Sul. *Rev. Bras. Parasitol. Vet.* **2018**, *27*, 112–117. [CrossRef]
52. Simonato, G.; Frangipane di Regalbono, A.; Cassini, R.; Traversa, D.; Tessarin, C.; Cesare, A.D.; Pietrobelli, M. Molecular detection of *Giardia duodenalis* and *Cryptosporidium* spp. in canine faecal samples contaminating public areas in Northern Italy. *Parasitol. Res.* **2017**, *116*, 3411–3418. [CrossRef]
53. Overgaauw, P.A.; van Zutphen, L.; Hoek, D.; Yaya, F.O.; Roelfsema, J.; Pinelli, E.; van Knapen, F.; Kortbeek, L.M. Zoonotic parasites in fecal samples and fur from dogs and cats in the Netherlands. *Vet. Parasitol.* **2009**, *163*, 115–122. [CrossRef]
54. Colombo, M.; Morelli, S.; Simonato, G.; Cesare, A.D.; Veronesi, F.; di Regalbono, A.F.; Grassi, L.; Russi, I.; Tiscar, P.G.; Morganti, G.; et al. Exposure to major Vector-Borne Diseases in dogs subjected to different preventative regimens in endemic areas of Italy. *Pathogens* **2021**, *10*, 507. [CrossRef]
55. Morelli, S.; Colombo, M.; Dimzas, D.; Barlaam, A.; Traversa, D.; Cesare, A.D.; Russi, I.; Spoletini, R.; Paoletti, B.; Diakou, A. *Leishmania infantum* seroprevalence in cats from touristic areas of Italy and Greece. *Front. Vet. Sci.* **2020**, *7*, 616566. [CrossRef] [PubMed]
56. Morelli, S.; Crisi, P.E.; Cesare, A.D.; Santis, F.D.; Barlaam, A.; Santoprete, G.; Parrinello, C.; Palermo, S.; Mancini, P.; Traversa, D. Exposure of client-owned cats to zoonotic vector-borne pathogens: Clinicpathological alterations and infection risk analysis. *Comp. Immunol. Microbiol. Infect. Dis.* **2019**, *66*, 101344. [CrossRef] [PubMed]
57. Ortuño, A.; Scorza, V.; Castellà, J.; Lappin, M. Prevalence of intestinal parasites in shelter and hunting dogs in Catalonia, Northeastern Spain. *Vet. J.* **2014**, *199*, 465–467. [CrossRef]

58. Berrilli, F.; D'Alfonso, R.; Giangaspero, A.; Marangi, M.; Brandonisio, O.; Kaboré, Y.; Glé, C.; Cianfanelli, C.; Lauro, R.; Cave, D.D. *Giardia duodenalis* genotypes and *Cryptosporidium species* in humans and domestic animals in Cote d'Ivoire: Occurrence and evidence for environmental contamination. *Trans. R. Soc. Trop. Med. Hyg.* **2012**, *106*, 191–195. [CrossRef] [PubMed]

59. Wang, S.S.; Wang, R.J.; Fan, X.C.; Liu, T.L.; Zhang, L.X.; Zhao, G.H. Prevalence and genotypes of *Enterocytozoon bieneusi* in China. *Acta Trop.* **2018**, *183*, 142–152. [CrossRef]

60. Ballweber, L.R.; Panuska, C.; Huston, C.L.; Vasilopulos, R.; Pharr, G.T.; Mackin, A. Prevalence of and risk factors associated with shedding of *Cryptosporidium felis* in domestic cats of Mississippi and Alabama. *Vet. Parasitol.* **2009**, *160*, 306–310. [CrossRef]

61. Cacciò, S.M.; Beck, R.; Lalle, M.; Marinculic, A.; Pozio, E. Multilocus genotyping of *Giardia duodenalis* reveals striking differences between assemblages A and B. *Int. J. Parasitol.* **2008**, *38*, 1523–1531. [CrossRef]

62. Xiao, L.; Morgan, U.M.; Limor, J.; Escalante, A.; Arrowood, M.; Shulaw, W.; Thompson, R.C.; Fayer, R.; Lal, A.A. Genetic diversity within *Cryptosporidium parvum* and related *Cryptosporidium* species. *Appl. Environ. Microbiol.* **1999**, *65*, 3386–3391. [CrossRef]

63. Ma, Y.T.; Zou, Y.; Liu, Q.; Xie, S.C.; Li, R.L.; Zhu, X.Q.; Gao, W.W. Prevalence and multilocus genotypes of *Enterocytozoon bieneusi* in alpacas (Vicugna pacos) in Shanxi Province, northern China. *Parasitol. Res.* **2019**, *118*, 1–5. [CrossRef] [PubMed]

64. Zhao, W.; Zhou, H.H.; Ren, G.X.; Qiang, Y.; Huang, H.C.; Lu, G.; Tan, F. Occurrence and potentially zoonotic genotypes of *Enterocytozoon bieneusi* in wild rhesus macaques (Macaca mulatta) living in Nanwan Monkey Island, Hainan, China: A public health concern. *BMC Vet. Res.* **2021**, *17*, 213. [CrossRef] [PubMed]

65. Santín, M.; Fayer, R. *Enterocytozoon bieneusi* genotype nomenclature based on the internal transcribed spacer sequence: A consensus. *J. Eukaryot. Microbiol.* **2009**, *56*, 34–38. [CrossRef] [PubMed]

Article

Genetic Diversity of *Cryptosporidium* in Bactrian Camels (*Camelus bactrianus*) in Xinjiang, Northwestern China

Yangwenna Cao [1,2,†], Zhaohui Cui [3,4,†], Qiang Zhou [1,2], Bo Jing [1,2], Chunyan Xu [1,2], Tian Wang [1,2], Meng Qi [1,2,*] and Longxian Zhang [4,*]

1   College of Animal Science, Tarim University, Alar 843300, Xinjiang, China; cyangwenna@gmail.com (Y.C.); zqiang1016@gmail.com (Q.Z.); 120050010@taru.edu.cn (B.J.); 120170015@taru.edu.cn (C.X.); 120180029@taru.edu.cn (T.W.)
2   Engineering Laboratory of Tarim Animal Diseases Diagnosis and Control, Xinjiang Production & Construction Corps, Alar 843300, Xinjiang, China
3   Key Laboratory of Biomarker Based Rapid-detection Technology for Food Safety of Henan Province, Xuchang University, Xuchang 461000, China; czh9008@xcu.edu.cn
4   College of Veterinary Medicine, Henan Agricultural University, Zhengzhou 450002, China
*   Correspondence: 120150006@taru.edu.cn (M.Q.); zhanglx8999@henau.edu.cn (L.Z.); Tel.: +86-371-5699-0163 (L.Z.)
†   These authors contributed equally to this work.

Received: 23 October 2020; Accepted: 12 November 2020; Published: 13 November 2020

Abstract: *Cryptosporidium* species are ubiquitous enteric protozoan pathogens of vertebrates distributed worldwide. The purpose of this study was to gain insight into the zoonotic potential and genetic diversity of *Cryptosporidium* spp. in Bactrian camels in Xinjiang, northwestern China. A total of 476 fecal samples were collected from 16 collection sites in Xinjiang and screened for *Cryptosporidium* by PCR. The prevalence of *Cryptosporidium* was 7.6% (36/476). Six *Cryptosporidium* species, *C. andersoni* ($n = 24$), *C. parvum* ($n = 6$), *C. occultus* ($n = 2$), *C. ubiquitum* ($n = 2$), *C. hominis* ($n = 1$), and *C. bovis* ($n = 1$), were identified based on sequence analysis of the small subunit (SSU) rRNA gene. Sequence analysis of the *gp60* gene identified six *C. parvum* isolates as subtypes, such as IIf-like-A15G2 ($n = 5$) and IIdA15G1 ($n = 1$), two *C. ubiquitum* isolates, such as subtype XIIa ($n = 2$), and one *C. hominis* isolate, such as IxiAs IkA19G1 ($n = 1$). This is the first report of *C. parvum*, *C. hominis*, *C. ubiquitum*, and *C. occultus* in Bactrian camels in China. These results indicated that the Bactrian camel may be an important reservoir for zoonotic *Cryptosporidium* spp. and these infections may be a public health threat in this region.

Keywords: *Cryptosporidium*; genotype; Bactrian camels; zoonotic potential; public health

---

1. Introduction

Cryptosporidium is a significant cause of diarrheal disease worldwide, with broad host ranges and the ability to infect all vertebrate groups, including humans [1]. As commonly seen, the transmission of enteric pathogens through contaminated surface water, such as *Cryptosporidium* spp. potentially cause large outbreaks of water- and food-borne infections in human populations [2]. Cryptosporidiosis is a global disease and is considered an important opportunistic disease in immunocompromised patients due to its high association with mortality in AIDS patients [3].

Characterization of pathogens at the species or genotype level is mandatory when assessing the potential sources of infection, pathogen load in animals, the environment, transmission routes in human populations, and public health relevance [1,4]. Currently, *Cryptosporidium* genotyping is mostly based on PCR and sequencing of the small subunit (SSU) rRNA gene, which has revealed no less than 40 valid species and more than 70 genotypes of *Cryptosporidium* in humans and animals [1,5,6]. Humans infected by approximately 20 *Cryptosporidium* species and genotypes, with *C. hominis* and *C. parvum*, are responsible for the highest proportion (~90%) of human *Cryptosporidium* infections globally [7]; nevertheless, several primarily animal pathogens, such as *C. meleagridis*, *C. felis*, *C. canis*, and *C. cuniculus* are less commonly found in humans [7].

In northwestern China, Bactrian camels (*Camelus bactrianus*) represent the major livestock species, especially in Xinjiang Uygur Autonomous Region (hereinafter referred to as Xinjiang), because they are well adapted to desert and semi-desert areas and provide milk, meat, and camel hair. *Cryptosporidium* infection of camel calves resulted in diarrhea and debility, while infected adult camels showed no symptoms [8]. Camels infected with *Cryptosporidium* have been reported in many countries, such as the United States, Australia, Czech Republic, Algeria, Iran, Egypt, and China [9–20]. However, compared with other livestock animals, information on prevalence, species, genotype, and zoonotic potential of *Cryptosporidium* spp. in Bactrian camels is still limited in China.

The main focus of the current study was to investigate the prevalence of *Cryptosporidium* and identify the species and subtypes of Bactrian camels in Xinjiang, China (Figure 1). The data will contribute to an improved understanding of *Cryptosporidium* spp. in Bactrian camels and assessment of their zoonotic potential.

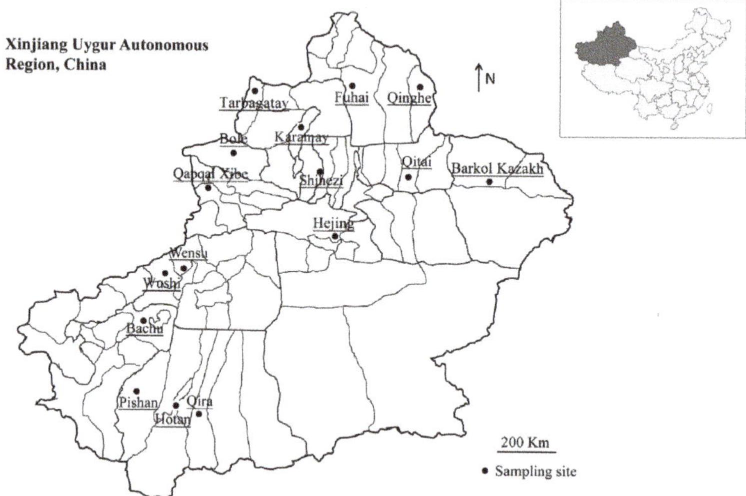

Figure 1. Bactrian camels fecal sampling locations in Xinjiang, northwestern China. No copyright permission was required. The figure was designed with the software ArcGIS 10.2. The map has been originally modified and assembled according to permission and attribution guidelines of the National Geomatics Center of China (http://www.ngcc.cn).

2. Results

2.1. Occurrence of Cryptosporidium

All fecal samples were screened for *Cryptosporidium* by nested PCR targeting of the SSU rRNA gene. In total, 36 samples were *Cryptosporidium*-positive, resulting in an overall infection rate of 7.6% (36/476). In total, 11 of 16 Bactrian camel herds tested contained individuals positive for *Cryptosporidium* spp., and the infection rate at the different collection sites ranged from 0–33.3%; the highest infection rate was observed in Qapqal Xibe County (Table 1).

Table 1. Occurrence of *Cryptosporidium* species/subtypes in Bactrian camels in Xinjiang, China.

Collection Sites	N/T (%) [95% CI]	*Cryptosporidium* Species/Subtypes (No.)
Barkol Kazakh	0/58 (0)	None
Qinghe	0/57 (0)	None
Qitai	2/18 (11.1) [0–27.2]	*C. andersoni* (1); *C. hominis* (1)/IkA19G1 (1)
Fuhai	1/26 (3.8) [0–11.8]	*C. parvum* (1)/If-like-A15G2 (1)
Karamay	3/45 (6.7) [0–14.2]	*C. andersoni* (3)
Shihezi	5/60 (8.3) [1.1–15.5]	*C. parvum* (5)/If-like-A15G2 (4), IIdA15G1 (1)
Hejing	5/16 (31.3) [5.7–56.8]	*C. andersoni* (5)
Tarbagatay	3/16 (18.8) [0–40.2]	*C. andersoni* (3)
Bole	0/61 (0)	None
Qapqal Xibe	4/12 (33.3) [2–64.6]	*C. andersoni* (4)
Wensu	1/24 (4.2) [0–12.8]	*C. bovis* (1)
Wushi	2/10 (20.0) [0–50.2]	*C. andersoni* (2)
Bachu	0/17 (0)	None
Pishan	3/17 (17.6) [0–37.9]	*C. andersoni* (1); *C. ubiquitum* (2)/XIIa (2)
Hotan	0/17 (0)	None
Qira	7/22 (31.8) [10.7–53.0]	*C. andersoni* (5); *C. occultus* (2)
Total	36/476 (7.6) [5.2–9.9]	*C. andersoni* (24); *C. bovis* (1); *C. occultus* (2); *C. parvum* (6)/If-like-A15G2 (5), IIdA15G1 (1); *C. hominis* (1)/IkA19G1 (1); *C. ubiquitum* (2)/XIIa (2)

N = Number of positives for *Cryptosporidium*; T = Total analyzed samples.

2.2. Cryptosporidium Species and Subtypes

Six species were detected from the 36 *Cryptosporidium*-positive samples. *C. andersoni* ($n = 24$) was the predominant species, followed by *C. parvum* ($n = 6$), *C. ubiquitum* ($n = 2$), *C. occultus* ($n = 2$), *C. hominis* ($n = 1$), and *C. bovis* ($n = 1$) (Table 1). The six *C. parvum*-positive samples were identified once again by restriction fragment length polymorphism (RFLP) analysis, and no mixed infections were found. Phylogenetic analysis revealed that all *C. andersoni* sequences were identical to the GenBank sequence KX710084, derived from Bactrian camels in China. Two types of sequences were identified from the six *C. parvum* isolates: *C. parvum* type 1 ($n = 5$) and *C. parvum* type 2 ($n = 1$) were identical to Genbank sequences KX259139 and KX259140, respectively, derived from deer in China. The two sequences of *C. occultus* were identical to sequence MK982467, derived from calves in Bangladesh. Moreover, the sequence of *C. hominis* was identical to sequence KU200955, derived from horses, while the sequence of *C. bovis* was identical to sequence MF074602, derived from dairy cattle in China. Two sequence types were identified in the two *C. ubiquitum* isolates: *C. ubiquitum* type 1 was identical to sequence KT235697, derived from goats in China, while *C. ubiquitum* type 2 represented a new sequence, bearing two single-nucleotide polymorphism (SNP) deletions at positions 485 and 486 and one SNP substitution at position 298 (A to G), compared with KT235697.

Sequence and phylogenetic analysis of the *gp60* gene revealed two subtypes present in the five *C. parvum* isolates: If-like-A15G2 ($n = 5$) and IIdA15G1 ($n = 1$). The sequence of If-like-A15G2 was similar to an isolate derived from a Swedish patient infected in South Africa (JN867334), except for the copy number differences in the trinucleotide repeat (A15G2 versus A12G2). The sequence of IIdA15G1 was identical to sequence KT964798, derived from dairy cattle in China. The single *C. hominis* isolate was subtyped as IkA19G1 and was similar to sequence KU727290, derived from an infected human patient in Sweden (A19G1 versus A18G2). The two new sequences of *C. ubiquitum* identified were identical to one another and subtyped to family XIIa. All of the subtype sequences, If-like, IId, Ik, and XIIa, clustered with published sequences, If-like, IId, Ik, and XIIa, respectively (Figure 2).

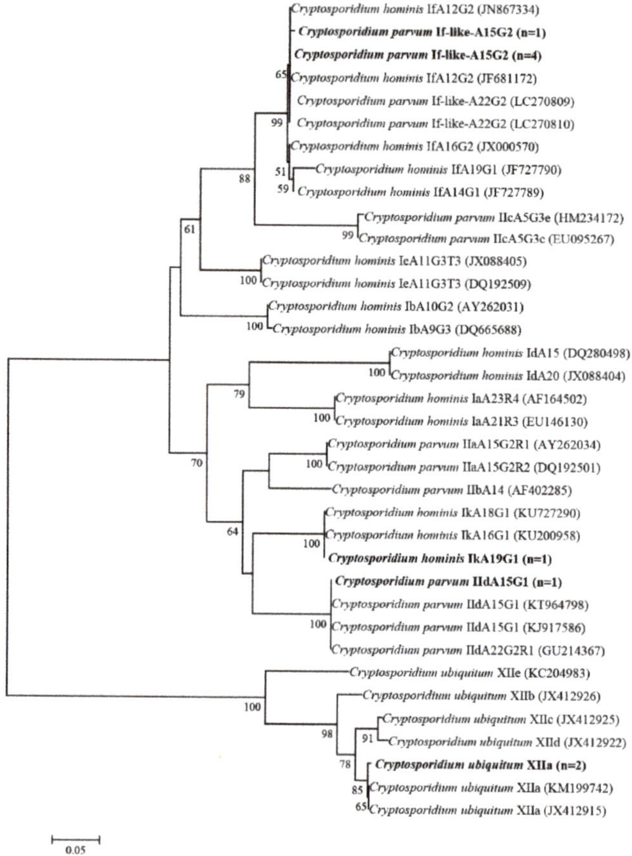

Figure 2. Phylogenetic relationships between *Cryptosporidium* spp. partial *gp60* sequences obtained in this study and sequences retrieved from the GenBank database. Phylogenetic trees were constructed using neighbor-joining methods based on genetic distance, calculated using the Kimura two-parameter model implemented in MEGA 7.0. Bootstrap values >50% from 1000 replicates are indicated at each node. Isolates from this study are shown in bold.

3. Discussion

Camels are well known as the ships of the desert and are famous as the beasts of the burden. Camels provide wool, milk, meat, leather, and even dung as fuel for the people in many semi-arid and arid zones, mainly in Africa and Asia [21]. Currently, camel husbandry has been transforming from nomadism to intensive production, resulting in the increase of the total population of camels, with an estimated global population of 35 million [21]. This intensive farming practice of camels has been posing an increased risk for zoonotic disease transmission to humans [22]. Many zoonotic parasites are reported to be transmitted from camels to humans globally [21]. However, there is scarce knowledge regarding camel parasites and their zoonotic importance in China. In this study, the overall *Cryptosporidium* prevalence was 7.6% (36/476), and six species of *Cryptosporidium* (*C. andersoni*, *C. parvum*, *C. hominis*, *C. ubiquitim*, *C. occultus*, and *C. bovis*) were identified, which indicated the genetic diversity of *Cryptosporidium* in Bactrian camels from Xinjiang, China.

From previously published studies, *C. andersoni*, *C. parvum*, *C. muris*, *C. bovis*, *Cryptosporidium* rat genotype IV, and camel genotype have been detected in camels [19]. Among them, only two species of *Cryptosporidium* have been reported in China, namely *C. andersoni* and *C. bovis* [10,12–14]. In the present study, both *C. andersoni* and *C. bovis* were identified, and *C. andersoni* was the dominant genotype detected in Bactrian camels. Although *C. andersoni* and *C. bovis* are commonly seen in calves and sheep, *C. andersoni* has also been found in several human cases [23,24].

Perhaps unsurprisingly, the most important zoonotic *Cryptosporidium* species., *C. parvum*, was previously reported in Dromedary camels in Algeria, Australia, and Egypt [11,15,19]. According to sequence analysis of the *gp60* gene, two subtypes of *C. parvum* were identified: IIdA15G1 and If-like-A15G2. In China, *C. parvum* isolates, including IIdA14G1, IIdA15G1, IIdA17G1, IIdA18G1, and IIdA19G1, mostly belong to the IId subtype family in goats, humans, cattle, donkeys, horses, rodents, monkeys, Golden takins, and yaks [1]. Previous studies have shown that IIdA15G1 was the predominant subtype in dairy calves and yaks in northwestern China [25,26]. In the present study, IIdA15G1 was identified in Bactrian camels in northwestern China, further confirming the dominance of the IIdA15G1 subtype in western China.

A unique *C. parvum* subtype If-like-A15G2 isolate was identified in Bactrian camels in the current research, which was similar to a previously observed If-like-A22G2 isolate found in Dromedary camels in Algeria [11]. Moreover, subtypes IIaA17G2R1, IIaA15G1R1, and IIdA19G1 were also identified in Dromedary camels in Australia and Egypt [15,19]. The *gp60* gene is highly polymorphic and can be used to categorize *C. parvum* and *C. hominis* into multiple subtypes according to nucleotide sequence differences [27]. However, it seems that *gp60* polymorphisms are ineffective for *C. parvum* subtype identification in camels. In the phylogenetic analysis of *gp60* sequences, *C. parvum* If-like genetically related to the *C. hominis* If subfamily and all If and If-like sequences formed a large clade (Figure 2). More extensive genetic characterization is needed to improve our understanding of the genetic similarity between *C. parvum* and *C. hominis* within the *gp60* gene.

Using *gp60* sequence analysis, *C. hominis* subtype IkA19G1 appeared to belong to subfamily Ik. Family Ik is commonly found in horses and donkeys [28,29] and has been also isolated from patients in Sweden and squirrel monkeys in China [30,31]. This is the first report of *C. hominis* in camels. Further studies should be carried out to expand the biological characterization of *C. hominis* subtype family Ik due to its potential for zoonotic transmission.

C. ubiquitum has a worldwide distribution, and six subtypes/families (XIIa–XIIf) have been identified [32]. Among these subtypes, subtype XIIa has been commonly observed in humans and a wide range of animals, especially domestic and wild ruminants [33,34]. In this study, *C. ubiquitum* and subtype XIIa were detected in Bactrian camels, which indicated that *C. ubiquitum* has a broad host range and high significance for zoonotic infection in this region. *C. occultus*, previously described as *Cryptosporidium* suis-like, was recognized as a valid species in 2018 and has been identified in cattle, yaks, alpacas, and wild rats in China [35–38]. Moreover, cases of human infection with *C. occultus* have also been found in Canada, China, and the UK [39–41]. The present study is the first report of *C. ubiquitum* and *C. occultus* in camels. Further studies into the epidemiology of *Cryptosporidium* infection in both human and livestock is essential.

4. Materials and Methods

4.1. Ethics Approval

The study was designed and conducted in accordance with the Guide for the Care and Use of Laboratory Animals of the Ministry of Health in China. The Research Ethics Committee of Tarim University critically reviewed this research protocol (approval no. ECTU 2016-0007) and then cleared it for performing. Finally, before fecal sample collection from Bactrian camels, appropriate permission was obtained from the farm owners.

4.2. Sample Collection

In total 476 fresh fecal samples were collected randomly from Bactrian camels grouped into 16 herds located at 16 collection sites in Xinjiang, from July 2016 to September 2019 (Figure 1). Each herd contained between 30 and 300 Bactrian camels. The Bactrian camels were free grazing in desert and semi-desert areas, so their age could not be accurately divided. In some of these areas, cattle, sheep, and horses in pastures also grazed freely. The Bactrian camels had access to pastures or areas where cattle, sheep, and horses had grazed. For each animal, the fresh fecal sample was collected from the ground immediately after defecation, and only one sample was collected per animal into a plastic container that was marked with the sample number and site. After shipping to the laboratory in a cool condition, the fecal samples were stored at 4 °C prior to DNA extraction.

4.3. DNA Extraction and PCR Amplification

The E.Z.N.A.® Stool DNA kit (Omega Biotek Inc., Norcross, GA, USA) was used to extract the total DNA from 200 mg of each precipitated sample, according to the manufacturer's recommendations. PCR analysis of the small subunit (SSU) rRNA gene was employed to screen the infection of *Cryptosporidium* spp. in fecal samples in Bactrian camels [42]. Furthermore, PCR amplification and subsequent sequencing of the 60-kDa glycoprotein (*gp60*) gene were used to subtype *C. parvum*, *C. ubiquitum*, and *C. hominis* [32,43]. The PCR reactions for the SSU rRNA and *gp60* genes conducted in 25 μL reaction mixtures consisted of 12.5 μL of 2 × EasyTaq PCR SuperMix (TransGen Biotech, Beijing, China), 0.3 μM of each primer, 1 μL of DNA sample, and 10.9 μL double-distilled water. *C. parvum* was also determined using restriction fragment length polymorphism (RFLP) analysis, as previously described [44].

4.4. Sequencing and Phylogenetic Analysis

Positive PCR amplicons were two-directionally sequenced at GENEWIZ (Suzhou, China). Sequences were assembled and edited using DNAstar Lasergene Editseq 7.1.0 (http://www.dnastar.com/), and reference sequences downloaded from the GenBank database were compared to determine the genotype and subtype of *Cryptosporidium* using ClustalX 2.1 (http://www.clustal.org/). The established nomenclature system was used in the naming subtype of *C. parvum* [11]. Phylogenetic analyses were conducted using neighbor-joining methods based on the Kimura-2 parameter model in MEGA 7.0 (http://www.megasoftware.net/). Seven presentative nucleotide sequences obtained in this study were submitted in the GenBank database (https://www.ncbi.nlm.nih.gov/) under the accession numbers: MH442993–MH442996, MT703861, MT703862, and MT724047.

5. Conclusions

Ultimately, Bactrian camels were infected with diverse *Cryptosporidium* species in Xinjiang, northwestern China. Most of these microorganisms have been reported in humans, showing their potential public health relevance and requiring the attention of public health authorities. More molecular studies may be helpful to assess the importance and genetic diversity of *Cryptosporidium* in this region.

Author Contributions: Conceptualization, M.Q. and L.Z.; methodology, M.Q. and L.Z.; validation, M.Q. and L.Z.; formal analysis, Y.C. and Z.C.; investigation, Q.Z., B.J., C.X., and T.W.; software, C.X.; resources, M.Q.; data curation, Z.C.; writing—original draft preparation, Y.C. and Z.C.; writing—review and editing, Y.C. and Z.C.; visualization, M.Q. and L.Z.; supervision, M.Q. and L.Z.; project administration, M.Q. and L.Z.; funding acquisition, M.Q. and L.Z. All authors have read and agreed to the published version of the manuscript.

Funding: This study was supported in part by the National Natural Science Foundation of China (31660712, 31702227), the Program for Young and Middle-aged Leading Science, Technology, and Innovation of Xinjiang Production & Construction Corps (2018CB034), and the Key Technologies R&D Programme of Xinjiang Production & Construction Corps (2020AB025).

Acknowledgments: We are grateful to Md Robiul Karim from Bangabandhu Sheikh Mujibur Rahman Agricultural University for his cordial help in editing the English text of a draft of this manuscript.

Conflicts of Interest: The authors declare no conflict of interest.

References

1. Xiao, L.; Feng, Y. Molecular epidemiologic tools for waterborne pathogens *Cryptosporidium* spp. and *Giardia duodenalis*. *Food Waterborne Parasitol.* **2017**, *8–9*, 14–32. [CrossRef] [PubMed]

2. Efstratiou, A.; Ongerth, J.E.; Karanis, P. Waterborne transmission of protozoan parasites: Review of worldwide outbreaks—An update 2011–2016. *Water Res.* **2017**, *114*, 14–22. [CrossRef] [PubMed]

3. Wang, R.J.; Li, J.Q.; Chen, Y.C.; Zhang, L.X.; Xiao, L.H. Widespread occurrence of *Cryptosporidium* infections in patients with HIV/AIDS: Epidemiology, clinical feature, diagnosis, and therapy. *Acta Trop.* **2018**, *187*, 257–263. [CrossRef] [PubMed]

4. Khan, A.; Shaik, J.S.; Grigg, M.E. Genomics and molecular epidemiology of *Cryptosporidium* species. *Acta Trop.* **2018**, *184*, 1–14. [CrossRef] [PubMed]

5. Liang, N.; Wu, Y.; Sun, M.; Chang, Y.; Lin, X.; Yu, L.; Hu, S.; Zhang, X.; Zheng, S.; Cui, Z.; et al. Molecular epidemiology of *Cryptosporidium* spp. in dairy cattle in Guangdong Province, South China. *Parasitology* **2019**, *146*, 28–32. [CrossRef]

6. Ryan, U.; Fayer, R.; Xiao, L. *Cryptosporidium* species in humans and animals: Current understanding and research needs. *Parasitology* **2014**, *141*, 1667–1685. [CrossRef]

7. Feng, Y.; Ryan, U.M.; Xiao, L. Genetic diversity and population structure of *Cryptosporidium*. *Trends Parasitol.* **2018**, *34*, 997–1011. [CrossRef]

8. Sazmand, A.; Joachim, A. Parasitic diseases of camels in Iran (1931–2017)—A literature review. *Parasite* **2017**, *24*, 21. [CrossRef]

9. Sazmand, A.; Rasooli, A.; Nouri, M.; Hamidinejat, H.; Hekmatimoghaddam, S. Prevalence of *Cryptosporidium* spp. in Camels and involved people in Yazd Province, Iran. *Iran. J. Parasitol.* **2012**, *7*, 80–84.

10. Zhang, Q.; Zhang, Z.; Ai, S.; Wang, X.; Zhang, R.; Duan, Z. *Cryptosporidium* spp., *Enterocytozoon bieneusi*, and *Giardia duodenalis* from animal sources in the Qinghai-Tibetan Plateau Area (QTPA) in China. *Comp. Immunol. Microbiol. Infect. Dis.* **2019**, *67*, 101346. [CrossRef]

11. Baroudi, D.; Zhang, H.; Amer, S.; Khelef, D.; Roellig, D.M.; Wang, Y.; Feng, Y.; Xiao, L. Divergent *Cryptosporidium parvum* subtype and *Enterocytozoon bieneusi* genotypes in dromedary camels in Algeria. *Parasitol. Res.* **2018**, *117*, 905–910. [CrossRef] [PubMed]

12. Gu, Y.; Wang, X.; Zhou, C.; Li, P.; Xu, Q.; Zhao, C.; Liu, W.; Xu, W. Investigation on *Cryptosporidium* infections in wild animals in zoo in Anhui Province. *J. Zoo Wildl. Med.* **2016**, *47*, 846–854. [CrossRef] [PubMed]

13. Liu, X.; Zhou, X.; Zhong, Z.; Deng, J.; Chen, W.; Cao, S.; Fu, H.; Zuo, Z.; Hu, Y.; Peng, G. Multilocus genotype and subtype analysis of *Cryptosporidium andersoni* derived from a Bactrian camel (Camelus bactrianus) in China. *Parasitol. Res.* **2014**, *113*, 2129–2136. [CrossRef] [PubMed]

14. Wang, R.; Zhang, L.; Ning, C.; Feng, Y.; Jian, F.; Xiao, L.; Lu, B.; Ai, W.; Dong, H. Multilocus phylogenetic analysis of *Cryptosporidium andersoni* (Apicomplexa) isolated from a bactrian camel (*Camelus bactrianus*) in China. *Parasitol. Res.* **2008**, *102*, 915–920. [CrossRef]

15. Zahedi, A.; Lee, G.K.C.; Greay, T.L.; Walsh, A.L.; Blignaut, D.J.C.; Ryan, U.M. First report of *Cryptosporidium parvum* in a dromedary camel calf from Western Australia. *Acta Parasitol.* **2018**, *63*, 422–427. [CrossRef]

16. Xiao, L.; Escalante, L.; Yang, C.; Sulaiman, I.; Escalante, A.A.; Montali, R.J.; Fayer, R.; Lal, A.A. Phylogenetic analysis of *Cryptosporidium* parasites based on the small-subunit rRNA gene locus. *Appl. Environ. Microbiol.* **1999**, *65*, 1578–1583. [CrossRef]

17. Kvác, M.; Sak, B.; Kvetonová, D.; Ditrich, O.; Hofmannová, L.; Modrý, D.; Vítovec, J.; Xiao, L. Infectivity, pathogenicity, and genetic characteristics of mammalian gastric *Cryptosporidium* spp. in domestic ruminants. *Vet. Parasitol.* **2008**, *153*, 363–367. [CrossRef]

18. Morgan, U.M.; Xiao, L.; Monis, P.; Sulaiman, I.; Pavlasek, I.; Blagburn, B.; Olson, M.; Upton, S.J.; Khramtsov, N.V.; Lal, A. Molecular and phylogenetic analysis of *Cryptosporidium muris* from various hosts. *Parasitology* **2000**, *120*, 457–464. [CrossRef]

19. El-Alfy, E.S.; Abu-Elwafa, S.; Abbas, I.; Al-Araby, M.; Al-Kappany, Y.; Umeda, K.; Nishikawa, Y. Molecular screening approach to identify protozoan and trichostrongylid parasites infecting one-humped camels (*Camelus dromedarius*). *Acta Trop.* **2019**, *197*, 105060. [CrossRef]

20. Abdelwahab, A.; Abdelmaogood, S. Identification of *Cryptosporidium* species infecting camels (*Camelus dromedarius*) in Egypt. *J. Am. Sci.* **2011**, *7*, 609–612.

21. Sazmand, A.; Joachim, A.; Otranto, D. Zoonotic parasites of dromedary camels: So important, so ignored. *Parasites Vectors* **2019**, *12*, 610. [CrossRef] [PubMed]

22. Zhu, S.; Zimmerman, D.; Deem, L. A review of zoonotic pathogens of Dromedary Camels. *Ecohealth* **2019**, *16*, 356–377. [CrossRef] [PubMed]

23. Leoni, F.; Amar, C.; Nichols, G.; Pedraza-Díaz, S.; McLauchlin, J. Genetic analysis of *Cryptosporidium* from 2414 humans with diarrhoea in England between 1985 and 2000. *J. Med. Microbiol.* **2006**, *55*, 703–707. [CrossRef] [PubMed]

24. Jiang, Y.; Ren, J.; Yuan, Z.; Liu, A.; Zhao, H.; Liu, H.; Chu, L.; Pan, W.; Cao, J.; Lin, Y.; et al. *Cryptosporidium andersoni* as a novel predominant *Cryptosporidium* species in outpatients with diarrhea in Jiangsu Province, China. *BMC Infect. Dis.* **2014**, *14*, 555. [CrossRef] [PubMed]

25. Cui, Z.; Wang, R.; Huang, J.; Wang, H.; Zhao, J.; Luo, N.; Li, J.; Zhang, Z.; Zhang, L. Cryptosporidiosis caused by *Cryptosporidium parvum* subtype IIdA15G1 at a dairy farm in Northwestern China. *Parasites Vectors* **2014**, *7*, 529. [CrossRef] [PubMed]

26. Qi, M.; Wang, H.; Jing, B.; Wang, D.; Wang, R.; Zhang, L. Occurrence and molecular identification of *Cryptosporidium* spp. in dairy calves in Xinjiang, Northwestern China. *Vet. Parasitol.* **2015**, *212*, 404–407. [CrossRef]

27. Thompson, R.C.A.; Ash, A. Molecular epidemiology of *Giardia* and *Cryptosporidium* infections. *Infect. Genet. Evol.* **2016**, *40*, 315–323. [CrossRef]

28. Jian, F.; Liu, A.; Wang, R.; Zhang, S.; Qi, M.; Zhao, W.; Shi, Y.; Wang, J.; Wei, J.; Zhang, L.; et al. Common occurrence of *Cryptosporidium hominis* in horses and donkeys. *Infect. Genet. Evol.* **2016**, *43*, 261–266. [CrossRef]

29. Laatamna, A.E.; Wagnerová, P.; Sak, B.; Květoňová, D.; Xiao, L.; Rost, M.; McEvoy, J.; Saadi, A.R.; Aissi, M.; Kváč, M. Microsporidia and *Cryptosporidium* in horses and donkeys in Algeria: Detection of a novel *Cryptosporidium hominis* subtype family (Ik) in a horse. *Vet. Parasitol.* **2015**, *208*, 135–142. [CrossRef]

30. Lebbad, M.; Winiecka-Krusnell, J.; Insulander, M.; Beser, J. Molecular characterization and epidemiological investigation of *Cryptosporidium hominis* IkA18G1 and *C. hominis* monkey genotype IiA17, two unusual subtypes diagnosed in Swedish patients. *Exp. Parasitol.* **2018**, *188*, 50–57. [CrossRef]

31. Liu, X.; Xie, N.; Li, W.; Zhou, Z.; Zhong, Z.; Shen, L.; Cao, S.; Yu, X.; Hu, Y.; Chen, W.; et al. Emergence of *Cryptosporidium hominis* monkey genotype II and novel subtype family Ik in the Squirrel Monkey (*Saimiri sciureus*) in China. *PLoS ONE* **2015**, *10*, e0141450. [CrossRef]

32. Li, N.; Xiao, L.; Alderisio, K.; Elwin, K.; Cebelinski, E.; Chalmers, R.; Santin, M.; Fayer, R.; Kvac, M.; Ryan, U. Subtyping *Cryptosporidium ubiquitum,* a zoonotic pathogen emerging in humans. *Emerg. Infect. Dis.* **2014**, *20*, 217–224. [CrossRef] [PubMed]

33. Huang, J.; Zhang, Z.; Zhang, Y.; Yang, Y.; Zhao, J.; Wang, R.; Jian, F.; Ning, C.; Zhang, W.; Zhang, L. Prevalence and molecular characterization of *Cryptosporidium* spp. and *Giardia duodenalis* in deer in Henan and Jilin, China. *Parasites Vectors* **2018**, *11*, 239. [CrossRef]

34. Majeed, Q.A.H.; El-Azazy, O.M.E.; Abdou, N.M.I.; Al-Aal, Z.A.; El-Kabbany, A.I.; Tahrani, L.M.A.; AlAzemi, M.S.; Wang, Y.; Feng, Y.; Xiao, L. Epidemiological observations on cryptosporidiosis and molecular characterization of *Cryptosporidium* spp. in sheep and goats in Kuwait. *Parasitol. Res.* **2018**, *117*, 1631–1636. [CrossRef] [PubMed]

35. Kváč, M.; Vlnatá, G.; Ježková, J.; Horčičková, M.; Konečný, R.; Hlásková, L.; McEvoy, J.; Sak, B. *Cryptosporidium occultus* sp. n. (*Apicomplexa*: *Cryptosporidiidae*) in rats. *Eur. J. Protistol.* **2018**, *63*, 96–104. [CrossRef] [PubMed]

36. Li, F.; Zhang, Z.; Hu, S.; Zhao, W.; Zhao, J.; Kváč, M.; Guo, Y.; Li, N.; Feng, Y.; Xiao, L. Common occurrence of divergent *Cryptosporidium* species and *Cryptosporidium parvum* subtypes in farmed bamboo rats (*Rhizomys sinensis*). *Parasites Vectors* **2020**, *13*, 149. [CrossRef]

37. Ma, J.; Li, P.; Zhao, X.; Xu, H.; Wu, W.; Wang, Y.; Guo, Y.; Wang, L.; Feng, Y.; Xiao, L. Occurrence and molecular characterization of *Cryptosporidium* spp. and *Enterocytozoon bieneusi* in dairy cattle, beef cattle and water buffaloes in China. *Vet. Parasitol.* **2015**, *207*, 220–227. [CrossRef]

38. Zhang, Q.; Li, J.; Li, Z.; Xu, C.; Hou, M.; Qi, M. Molecular identification of *Cryptosporidium* spp. in alpacas (*Vicugna pacos*) in China. *Int. J. Parasitol. Parasites Wildl.* **2020**, *12*, 181–184. [CrossRef]

39. Xu, N.; Liu, H.; Jiang, Y.; Yin, J.; Yuan, Z.; Shen, Y.; Cao, J. First report of *Cryptosporidium viatorum* and *Cryptosporidium occultus* in humans in China, and of the unique novel *C. viatorum* subtype XVaA3h. *BMC Infect. Dis.* **2020**, *20*, 16. [CrossRef]

40. Ong, C.S.; Eisler, D.L.; Alikhani, A.; Fung, V.W.; Tomblin, J.; Bowie, W.R.; Isaac-Renton, J.L. Novel *Cryptosporidium* genotypes in sporadic cryptosporidiosis cases: First report of human infections with a cervine genotype. *Emerg. Infect. Dis.* **2002**, *8*, 263–268. [CrossRef]

41. Robinson, G.; Chalmers, R.M.; Stapleton, C.; Palmer, S.R.; Watkins, J.; Francis, C.; Kay, D. A whole water catchment approach to investigating the origin and distribution of *Cryptosporidium* species. *J. Appl. Microbiol.* **2011**, *111*, 717–730. [CrossRef] [PubMed]

42. Jiang, J.; Alderisio, K.A.; Xiao, L. Distribution of *Cryptosporidium* genotypes in storm event water samples from three watersheds in New York. *Appl. Environ. Microbiol.* **2005**, *71*, 4446–4454. [CrossRef] [PubMed]

43. Alves, M.; Xiao, L.; Sulaiman, I.; Lal, A.A.; Matos, O.; Antunes, F. Subgenotype analysis of *Cryptosporidium* isolates from humans, cattle, and zoo ruminants in Portugal. *J. Clin. Microbiol.* **2003**, *41*, 2744–2747. [CrossRef]

44. Xiao, L.; Morgan, U.M.; Limor, J.; Escalante, A.; Arrowood, M.; Shulaw, W.; Thompson, R.C.; Fayer, R.; Lal, A.A. Genetic diversity within *Cryptosporidium parvum* and related *Cryptosporidium* species. *Appl. Environ. Microbiol.* **1999**, *65*, 3386–3391. [CrossRef] [PubMed]

Publisher's Note: MDPI stays neutral with regard to jurisdictional claims in published maps and institutional affiliations.

Article

Occurrence and Multi-Locus Analysis of *Giardia duodenalis* in Coypus (*Myocastor coypus*) in China

Zhaohui Cui [1,2], Deguo Wang [1], Wen Wang [2], Ying Zhang [2], Bo Jing [2], Chunyan Xu [2], Yuanchai Chen [3], Meng Qi [2,*] and Longxian Zhang [3,*]

[1] Key Laboratory of Biomarker Based Rapid-Detection Technology for Food Safety of Henan Province, Food and Pharmacy College, Xuchang University, Xuchang 461000, China; czh9008@xcu.edu.cn (Z.C.); wangdg666@126.com (D.W.)

[2] College of Animal Science, Tarim University, Alar 843300, China; 18699708588@163.com (W.W.); yingzhang782@gmail.com (Y.Z.); 120050010@taru.edu.cn (B.J.); 120170015@taru.edu.cn (C.X.)

[3] College of Veterinary Medicine, Henan Agricultural University, Zhengzhou 450046, China; chenyc217@gmail.com

* Correspondence: qimengdz@163.com (M.Q.); zhanglx8999@henau.edu.cn (L.Z.)

Abstract: *Giardia duodenalis* is a major gastrointestinal parasite found globally in both humans and animals. This work examined the occurrence of *G. duodenalis* in coypus (*Myocastor coypus*) in China. Multi-locus analysis was conducted to evaluate the level of genetic variation and the potential zoonotic role of the isolates. In total, 308 fecal samples were collected from seven farms in China and subjected to PCR screening to reveal *G. duodenalis*. Notably, *G. duodenalis* was detected in 38 (12.3%) specimens from assemblages A (n = 2) and B (n = 36). Positive samples were further characterized by PCR and nucleotide sequencing of the triose phosphate isomerase (*tpi*), beta giardin (*bg*), and glutamate dehydrogenase (*gdh*) genes. Multi-locus genotyping yielded 10 novel multi-locus genotypes (MLGs) (one MLG and nine MLGs for assemblages A and B, respectively). Based on the generated phylogenetic tree, AI–novel 1 clustered more closely with MLG AI-2. Furthermore, within the assemblage B phylogenetic analysis, the novel assemblage B MLGs were identified as BIV and clustered in the MLG BIV branch. This is the first report of *G. duodenalis* in coypus in China. The presence of zoonotic genotypes and subtypes of *G. duodenalis* in coypus suggests that these animals can transmit human giardiasis.

Keywords: *Giardia duodenalis*; coypus (*Myocastor coypus*); multi-locus genotype; genetic variation; zoonotic genotypes; PCR (polymerase chain reaction); China

Citation: Cui, Z.; Wang, D.; Wang, W.; Zhang, Y.; Jing, B.; Xu, C.; Chen, Y.; Qi, M.; Zhang, L. Occurrence and Multi-Locus Analysis of *Giardia duodenalis* in Coypus (*Myocastor coypus*) in China. *Pathogens* **2021**, *10*, 179. https://doi.org/10.3390/pathogens10020179

Academic Editors: David Carmena, David González-Barrio and Pamela Carolina Köster

Received: 29 December 2020
Accepted: 4 February 2021
Published: 7 February 2021

Publisher's Note: MDPI stays neutral with regard to jurisdictional claims in published maps and institutional affiliations.

1. Introduction

Giardia duodenalis (syn. *G. lamblia*, *G. intestinalis*) is a flagellate protozoan parasite recognized as a significant global contributor to diarrheal disease, affecting humans, domestic animals, and wildlife across the globe [1,2]. The majority of *G. duodenalis* infections are asymptomatic; however, in rare cases, some patients may experience severe gastrointestinal disturbances for several weeks [3]. As *G. duodenalis* utilizes the fecal–oral route for lifecycle maintenance, projections indicate that this parasite causes ~28.2 million foodborne disease cases [4,5]. Based on the above data, the United Nations Food and Agriculture Organization (FAO) and the World Health Organization (WHO) in 2014 ranked *Giardia* 11th of 24 food-borne parasites [4].

The current wide use of genotyping tools has immensely improved our understanding of *G. duodenalis* transmission in humans and animals [5,6]. At least eight genotypes or assemblages have been described, including assemblages A and B containing zoonotic isolates potentially infecting humans and animals, and assemblages C–H, which exhibit specificity to particular animal hosts [7]. Moreover, several molecular markers (triosephosphate isomerase, *tpi*; glutamate dehydrogenase, *gdh* and beta giardin, *bg*) have been developed to

175

create a multi-locus genotyping (MLG) tool for subtyping assemblages A, B, and E and to explore the population genetic characterizations of *G. duodenalis* [6,8]. Subsequently, the MLG tool subdivided assemblage A into sub-assemblages AI, AII, and AIII and assemblage B into sub-assemblages BIII and BIV [9].

In recent years, studies on the epidemiology of *G. duodenalis* have been conducted for humans, non-human primates, ruminants, companion animals, domestic animals, wildlife, and in the environment in China [10]. However, limited information has been provided on the infection rate and genotype characteristics in rodents in China. The coypus (*Myocastor coypus*) is a large, amphibious rodent native to South America, which has become invasive in Europe and other parts of the world except Oceania and Antarctica [11]. Coypus were first introduced to China in 1956, then later widely reared in farms as important fur-bearing animals. The climate of China is very suitable for the growth of coypus, and 16 color-type strains have been bred. The number of the national stock reached more than 400,000 in 2000. To date, little is known about the genetic characteristics of *G. duodenalis* in coypus globally. Only two studies in Italy and the USA reported *Giardia* spp. prevalence in coypus, but neither identified the species [12,13]. Thus, the present study aimed to explore the distribution and genetic diversity of *G. duodenalis* in coypus in China and assess its zoonotic potential based on MLG analysis.

2. Results

2.1. Occurrence of G. duodenalis

All samples were initially tested using nested PCR amplification of the small subunit ribosomal RNA (SSU rRNA) gene. Of the 308 samples, *G. duodenalis* was present in 38 samples (12.3%, 95% CI: 8.5–16.2%) (Table 1). Each examined farm had infected animals. Notably, the highest infection rate of *G. duodenalis* in coypus was detected in Baoding (28.6%, 10/35), followed by Ganzhou (25.7%, 9/35), Chengdu (15.0%, 6/40), Laibin (13.6%, 3/22), Yongzhou (13.0%, 3/23), Kaifeng (11.5%, 6/52), and Anyang (1.0%, 1/101) (Table 1).

Table 1. Distribution of *G. duodenalis* assemblages in coypus from different farms in China.

Location	Age (Month)	N/T (%; 95% CI)	Assemblage (No.)	SSU rRNA (No.)	*tpi* (No.)	*gdh* (No.)	*bg* (No.)
Baoding, Hebei	<3	2/13 (15.4; 0–38.8)	B (2)	B (2)	B (1)	B (1)	B (1)
	3–6	5/10 (50.0; 14.0–85.9)	B (5)	B (5)	B (4)	B (2)	B (3)
	>6	3/12 (25.0; 0–53.7)	B (3)	B (3)	B (1)	B (1)	B (1)
	Subtotal	10/35 (28.6; 12.2–44.9)	B (10)				
Anyang, Henan	<3	0/52					
	3–6	0/10					
	>6	1/39 (2.6; 0–8.8)	B (1)	B (1)			
	Subtotal	1/101 (1.0; 0–3.4)	B (1)				
Kaifeng, Henan	3–6	2/27 (7.4; 0–19.1)	B (2)	B (2)	B (1)		
	>6	4/25 (16.0; 0–32.4)	B (4)	B (4)	B (2)		
	Subtotal	6/52 (11.5; 1.9–21.2)	B (6)				
Chengdu, Sichuan	>6	6/40 (15.0; 2.7–27.3)	B (6)	B (6)	B (6)	B (3)	B (3)
Yongzhou, Hunan	>6	3/23 (13.0; 0–28.9)	B (3)	B (3)	B (1)		
Ganzhou, Jiangxi	>6	9/35 (25.7; 9.8–41.6)	B (9)	B (9)	B (6)	B (3)	B (3)
Laibin, Guangxi	>6	3/22 (13.6; 0–30.2)	A (2), A + B (1)	A (2), B (1)	A (3)	A (1)	A (1)
Total		38/308 (12.3; 8.5–16.2)	A (2), B (35), A + B (1)				

N = number of positives for *G. duodenalis*; No. = number of samples; T = total analyzed samples.

By age, the highest infection rate was reported in the 3–6-month-old group (14.9%, 7/47), followed by >6-month-old group (14.8%, 29/196), and <3-month-old group (3.1%, 2/65) (Table 2). Furthermore, the correlation of age with the infection rates was evaluated based on the calculated ORs and 95% CI values (Table 2). There was a strong positive correlation between the infection rate and age, with an OR of 5.51 (95% CI: 1.09–27.87, $p = 0.023$) associated with the 3–6-month-old group and an OR of 5.47 (95% CI: 1.27–23.59, $p = 0.011$) associated with the >6-month-old group.

Table 2. Distribution of *G. duodenalis* assemblages in coypus of different ages.

Age (Month)	N/T (%; 95% CI)	Assemblage (No.)	*p*-Value	OR (95% CI)
<3	2/65 (3.1; 0–8.0)	B (2)		1
3–6	7/47 (14.9; 3.7–26.1)	B (7)	0.023	5.51 (1.09–27.87)
>6	29/196 (14.8; 9.6–20.0)	A (2), B (26), A + B (1)	0.011	5.47 (1.27–23.59)

N = number of positives for *G. duodenalis*; OR: odds ratio; T = total of analyzed samples.

2.2. Assemblage A and B Subtypes

Here, two genotypes, assemblages A (2) and B (35), were identified based on sequence analysis of the SSU rRNA, *tpi*, *gdh*, and *bg* loci (Table 1). Notably, assemblage B was the dominant genotype (92.1%, 35/38). Mixed infection was found in one sample. To reveal the genetic diversity of the *G. duodenalis*-positive samples, we sequenced the *tpi*, *gdh*, and *bg* genes, from which 25, 11, and 12 sequences were obtained, respectively (Table 3).

Table 3. Multi-locus characterization of *G. duodenalis* isolates in coypus in China based on *bg*, *gdh* and *tpi* genes.

Isolate Code	*tpi*	*gdh*	*bg*	MLG Type
118	B1 (MW321619)	PN	PN	
126	A * (MW321620)	A5 (MW321623)	A5 (MW321613)	AI-novel 1
127	A *	PN	PN	
128	A *	PN	PN	
147	B1	B3 (MW321626)	B1 (MW321614)	BIV-novel 1
150	B1	PN	B1	
151	B2 (MW321621)	PN	PN	
155, 256	B1	B2 (MW321625)	B2 (MW321615)	BIV-novel 2
157	B2	B1 (MW321624)	B1	BIV-novel 3
171	B1	B3	B2	BIV-novel 4
182, 184	B1	B1	B3 (MW321616)	BIV-novel 5
189	B1	PN	PN	
197	B3 (MW321622)	PN	B3	
210	B3	B1	PN	
217	B1	PN	PN	
225	B1	PN	PN	
233	B2	B1	B4 (MW321617)	BIV-novel 6
245	B1	PN	PN	
248	B1	B2	B5 (MW321618)	BIV-novel 7
249	B1	PN	PN	
283	B1	PN	PN	
287	B1	PN	PN	
302	B1	PN	PN	

* New variants without heterogeneous positions. MLG: multi-locus genotypes; PN: PCR negative.

Of the *tpi* sequences, 3 and 22 belonged to assemblages A and B, respectively. Sequences from all isolates from assemblage A exhibited two single-nucleotide polymorphisms (G155T and G186A) relative to the MN174855 sequence. Within the assemblage B isolates, three subtypes were formed, which were designated as B1 ($n = 17$), B2 ($n = 3$), and

B3 (*n* = 2) for convenience. Of note, B1, B2, and B3 sequences were identical to MH644772, KM977653, and HM140711, respectively. At the *gdh* locus, 1 and 10 samples were found positive and identified as assemblages A and B, respectively. The one assemblage A sequence was identical to MN174853. Among the assemblage B sequences, B1 (*n* = 5), B2 (*n* = 3), and B3 (*n* = 2) sequences were identical to KM977648, MK952603, and MK982476, respectively. Sequence analysis demonstrated high genetic diversity in assemblage B at the *bg* locus. Ultimately, five subtypes (B1–B5) were formed in the 11 assemblage B sequences. The subtypes B1 (*n* = 4), B2 (*n* = 2), B3 (*n* = 3), B4 (*n* = 1), and B5 (*n* = 1) exhibited consistency with MT487587, MK982544, MN174847, MT487587, and KY696837, respectively. Moreover, one assemblage A sequence was identical to MN704938.

2.3. MLG and Phylogenetic Analysis

Collectively, sequence data sets from the three loci were available from 10 isolates. Multi-locus genotyping yielded 10 novel MLGs (one MLG for assemblage A and nine MLGs for assemblage B) (Table 3). The single MLG in assemblage A was identified as AI–novel 1, whereas the nine MLGs in assemblage B were designated from BIV–novel 1 to BIV–novel 7. Phylogenetic relationships of MLGs in assemblages A and B with the reference genotypes are illustrated in Figure 1. Based on the phylogenetic tree, AI–novel 1 identified in the present study clustered more closely to MLG AI-2 (Figure 1A). Within the assemblage B phylogenetic analysis, all MLGs in assemblage B were identified as BIV and clustered in the MLG BIV branch (Figure 1B).

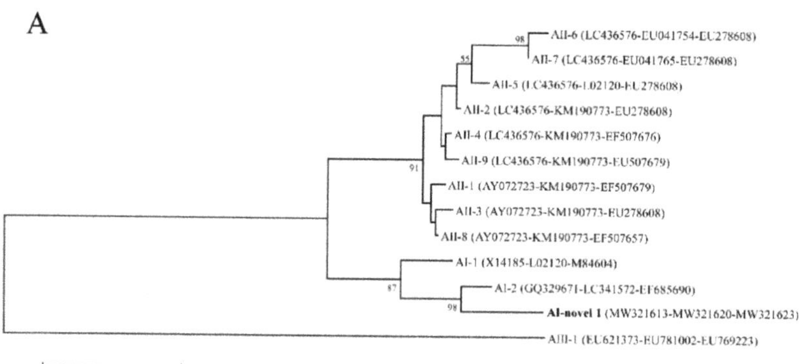

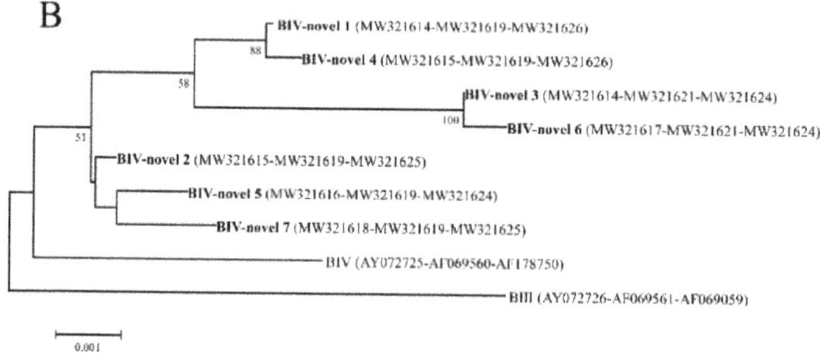

Figure 1. Nucleotide neighbor-joining trees based on concatenated datasets for *bg*, *tpi*, and *gdh* gene sequences of *G. duodenalis* assemblage (**A**,**B**) isolates obtained in this study and sequences retrieved from the GenBank database. Bootstrap values greater than 50% from 1000 replicates are shown on nodes. The bold texts represent the isolates of this study.

3. Discussion

G. duodenalis is very commonly found in humans and domestic animals, as revealed by numerous prevalence studies across the globe [7,10,14]. The molecular epidemiology of *G. duodenalis* has been widely studied in livestock, which revealed its transmission dynamics and zoonotic significance in these animals. However, the parasite has not been extensively investigated in rodents; therefore, very little knowledge on the distribution, genetic diversity, and zoonotic potential of *Giardia* spp. in these animal hosts has been published [6,7,15,16]. Reports have demonstrated that *G. duodenalis* infections in rodents in Australia, Belgium, China, Croatia, Germany, Malaysia, Poland, Romania, Spain, Sweden, and the USA (Table 4) have prevalence rates ranging from 1.2% to 100% [12–34]. Herein, molecular analysis of 308 fecal samples collected from coypus in six provinces in China confirmed a *G. duodenalis* prevalence of 12.3% (38/308). To the best of our knowledge, this is the first molecular study of *G. duodenalis* infections in coypus, except for two studies in Italy and the USA that assessed *Giardia* spp. infection in coypus with the prevalence of 0% (0/153) and 73.3% (22/30), respectively, using an immunoenzymatic assay [12,13].

Table 4. *Giardia duodenalis* infection rates and genotypes in rodents worldwide.

Animal	Location	Positive % (N/T)	Assemblage	Sub-Assemblage	Reference
Ash-grey mouse (*Pseudomys albocinereus*)	Australia	-	E	-	[17]
Asian house rats (*Rattus tanezumi*)	China	6.1 (2/33)	G	-	[18]
Bamboo rat (*Rhizomys sinensis*)	China	10.8 (42/480)	B	-	[19]
Bank vole (*Myodes glareolus*)	Germany	1.3 (4/301)	A, B	-	[20]
Bank vole (*Myodes glareolus*)	Poland	58.3 (849/1457)	-	-	[21]
Beaver (*Castor canadensis*)	USA	33.3 (30/100)	-	-	[13]
Beaver (*Castor canadensis*)	USA	-	B	-	[22]
Black rat (*Rattus rattus*)	Spain	36.2 (42/116)	G	GI, GII	[23]
Brown rats (*Rattus norvegicus*)	Malaysia	3.0 (4/134)	B	-	[24]
Brown rats (*Rattus norvegicus*)	China	6.6 (11/168)	G	-	[18]
Bush rat (*Rattus fuscipes*)	Australia	-	F + C	-	[17]
Chinchillas (*Chinchilla lanigera*)	Belgium	27.5 (22/80)	A, B, C, E	AI, AII, BIV	[25]
Chinchillas (*Chinchilla lanigera*)	China	27.1 (38/140)	A, B	AI, BIV	[26]
Chinchillas (*Chinchilla lanigera*)	Germany	61.4 (326/531)	A, B, D	AI, BIV	[27]
Chinchillas (*Chinchilla lanigera*)	Italy	29.8 (31/104)	B, C	-	[28]
Chinchillas (*Chinchilla lanigera*)	Romania	55.7 (190/341)	B, D, E	BIII, BIV	[29]
Chipmunks (*Eutamias asiaticus*)	China	8.6 (24/279)	A, G	AI, GI, GII	[30]
Common vole (*Microtus arvalis*)	Poland	74.2 (302/407)	-	-	[21]
Coypus (*Myocastor coypus*)	China	12.3 (38/308)	A, B	AI, BIV	This study
Coypus (*Myocastor coypus*)	Italy	0 (0/153)	-	-	[12]
Nutria (*Myocastor coypus*)	USA	73.3 (22/30)	-	-	[13]
Deer mice (*Peromyscus maniculatus*)	USA	25.5 (53/208)	-	-	[31]
Eurasian field mice (*Apodemus* sp.)	Germany	1.2 (1/82)	A	-	[20]
House mice (*Mus musculus*)	China	3.2 (1/31)	G	-	[18]
House mouse (*Mus musculus domesticus*)	Spain	17.6 (29/165)	G	-	[23]
Muskrat (*Ondatra zibethicus*)	Romania	100 (1/1)	C	-	[32]
Muskrat (*Ondatra zibethicus*)	USA	-	B	-	[22]
Norway rat (*Rattus norvegicus*)	Spain	66.7 (2/3)	B	-	[23]
Prevost's squirrel (*Callosciurus prevosti*)	Croatia	-	B	-	[33]
Patagonian cavy (*Docilchotis patagonum*)	Croatia	-	B	BIV	[33]
Rat	Sweden	-	G	-	[34]
Yellow-necked mouse (*Apodemus flavicollis*)	Poland	24.4 (150/616)	-	-	[21]

N: number of positives for *G. duodenalis*; T: total of analyzed samples; "-" indicates not available.

Based on the current knowledge, eight *Giardia* species are considered valid, including *G. duodenalis*, *Giardia agilis*, *Giardia ardeae*, *Giardia psittaci*, *Giardia muris*, *Giardia microti*, *Giardia peramelis*, and *Giardia cricetidarum* [4,7]. In particular, two host-adapted species of *Giardia* spp. have been detected in rodents, including *G. muris* and *G. microti* [20,35]. To date, in China, *G. duodenalis* has been reported in rodents of several genera, including Chinchillas (*Chinchilla lanigera*), Asian house rats (*Rattus tanezumi*), brown rats (*Rattus norvegicus*), house mice (*Mus musculus*), chipmunks (*Eutamias asiaticus*), and bamboo rats (*Rhizomys sinensis*) [18,19,26,30]. In previous work, seven assemblages of *G. duodenalis* were found in rodents, including A, B, C, D, E, F, and G (Table 4). Both zoonotic and rodent-specific assemblages A, B, and G of *G. duodenalis* have been detected in rodents in China. Herein, the assemblages A and B were identified; assemblage B was the predominant genotype in coypus. It is worth noting that *G. duodenalis* assemblages A and B are important human pathogens; of them, assemblage B is more commonly reported in Asia, Oceania, Europe, and Africa than assemblage A [5,15,36]. We also reported mixed infections in the present study. Of note, the use of assemblage-specific primers in addition to MLST data is advocated in molecular epidemiological surveys, as mixed infections are likely to be underestimated [15,37]. Mixed infections are likely to be one of the main reasons for the adoption of multiple genetic markers in identifying distinct assemblages in the same sample [8,15]. Consequently, further studies on the molecular epidemiology of *G. duodenalis* is warranted to improve our understanding of the genetic diversity in rodents in China.

In recent studies, the MLG tool based on sequence analysis of the *gdh*, *bg*, and *tpi* loci represents a more informative approach for genotyping and elucidating the characteristics of *G. duodenalis* in different hosts, in addition to its zoonotic potential [9,15,19]. This has permitted the identification of sub-assemblages within assemblage A and B: AI to AIV and BI to BIV. Moreover, there are differences in the distribution of these sub-assemblages among hosts; for instance, human *Giardia* isolates mainly belong to sub-assemblage AII but also AI, whereas animals harbor AI, AII, and AIII [15]. In this study, 10 *G. duodenalis* isolates from coypus were successfully sequenced at all three gene loci; the sequence analysis of which resulted in one MLG in assemblage A and nine MLGs in assemblage B. Based on the phylogenetic analysis, the one MLG in assemblage A (AI–novel 1) was clustered with sub-assemblage AI-2, whereas the nine remaining MLGs in assemblage B (BIV–novel 1 to 7) were clustered with sub-assemblage BIV. Sub-assemblage AI was previously reported in various hosts in China, including humans, cattle, goats, sheep, dogs, cats, pigs, chinchillas, and chipmunks [6,10]. Other researchers also found sub-assemblage BIV in humans and non-human primates [6,10]. Additionally, compelling evidence on zoonotic transmission of *G. duodenalis* assemblage AI was revealed by epidemiological data showing a highly significant association between *Giardia* infection of schoolchildren and the presence of *Giardia*-positive dogs in the same household in Mexico [38]. Therefore, further studies into the epidemiology of *G. duodenalis* in handlers/workers on farms should be initiated to address the zoonotic potential of *G. duodenalis* in coypus.

4. Materials and Methods

4.1. Ethics Statement

This study was performed with strict adherence to the recommendations of the Guide for the Care and Use of Laboratory Animals of the Ministry of Health, China. The research protocol was reviewed and approved by the Research Ethics Committee of Tarim University (approval no. ECTU 2018-0026). Farm owners gave permission before we commenced fecal sample collection.

4.2. Study Area and Sample Collection

A total of 308 fecal samples were collected during autumn, winter, and spring of 2018 and 2019 from seven farms in Hebei Province (Baoding), Henan Province (Anyang and Kaifeng), Sichuan Province (Chengdu), Hunan Province (Yongzhou), Jiangxi Province

(Ganzhou) and Guangxi Zhuang Autonomous Region (Laibin) in China (Table 1). The breeding scale of coypus was about 500–2000 in each farm, and the breeding conditions of each farm were basically the same. Approximately 10–15% of coypus representing each age group were investigated at each farm. All the fecal samples were collected immediately after excretion and placed in a plastic container; we recorded the date, site, age, and health condition at collection time. All samples were shipped to the laboratory in a cooler with ice packs within 48 h and stored at 4 °C.

4.3. DNA Extraction and Nested PCR Analysis

The genomic DNA sample was extracted from approximately 200 mg of each fecal specimen using a commercial E.Z.N.A® Stool DNA Kit (Omega Bio-Tek Inc., Norcross, GA, USA) following the protocol stipulated by the manufacturer. The quantity of nucleic acid in samples was photometrically estimated at OD_{260} and stored at −20 °C for subsequent molecular analysis.

Nested PCR amplification of the SSU rRNA gene was employed to screen *G. duodenalis* infection [39]. The *G. duodenalis*-positive samples were further analyzed via nested PCRs of gene loci *tpi*, *gdh*, and *bg* (see Table S1 and Figure S1) [8,22,40]. Samples positive for all four loci were used to assess the MLGs of *G. duodenalis*.

4.4. Sequencing and Phylogenetic Analysis

Bi-directional sequencing of all the secondary PCR products from SSU rRNA, *tpi*, *gdh*, and *bg* loci was performed at the biotechnology company, GENEWIZ (Suzhou, China). The sequence assembly and editing were performed with the DNAstar Lasergene Editseq 7.1.0. Multiple-sequence alignment analysis was employed on the obtained and GenBank reference sequences using software ClustalX 2.1 to ascertain the genotypes and subtypes of *G. duodenalis*. We concatenated sequences for each positive isolate to obtain a multi-locus sequence (*bg*, *tpi*, *gdh*) in accordance with a previous report [41]. Phylogenetic analyses of the concatenated MLG sequences were achieved using neighbor-joining methods based on the Kimura-2 parameter model in MEGA 7.0.

4.5. Statistical Analysis

The differences in infection rates among locations and age groups were compared with the $\chi 2$ test performed in SPSS 18. At the level of *p* value < 0.05, the differences were appraised as significant.

4.6. Nucleotide Sequence Accession Numbers

Here, the representative nucleotide sequences of this study were deposited in the GenBank database under the following accession numbers: MW322754–MW322755, MW321613–MW321626.

5. Conclusions

This work presents the first report of *G. duodenalis* infections in coypus in China based on MLG analysis. Multi-locus genotyping yielded 10 novel MLGs, including one assemblage A MLG (AI–novel 1) and nine assemblage B MLGs (BIV–novel 1 to 7). Of note, the presence of zoonotic assemblages and sub-assemblages of *G. duodenalis* in coypus suggests the potential contribution of these animals to human giardiasis transmission.

Supplementary Materials: The following are available online at https://www.mdpi.com/2076-081 7/10/2/179/s1, Table S1 Primer sequences and reaction conditions used in nested PCR amplifications. Figure S1. Genomic map of the positions of *tpi*, *gdh* and *bg* genes.

Author Contributions: Conceptualization, M.Q. and L.Z.; methodology, M.Q. and L.Z.; validation, M.Q. and L.Z.; formal analysis, Z.C. and D.W.; investigation, W.W., Y.Z., B.J. and C.X.; software, C.X. and Y.C.; resources, M.Q.; data curation, Z.C.; writing—original draft preparation, Z.C.; writing— review and editing, M.Q. and Z.C.; visualization, M.Q. and L.Z.; supervision, M.Q. and L.Z.; project administration, M.Q. and L.Z.; funding acquisition, M.Q. All authors have read and agreed to the published version of the manuscript.

Funding: This research was funded by the Program for Young and Middle-Aged Leading Science, Technology, and Innovation of Xinjiang Production & Construction Corps (2018CB034).

Institutional Review Board Statement: Not applicable.

Informed Consent Statement: Not applicable.

Acknowledgments: We are grateful to Robiul Karim from Bangabandhu Sheikh Mujibur Rahman Agricultural University for his cordial help in editing the English text of a draft of this manuscript.

Conflicts of Interest: The authors declare no conflict of interest.

References

1. Simonato, G.; Danesi, P.; Frangipane di Regalbono, A.; Dotto, G.; Tessarin, C.; Pietrobelli, M.; Pasotto, D. Surveillance of zoonotic parasites in animals involved in animal-assisted interventions (AAIs). *Int. J. Environ. Res. Public Health* **2020**, *17*, 7914. [CrossRef]
2. Dixon, B. *Giardia duodenalis* in humans and animals—Transmission and disease. *Res. Vet. Sci.* **2020**. [CrossRef]
3. Fink, M.Y.; Singer, S.M. The intersection of immune responses, microbiota, and pathogenesis in giardiasis. *Trends Parasitol.* **2017**, *33*, 901–913. [CrossRef]
4. Ryan, U.; Hijjawi, N.; Feng, Y.; Xiao, L. *Giardia*: An under-reported foodborne parasite. *Int. J. Parasitol.* **2019**, *49*, 1–11. [CrossRef] [PubMed]
5. Xiao, L.; Feng, Y. Molecular epidemiologic tools for waterborne pathogens *Cryptosporidium* spp. and *Giardia duodenalis*. *Food Waterborne Parasitol.* **2017**, *8–9*, 14–32. [CrossRef] [PubMed]
6. Feng, Y.; Xiao, L. Zoonotic potential and molecular epidemiology of *Giardia* species and giardiasis. *Clin. Rev.* **2011**, *24*, 110–140. [CrossRef] [PubMed]
7. Ryan, U.; Zahedi, A. Molecular epidemiology of giardiasis from a veterinary perspective. *Adv. Parasitol.* **2019**, *106*, 209–254. [PubMed]
8. Cacciò, S.M.; Beck, R.; Lalle, M.; Marinculic, A.; Pozio, E. Multi-locus genotyping of *Giardia duodenalis* reveals striking differences between assemblages A and B. *Int. J. Parasitol.* **2008**, *38*, 1523–1531. [CrossRef] [PubMed]
9. Ankarklev, J.; Lebbad, M.; Einarsson, E.; Franzén, O.; Ahola, H.; Troell, K.; Svärd, S.G. A novel high-resolution multi-locus sequence typing of *Giardia intestinalis* Assemblage a isolates reveals zoonotic transmission, clonal outbreaks and recombination. *Infect. Genet. Evol.* **2018**, *60*, 7–16. [CrossRef] [PubMed]
10. Li, J.; Wang, H.; Wang, R.; Zhang, L. *Giardia duodenalis* infections in humans and other animals in China. *Front. Microbiol.* **2017**, *8*, 2004. [CrossRef] [PubMed]
11. Murphy, W.J.; Eizirik, E.; Johnson, W.E.; Zhang, Y.P.; Ryder, O.A.; O'Brien, S.J. Molecular phylogenetics and the origins of placental mammals. *Nature* **2001**, *409*, 614–618. [CrossRef]
12. Zanzani, S.A.; Di Cerbo, A.; Gazzonis, A.L.; Epis, S.; Invernizzi, A.; Tagliabue, S.; Manfredi, M.T. Parasitic and bacterial infections of Myocastor coypus in a metropolitan area of northwestern Italy. *J. Wildl. Dis.* **2016**, *52*, 126–130. [CrossRef]
13. Dunlap, B.G.; Thies, M.L. *Giardia* in beaver (*Castor canadensis*) and nutria (*Myocastor coypus*) from east Texas. *J. Parasitol.* **2002**, *88*, 1254–1258. [CrossRef]
14. Heyworth, M.F. *Giardia duodenalis* genetic assemblages and hosts. *Parasite* **2016**, *23*, 13. [CrossRef] [PubMed]
15. Ryan, U.; Cacciò, S.M. Zoonotic potential of *Giardia*. *Int. J. Parasitol.* **2013**, *43*, 943–956. [CrossRef] [PubMed]
16. Einarsson, E.; Ma'ayeh, S.; Svärd, S.G. An up-date on *Giardia* and giardiasis. *Curr. Opin. Microbiol.* **2016**, *34*, 47–52. [CrossRef] [PubMed]
17. Thompson, R.C.; Smith, A.; Lymbery, A.J.; Averis, S.; Morris, K.D.; Wayne, A.F. *Giardia* in Western Australian wildlife. *Vet. Parasitol.* **2010**, *170*, 207–211. [CrossRef] [PubMed]
18. Zhao, Z.; Wang, R.; Zhao, W.; Qi, M.; Zhao, J.; Zhang, L.; Li, J.; Liu, A. Genotyping and subtyping of *Giardia* and *Cryptosporidium* isolates from commensal rodents in China. *Parasitology* **2015**, *142*, 800–806. [CrossRef] [PubMed]
19. Ma, X.; Wang, Y.; Zhang, H.J.; Wu, H.X.; Zhao, G.H. First report of *Giardia duodenalis* infection in bamboo rats. *Parasit. Vectors* **2018**, *11*, 520. [CrossRef] [PubMed]
20. Helmy, Y.A.; Spierling, N.G.; Schmidt, S.; Rosenfeld, U.M.; Reil, D.; Imholt, C.; Jacob, J.; Ulrich, R.G.; Aebischer, T.; Klotz, C. Occurrence and distribution of *Giardia* species in wild rodents in Germany. *Parasit. Vectors* **2018**, *11*, 213. [CrossRef] [PubMed]
21. Bajer, A. Between-year variation and spatial dynamics of *Cryptosporidium* spp. and *Giardia* spp. infections in naturally infected rodent populations. *Parasitology* **2008**, *135*, 1629–1649. [CrossRef] [PubMed]
22. Sulaiman, I.M.; Fayer, R.; Bern, C.; Gilman, R.H.; Trout, J.M.; Schantz, P.M.; Das, P.; Lal, A.A.; Xiao, L. Triosephosphate isomerase gene characterization and potential zoonotic transmission of *Giardia duodenalis*. *Emerg. Infect. Dis.* **2003**, *9*, 1444–1452. [CrossRef] [PubMed]

23. Fernandez-Alvarez, A.; Martin-Alonso, A.; Abreu-Acosta, N.; Feliu, C.; Hugot, J.P.; Valladares, B.; Foronda, P. Identification of a novel assemblage G subgenotype and a zoonotic assemblage B in rodent isolates of *Giardia duodenalis* in the Canary Islands, Spain. *Parasitology* **2014**, *141*, 206–215. [CrossRef]
24. Tan, T.K.; Low, V.L.; Ng, W.H.; Ibrahim, J.; Wang, D.; Tan, C.H.; Chellappan, S.; Lim, Y.A.L. Occurrence of zoonotic *Cryptosporidium* and *Giardia duodenalis* species/genotypes in urban rodents. *Parasitol. Int.* **2019**, *69*, 110–113. [CrossRef] [PubMed]
25. Levecke, B.; Meulemans, L.; Dalemans, T.; Casaert, S.; Claerebout, E.; Geurden, T. Mixed *Giardia duodenalis* assemblage A, B, C and E infections in pet chinchillas (*Chinchilla lanigera*) in Flanders (Belgium). *Vet. Parasitol.* **2011**, *177*, 166–170. [CrossRef] [PubMed]
26. Qi, M.; Yu, F.; Li, S.; Wang, H.; Luo, N.; Huang, J.; Zhang, L. Multi-locus genotyping of potentially zoonotic *Giardia duodenalis* in pet chinchillas (*Chinchilla lanigera*) in China. *Vet. Parasitol.* **2015**, *208*, 113–117. [CrossRef]
27. Pantchev, N.; Broglia, A.; Paoletti, B.; Globokar, M.; Bertram, A.; Nockler, K.; Caccio, S.M. Occurrence and molecular typing of *Giardia* isolates in pet rabbits, chinchillas, guinea pigs and ferrets collected in Europe during 2006–2012. *Vet. Rec.* **2014**, *175*, 18. [CrossRef] [PubMed]
28. Veronesi, F.; Piergili, D.; Morganti, G.; Bietta, A.; Moretta, I.; Moretti, A.; Traversa, D. Occurrence of *Giardia duodenalis* infection in chinchillas (*Chincilla lanigera*) from Italian breeding facilities. *Res. Vet. Sci.* **2012**, *93*, 807–810. [CrossRef] [PubMed]
29. Gherman, C.M.; Kalmar, Z.; Gyorke, A.; Mircean, V. Occurrence of *Giardia duodenalis* assemblages in farmed long-tailed chinchillas Chinchilla lanigera (*Rodentia*) from Romania. *Parasit. Vectors* **2018**, *11*, 86. [CrossRef] [PubMed]
30. Deng, L.; Luo, R.; Liu, H.; Zhou, Z.; Li, L.; Chai, Y.; Yang, L.; Wang, W.; Fu, H.; Zhong, Z.; et al. First identification and multi-locus genotyping of *Giardia duodenalis* in pet chipmunks (*Eutamias asiaticus*) in Sichuan Province, southwestern China. *Parasit. Vectors* **2018**, *11*, 199. [CrossRef]
31. Kilonzo, C.; Li, X.; Vodoz, T.; Xiao, C.; Chase, J.A.; Jay-Russell, M.T.; Vivas, E.J.; Atwill, E.R. Quantitative shedding of multiple genotypes of *Cryptosporidium* and *Giardia* by deer mice (*Peromyscus maniculatus*) in a major agricultural region on the california central coast. *J. Food. Prot.* **2017**, *80*, 819–828. [CrossRef] [PubMed]
32. Adriana, G.; Zsuzsa, K.; Mirabela, D.; Mircea, G.C.; Viorica, M. *Giardia duodenalis* genotypes in domestic and wild animals from Romania identified by PCR-RFLP targeting the *gdh* gene. *Vet. Parasitol.* **2016**, *217*, 71–75. [CrossRef] [PubMed]
33. Beck, R.; Sprong, H.; Bata, I.; Lucinger, S.; Pozio, E.; Caccio, S.M. Prevalence and molecular typing of *Giardia* spp. in captive mammals at the zoo of Zagreb, Croatia. *Vet. Parasitol.* **2011**, *175*, 40–46. [CrossRef] [PubMed]
34. Lebbad, M.; Mattsson, J.G.; Christensson, B.; Ljungström, B.; Backhans, A.; Andersson, J.O.; Svärd, S.G. From mouse to moose: Multi-locus genotyping of *Giardia* isolates from various animal species. *Vet. Parasitol.* **2010**, *168*, 231–239. [CrossRef] [PubMed]
35. Thompson, R.C.; Monis, P. *Giardia*–from genome to proteome. *Adv. Parasitol.* **2012**, *78*, 57–95. [PubMed]
36. Squire, S.A.; Ryan, U. *Cryptosporidium* and *Giardia* in Africa: Current and future challenges. *Parasit. Vectors* **2017**, *10*, 195. [CrossRef] [PubMed]
37. Levecke, B.; Dorny, P.; Geurden, T.; Vercammen, F.; Vercruysse, J. Gastrointestinal protozoa in non-human primates of four zoological gardens in Belgium. *Vet. Parasitol.* **2007**, *148*, 236–246. [CrossRef] [PubMed]
38. García-Cervantes, P.C.; Báez-Flores, M.E.; Delgado-Vargas, F.; Ponce-Macotela, M.; Nawa, Y.; De-la-Cruz-Otero, M.D.; Martínez-Gordillo, M.N.; Díaz-Camacho, S.P. *Giardia duodenalis* genotypes among schoolchildren and their families and pets in urban and rural areas of Sinaloa, Mexico. *J. Infect. Dev. Ctries.* **2017**, *11*, 180–187. [CrossRef] [PubMed]
39. Appelbee, A.J.; Frederick, L.M.; Heitman, T.L.; Olson, M.E. Prevalence and genotyping of *Giardia duodenalis* from beef calves in Alberta, Canada. *Vet. Parasitol.* **2003**, *112*, 289–294. [CrossRef]
40. Lalle, M.; Pozio, E.; Capelli, G.; Bruschi, F.; Crotti, D.; Cacciò, S.M. Genetic heterogeneity at the beta-giardin locus among human and animal isolates of *Giardia duodenalis* and identification of potentially zoonotic subgenotypes. *Int. J. Parasitol.* **2005**, *35*, 207–213. [CrossRef]
41. Yu, F.; Amer, S.; Qi, M.; Wang, R.; Wang, Y.; Zhang, S.; Jian, F.; Ning, C.; El Batae, H.; Zhang, L. Multi-locus genotyping of *Giardia duodenalis* isolated from patients in Egypt. *Acta Trop.* **2019**, *196*, 66–71. [CrossRef] [PubMed]

Article

Cryptosporidium Species and *C. parvum* Subtypes in Farmed Bamboo Rats

Falei Li [1,2], Wentao Zhao [1], Chenyuan Zhang [1], Yaqiong Guo [1], Na Li [1], Lihua Xiao [1,2,*] and Yaoyu Feng [1,2,*]

1   Center for Emerging and Zoonotic Diseases, South China Agricultural University, Wushan Road, Guangzhou 510642, China; falei0316@126.com (F.L.); 17818522213@163.com (W.Z.); chenyuanzhang@gdaib.edu.cn (C.Z.); guoyq@scau.edu.cn (Y.G.); nli@scau.edu.cn (N.L.)
2   Guangdong Laboratory for Lingnan Modern Agriculture, Wushan Road, Guangzhou 510642, China
*   Correspondence: lxiao@scau.edu.cn (L.X.); yyfeng@scau.edu.cn (Y.F.)

Received: 6 November 2020; Accepted: 1 December 2020; Published: 2 December 2020

Abstract: Bamboo rats (*Rhizomys sinensis*) are widely farmed in Guangdong, China, but the distribution and public health potential of *Cryptosporidium* spp. in them are unclear. In this study, 724 fecal specimens were collected from bamboo rats in Guangdong Province and analyzed for *Cryptosporidium* spp. using PCR and sequence analyses of the small subunit rRNA gene. The overall detection rate of *Cryptosporidium* spp. was 12.2% (88/724). By age, the detection rate in animals under 2 months (23.2% or 13/56) was significantly higher than in animals over 2 months (11.2% or 75/668; $\chi^2 = 6.95$, $df = 1$, $p = 0.0084$). By reproduction status, the detection rate of *Cryptosporidium* spp. in nursing animals (23.1% or 27/117) was significantly higher than in other reproduction statuses (6.8% or 4/59; $\chi^2 = 7.18$, $df = 1$, $p = 0.0074$). Five *Cryptosporidium* species and genotypes were detected, including *Cryptosporidium* bamboo rat genotype I ($n = 49$), *C. parvum* ($n = 31$), *Cryptosporidium* bamboo rat genotype III ($n = 5$), *C. occultus* ($n = 2$), and *C. muris* ($n = 1$). The average numbers of oocysts per gram of feces for these *Cryptosporidium* spp. were 14,074, 494,636, 9239, 394, and 323, respectively. The genetic uniqueness of bamboo rat genotypes I and III was confirmed by sequence analyses of the 70 kDa heat shock protein and actin genes. Subtyping *C. parvum* by sequence analysis of the 60 kDa glycoprotein gene identified the presence of IIoA15G1 ($n = 20$) and IIpA6 ($n = 2$) subtypes. The results of this study indicated that *Cryptosporidium* spp. are common in bamboo rats in Guangdong, and some of the *Cryptosporidium* spp. in these animals are known human pathogens.

Keywords: *Cryptosporidium parvum*; subtype; bamboo rat; human pathogen

---

1. Introduction

Cryptosporidium spp. are major pathogens that mainly parasitize the gastrointestinal epithelium, causing moderate-to-severe diarrhea in humans and animals [1,2]. They are responsible for significant mortality in both young children [1,3] and neonatal farm animals [4–8]. In addition, cryptosporidiosis has been associated with retarded growth in humans and farm animals [9–12].

Cryptosporidium spp. are especially common in rodents [13]. Among the >40 established *Cryptosporidium* species and an equal number of genotypes of unknown species status [14], *C. parvum* is commonly found in various rodents in China [15]. Other *Cryptosporidium* spp. from rodents have more limited host ranges, such as *C. meleagridis* and *Cryptosporidium* deer mouse genotypes I, II, III, and IV in deer mice [16–19]; *Cryptosporidium* chipmunk genotypes I and II in chipmunks [16]; and *Cryptosporidium* ferret genotype, *C. rubeyi* and several squirrel genotypes in squirrels [20–23].

Bamboo rats are widely farmed in southern China. They are commonly infected with *Cryptosporidium* spp. Several *Cryptosporidium* species and genotypes have been identified in these animals, including *C. parvum*, *C. parvum*-like genotype, *C. occultus*, and *Cryptosporidium* bamboo rat genotypes I and II [24,25]. Thus, the distribution of *Cryptosporidium* spp. in bamboo rats appears to be different from other rodents. At the subtype level, *C. parvum* subtypes found in bamboo rats also differ from those in other rodents. In China, rodents such as hamsters, chipmunks, and rats are mostly infected with the IId subtype family of *C. parvum*, with IIdA15G1 and IIdA19G1 as the most common subtypes [26]. In contrast, bamboo rats in southern China are seemingly infected with rare IIo and IIp subtype families of *C. parvum* [25,27]. Most of data on these two *C. parvum* subtype families in bamboo rats, however, were from a study of animals in Jiangxi, Guangxi, and Hainan [24].

In this study, we examined the occurrence of *Cryptosporidium* spp. and *C. parvum* subtypes in bamboo rats in Guangdong Province. The distributions of *Cryptosporidium* spp. and *C. parvum* subtypes were compared among farms, age groups and reproduction statuses. The oocyst shedding intensity was compared among *Cryptosporidium* species and genotypes for the first time.

2. Results

2.1. Occurrence of Cryptosporidium spp. in Bamboo Rats

The overall detection rate of *Cryptosporidium* spp. in bamboo rats in Guangdong was 12.2% (88/724). The detection rate of *Cryptosporidium* spp. on farm 1 (35.9% or 33/92) was significantly higher than on farm 2 (3.5% or 5/142; $\chi^2 = 42.95$, $df = 1$, $p < 0.0001$; df: degrees of freedom), farm 3 (1.0% or 2/205; $\chi^2 = 74.38$, $df = 1$, $p < 0.0001$), farm 5 (0% or 0/56; $\chi^2 = 25.85$, $df = 1$, $p < 0.0001$), and farm 6 (19.7% or 37/188; $\chi^2 = 8.63$, $df = 1$, $p = 0.0033$). The difference in detection rate between farms 4 (26.8% or 11/41) and 1 was not significant ($\chi^2 = 1.05$, $df = 1$, $p = 0.3062$; Table 1).

By age, the detection rates of *Cryptosporidium* spp. in bamboo rats ranged from 0.0% (0/8) in animals of 6–8 months to 23.2% (13/56) in animals under 2 months. The detection rate of *Cryptosporidium* spp. in animals under 2 months of age was significantly higher than in older animals (11.2% or 75/668; $\chi^2 = 6.95$, $df = 1$, $P = 0.0084$; Table 2). By reproduction status of adult animals (1–3 years in age), nursing animals had a significantly higher detection rate (23.1% or 27/117) than breeding animals (4.9% or 10/205; $\chi^2 = 24.26$, $df = 1$, $p < 0.0001$), pregnant animals (6.8% or 4/59; $\chi^2 = 7.18$, $df = 1$, $p = 0.0074$), and nonpregnant animals (10.6% or 21/198; $\chi^2 = 8.86$, $df = 1$, $p = 0.0029$; Table 3).

Table 1. Occurrence of *Cryptosporidium* spp. and *C. parvum* subtypes in bamboo rats in Guangdong, China.

Farm	Age Group	No. Specimens	No. Positive (%)	*Cryptosporidium* spp.					*C. parvum* Subtype
				Bamboo Rat Genotype I	Bamboo Rat Genotype III	*C. occultus*	*C. muris*	*C. parvum*	
1	<2 months	9	6 (66.7)	6	-	-	-	-	-
	1–3 years	83	27 (32.5)	27	-	-	-	-	-
	subtotal	92	33 (35.9)	33	-	-	-	-	-
2	4–6 months	1	0 (0.0)	-	-	-	-	-	-
	1–3 years	141	5 (3.5)	4	-	1	-	-	-
	subtotal	142	5 (3.5)	4	-	1	-	-	-
3	<2 months	36	0 (0.0)	-	-	-	-	-	-
	2–4 months	29	0 (0.0)	-	-	-	-	-	-
	4–6 months	18	1 (5.6)	-	-	1	-	-	-
	6–8 months	8	0 (0.0)	-	-	-	-	-	-
	1–3 years	114	1 (0.9)	-	-	-	-	1	-
	subtotal	205	2 (1.0)	-	-	1	-	1	-
4	2–4 months	16	6 (37.5)	3	-	-	-	3	IIoA15G1 (1)
	1–3 years	25	5 (20.0)	5	-	-	-	-	-
	subtotal	41	11 (26.8)	8	-	-	-	3	IIoA15G1 (1)
5	1–3 years	56	0 (0.0)	-	-	-	-	-	-
	subtotal	56	0 (0.0)	-	-	-	-	-	-
6	<2 months	11	7 (63.6)	-	-	-	-	7	IIoA15G1 (5)
	2–4 months	17	6 (35.3)	3	1	-	-	2	IIoA15G1 (1)
	1–3 years	160	24 (15.0)	1	4	-	1	18	IIoA15G1 (13), IIpA6 (2)
	subtotal	188	37 (19.7)	4	5	-	1	27	IIoA15G1 (19), IIpA6 (2)
	Total	724	88 (12.2)	49	5	2	1	31	IIoA15G1 (20), IIpA6 (2)

Table 2. Occurrence of *Cryptosporidium* spp. in farmed bamboo rats in Guangdong by age.

Age	No. Specimens	No. Positive (%)	*Cryptosporidium* spp.					*C. parvum* Subtype
			Bamboo Rat Genotype I	Bamboo Rat Genotype III	*C. muris*	*C. occultus*	*C. parvum*	
<2 months	56	13 (23.2)	6	-	-	-	7	IIoA15G1 (5)
2–4 months	62	12 (19.4)	6	1	-	-	5	IIoA15G1 (2)
4–6 months	19	1 (5.3)	-	-	-	1	-	-
6–8 months	8	0 (0.0)	-	-	-	-	-	-
1–3 years	579	62 (10.7)	37	4	1	1	19	IIoA15G1 (13), IIpA6 (2)
Total	724	88 (12.2)	49	5	1	2	31	IIoA15G1 (20), IIpA6 (2)

Table 3. Occurrence of *Cryptosporidium* spp. in farmed adult bamboo rats in Guangdong by reproduction status.

Animal Status	No. Specimens	No. Positive (%)	*Cryptosporidium* spp.					*C. parvum* Subtype
			Bamboo Rat Genotype I	Bamboo Rat Genotype III	*C. muris*	*C. occultus*	*C. parvum*	
Nursing	117	27 (23.1)	5	3	1	-	18	IIoA15G1(13), IIpA6 (2)
Breeding	205	10 (4.9)	8	1	-	-	1	-
Pregnancy	59	4 (6.8)	4	-	-	-	-	-
Nonpregnant	198	21 (10.6)	20	-	-	1	-	-
Total	579	62 (10.7)	37	4	1	1	19	IIoA15G1 (13), IIpA6 (2)

2.2. Cryptosporidium Species and Genotypes Identified

All 88 *Cryptosporidium*-positive specimens were successfully genotyped by sequence analysis of the *SSU* rRNA gene. Among them, five *Cryptosporidium* species and genotypes were detected, including bamboo rat genotype I in 49 specimens, *C. parvum* in 31 specimens, bamboo rat genotype III in 5 specimens, *C. occultus* in 2 specimens, and *C. muris* in 1 specimen. The *SSU* rRNA sequences from bamboo rat genotype I, bamboo rat genotype III, *C. parvum*, and *C. occultus* were identical to reference sequences MK956935 (from bamboo rats in China), MK956936 (from bamboo rats in China), MK956932 (from bamboo rats in China), and MK982467 (from calves in Bangladesh), respectively. The bamboo rat genotype III was previously named the *C. parvum*-like genotype. In contrast, the sequence from *C. muris* was similar to the reference sequence GU319781 from nonhuman primates in China, with three nucleotide substitutions. In phylogenetic analysis of the *SSU* rRNA sequences, these *Cryptosporidium* species and genotypes from bamboo rats clustered with their reference sequences (Figure 1a).

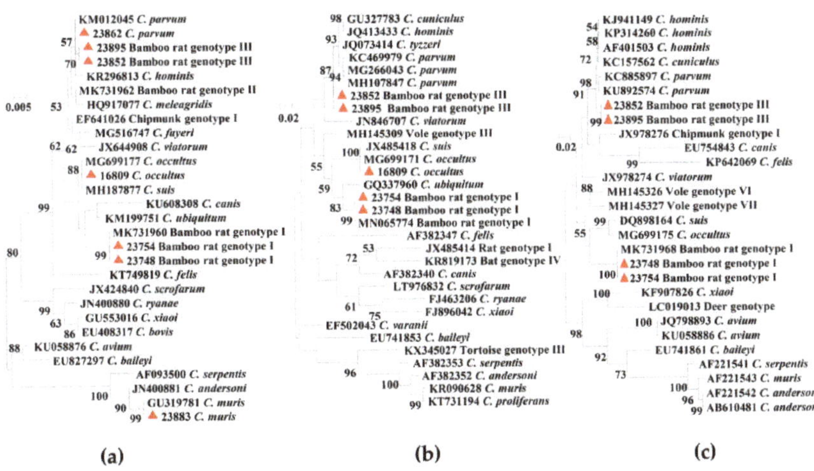

Figure 1. Phylogenetic relationship of *Cryptosporidium* spp. based on the maximum likelihood analyses of the *SSU* rRNA gene (**a**); actin gene (**b**) and 70 kDa heat shock protein gene (**c**). Bootstrap values greater than 50% from 1000 replicates are displayed. The red triangles indicate the *Cryptosporidium* spp. identified in the present study.

Representative isolates of bamboo rat genotypes I and III and *C. occultus* were characterized by sequence analysis of the actin gene. The sequences from bamboo rat genotype I and *C. occultus* were identical to reference sequences MN065774 (from bamboo rats in China) and MG699171 (from *Meriones unguiculatus* in Czech Republic), respectively. The sequence from bamboo rat genotype III, however, was similar to the *C. parvum* sequence MG266043 from *Apodemus agrarius* in Slovakia, with 16 nucleotide substitutions. As expected, these *Cryptosporidium* species and genotypes clustered with their reference sequences in phylogenetic analysis of the actin gene (Figure 1b).

Nucleotide sequences of the *hsp70* gene were also obtained from representative isolates of bamboo rat genotypes I and III; the two *C. occultus* isolates were PCR-negative at this locus. The sequences from bamboo rat genotype I was identical to reference sequences MK731968 from bamboo rats in China. In contrast, the sequences from bamboo rat genotype III had 14 single nucleotide polymorphisms (SNPs) and 1 nucleotide deletion compared with *C. parvum* sequences (AF401503 and KU892574). As expected, bamboo rat genotypes I and III clustered these reference sequences in phylogenetic analysis of the *hsp70* gene (Figure 1c).

2.3. Age Pattern of Cryptosporidium Species and Genotypes

Bamboo rat genotype I and *C. parvum* were the dominant *Cryptosporidium* spp. in bamboo rats, especially in animals under 2 months of age and in nursing adults. Among adult animals, both *Cryptosporidium* spp. were detected in nursing animals and breeding animals. In contrast, only bamboo rat genotype I was detected in pregnant and nonpregnant animals (Table 3). Among young animals, both bamboo rat genotype I and *C. parvum* were detected in animals under 4 months of age. They, however, were not detected in animals with an age of 4–8 months (Table 2).

2.4. Occurrence of C. parvum Subtypes

Twenty-two of the 31 *C. parvum*-positive specimens were successfully subtyped based on sequence analysis of the *gp60* gene. Subtypes IIoA15G1 ($n = 20$) and IIpA6 ($n = 2$) were identified from them. The sequences generated were identical to reference sequences MK956002 and MK955996, respectively.

2.5. Intensity of Oocyst Shedding among Cryptosporidium spp.

The oocyst shedding intensity, oocysts per gram of feces (OPG), in *Cryptosporidium*-infected bamboo rats was measured using qPCR. The average OPG values were 494,636 ± 1,892,289 for *C. parvum* ($n = 28$), 9239 ± 12,939 for bamboo rat genotype III ($n = 5$), 14,074 ± 46,621 for bamboo rat genotype I ($n = 39$), 394 ± 442 for *C. occultus* ($n = 3$), and 323 for *C. muris* ($n = 1$; Figure 2).

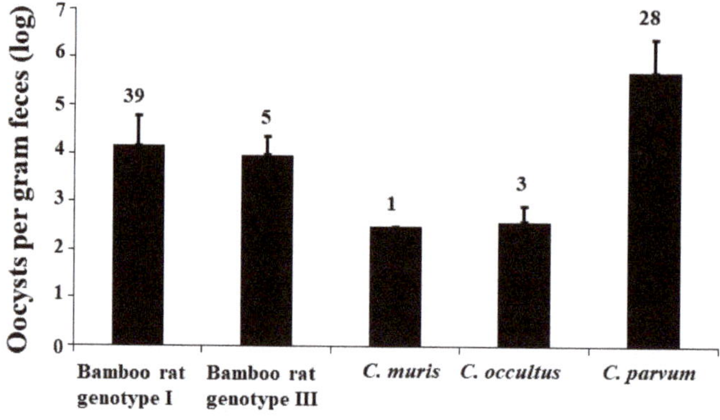

Figure 2. Oocyst shedding intensity (oocysts per gram of feces) of *Cryptosporidium* species and genotypes in bamboo rats (mean ± standard deviation).

3. Discussion

The results of this study suggest that *Cryptosporidium* spp. are common in bamboo rats in Guangdong, China. The detection rate of *Cryptosporidium* spp. in present study (12.2% or 88/724) is higher than in previous study (2.2% or 1/46) in the same area [25]. It is also higher than the 3.2% detection rate in bamboo rats from a pet market [27]. In contrast, it is lower than several reports of *Cryptosporidium* spp. in farmed bamboo rats sampled in Guangxi (20.9% or 100/477), Jiangxi (33.3% or 51/153), and Hainan (69.6% or 55/79) [24]. Among six farms, the detection rate of *Cryptosporidium* spp. on farm 1 (35.9%) was higher than other farms (0.0–26.8%), probably because of the higher animal density and poor hygiene conditions on the farm. The detection rates of *Cryptosporidium* spp. in animals under 2 months of age (23.2%) and nursing animals (23.1%) were higher than other age groups (0.0–19.4%) and reproduction statuses (4.9–10.6%). These differences might be due to the lower immunity of young animals and nursing animals, making them more susceptible to infection with *Cryptosporidium* spp. As the detection rates differed among age groups and reproduction statuses, these variations in detection rates among studies are expected.

Diverse *Cryptosporidium* species and genotypes are present in farmed bamboo rats in the study area. Five *Cryptosporidium* species and genotypes were identified in this study, including bamboo rat genotype I (55.7%, 49/88), *C. parvum* (35.2%, 31/88), bamboo rat genotype III (5.7%, 5/88), *C. occultus* (2.3%, 2/88), and *C. muris* (1.1%, 1/88). Among them, bamboo rat genotype I and *C. parvum* are dominant *Cryptosporidium* spp. in present study. This is similar to observations of two previous studies [24,25]. However, bamboo rat genotype III, commonly found in bamboo rats in Guangxi and Hainan [24], is rare in the present study. Among the five *Cryptosporidium* species and genotypes identified in the study, *C. parvum* is a well-known human pathogen and has been found in a wide range of animals. *Cryptosporidium muris* and *C. occultus*, in contrast, are mostly *Cryptosporidium* species of rats and have been only occasionally found in humans [14]. Although bamboo rat genotypes I and III are genetically related to *C. ubiquitum* and *C. parvum*, respectively, in sequence analysis of three genetic loci, it is still unclear whether they can infect humans [14,28].

Results of oocyst shedding intensity in this study suggest that bamboo rats could be natural hosts of bamboo rat genotype I, bamboo rat genotype III, and *C. parvum*. Among five *Cryptosporidium* spp. identified in this study, these three *Cryptosporidium* genotypes had higher oocyst shedding intensity than *C. occultus* and *C. muris*. In addition, numerous bamboo rats are known to be infected with bamboo rat genotypes I and III and *C. parvum*. In contrast, brown rats appear to be natural hosts of *C. occultus* and *C. muris* [29–31]. Infection with these two *Cryptosporidium* species in bamboo rats could be due to contact with *Cryptosporidium*-infected wild rats living in the same ecological niche.

The unique *C. parvum* subtypes identified in the study probably represent emerging pathogens in a broad range of hosts. Thus far, more than 20 *C. parvum* subtype families have been identified based on sequence analyses of the *gp60* gene [32]. Among them, the IIo and IIp were previously seen in a few bamboo rats in Sichuan, China and were considered rare *C. parvum* subtype families [27]. IIo subtypes, however, have been found in numerous bamboo rats from Jiangxi, Guangxi, and Hainan in southern China [24]. One IIo subtype, IIoA14, has also been widely found in crab-eating macaques in Hainan, China [33]. Human infections with IIo subtypes have been identified in Thailand and New Zealand [34–36]. IIp subtypes of *C. parvum* also appear to be common in bamboo rats, having been thus far found in Jiangxi, Sichuan, Guangxi, Guangdong, and Hainan [24,27]. As they are genetically related to IId and IIo subtypes, they could also have a broader host range and human-infective potential [33]. Interestingly, IIp subtypes were much more prevalent than IIo subtypes in the previous survey of *Cryptosporidium* spp. in bamboo rats in Jiangxi, Guangxi, and Hainan. In the present study in Guangdong, the IIoA15G1 subtype was more commonly detected than the IIpA6 subtype.

There appears to be some transmission of *C. parvum* between nursing bamboo rats and their babies. Most *C. parvum* infections in adult bamboo rats (18 of 19 *C. parvum*-positive) were detected in nursing mothers; only one of them was found in another adult bamboo rat. In contrast, all *C. parvum* infections in young bamboo rats (12 of 12 *C. parvum*-positive) were detected in animals under 4 months of age. The *C. parvum* infections in nursing animals were probably due to sharing cages between the nursing dams and their babies. This is supported by the result of subtype analysis, in which the IIoA15G1 subtype was commonly found in both nursing animals and young bamboo rats. This is also consistent with results of studies in sheep and cattle, in which *C. parvum* is mostly detected in pre-weaned animals [37–40]. As the number of animals examined in the present study is small, further studies are needed to support this hypothesis. In contrast, *Cryptosporidium* bamboo rat genotype I appeared to be transmitted differently from *C. parvum*, as it was commonly found in adult bamboo rats of all reproduction statuses.

4. Materials and Methods

4.1. Ethics Statement

All fecal specimens were collected from bamboo rats with the approval of the farmers. The animals were handled in compliance with the regulations of the Chinese Laboratory Animal Administration

Act of 2017. The study protocol was approved by the Research Ethics Committee of the South China Agricultural University (approval no. 2019g001).

4.2. Specimens

A total of 724 fecal specimens were collected during March to May 2019 from bamboo rats on six farms in Guangdong Province, China. On these farms, animals were kept in cages of 60 × 60 × 60 cm. Animals under six months of age were mostly kept in groups of 10, while older animals (between 7 months and 3 years) were kept in groups of 3. For the former, four specimens were collected from fresh feces in different corners of the cage, while for the latter, only one fecal specimen was collected from the cage (Table 1). These fecal specimens were stored in 2.5% potassium dichromate at 4 °C before DNA extraction.

The bamboo rats examined were divided into five age groups, including <2 months, 2–4 months, 4–6 months, 6–8 months, and 1–3 years (Table 2). The adult animals (1–3 years-old) were further divided into four reproduction statuses: nursing (female bamboo rats feeding babies), breeding (one female and 1–2 males kept together for mating), pregnancy, and nonpregnancy (Table 3). On the studies farms, pregnant bamboo rats were moved to new cages for delivery after for delivery after cleaning of the cages without any further disinfection.

4.3. Detection, Genotyping and Subtyping of Cryptosporidium spp.

Fecal specimens were washed off potassium dichromate with distilled water by centrifugation at 2000× *g* for 10 min. Genomic DNA was extracted from 200 mg washed fecal material using the Fast DNA Spin Kit for Soil (MP Biomedical, Santa Ana, CA, USA) as described [41]. *Cryptosporidium* spp. in the extracted DNA were detected using a nested PCR assay targeting a ~830-bp fragment of the small subunit rRNA (*SSU*) rRNA gene [42]. The *C. parvum* identified was subtyped by sequence analysis of a ~800-bp fragment of the 60 kDa (*gp60*) gene [43]. Representative isolates of the bamboo rat genotypes I and III and *C. occultus* were further characterized by sequence analyses of the 70 kDa heat shock protein (*hsp70*) and actin genes [44,45].

4.4. Measurement of Oocyst Shedding Intensity

The intensity of oocyst shedding in *Cryptosporidium*-positive specimens from naturally infected bamboo rats was measured by quantitative PCR (qPCR) targeting the *SSU* rRNA gene [33,46]. The qPCR was conducted on a LightCycler 480 II (Roche, Indianapolis, IN, USA). To calculate oocysts per gram of feces (OPG), Cq (quantitation cycle) values generated were analyzed based on a standard curve constructed with DNA preparations from fecal specimens spiked with known numbers of oocysts of the IOWA (USA) strain of *C. parvum* (Waterborne, Inc., New Orleans, LA, USA).

4.5. Sequence Analysis

All secondary PCR products from *Cryptosporidium*-positive specimens were sequenced bi-directionally on an ABI 3730 Autosequencer (Applied Biosystems, Foster City, CA, USA) for determining *Cryptosporidium* spp. and *C. parvum* subtypes. The nucleotide sequences generated were assembled using ChromasPro 2.1.5.0 (http://technelysium. com. au/ChromasPro. html), edited using BioEdit 7.1.3.0 (http://www.mbio.ncsu.edu/BioEdit/bioedit.html), and aligned with reference sequences downloaded from GenBank using ClustalX 2.0.11 (http://clustal.org). Phylogenetic analysis of sequence generated was performed by constructing maximum likelihood trees using Mega 6.0 (http://www.megasoftware.net/) based on substitution rates calculated using the general time reversible model. The robustness of clade formation was assessed using bootstrapping with 1000 replicates. Representative sequences generated in this study were submitted to the GenBank under accession numbers MW092529-MW092535 and MW117315-MW117325.

4.6. Statistical Analysis

The occurrence rates of *Cryptosporidium* spp. in bamboo rats were compared among farms, age groups, and reproduction statuses using the chi-square test implemented in SPSS v.20.0 (IBM Corp., New York, NY, USA). Differences were considered significant at $p \leq 0.05$.

5. Conclusions

The results of this study suggest that divergent *Cryptosporidium* spp., including *C. parvum*, bamboo rat genotypes I and III, *C. occultus* and *C. muris*, occur in farmed bamboo rats in Guangdong, China. The *C. parvum* identified belongs to the unique subtype families IIo and IIp. As most of the *Cryptosporidium* species and *C. parvum* subtypes are not commonly found in domestic animals, they could have wildlife origin and have maintained at high transmission intensity in some of the semi-domesticated exotic animals. Attention should be paid to their spillover to other farm animals as well as human populations.

Author Contributions: Conceptualization, L.X. and Y.F.; data curation, F.L.; formal analysis, F.L.; funding acquisition, L.X. and Y.F.; investigation, W.Z. and C.Z.; methodology, L.X. and Y.F.; project administration, N.L.; resources, Y.F.; software, L.X. and Y.F.; supervision, Y.G.; validation, L.X. and Y.F.; visualization, L.X. and Y.F.; writing—original draft, F.L.; writing—review and editing, L.X. All authors have read and agreed to the published version of the manuscript.

Funding: This research was funded by the National Natural Science Foundation of China (U1901208 and 31820103014), 111 Project (D20008), and Innovation Team Project of Guangdong University (2019KCXTD001).

Conflicts of Interest: The authors declare no conflict of interest.

References

1. Kotloff, K.L.; Nataro, J.P.; Blackwelder, W.C.; Nasrin, D.; Farag, T.H.; Panchalingam, S.; Wu, Y.; Sow, S.O.; Sur, D.; Breiman, R.F.; et al. Burden and aetiology of diarrhoeal disease in infants and young children in developing countries (the Global Enteric Multicenter Study, GEMS): A prospective, case-control study. *Lancet* **2013**, *382*, 209–222. [CrossRef]
2. Meganck, V.; Hoflack, G.; Opsomer, G. Advances in prevention and therapy of neonatal dairy calf diarrhoea: A systematical review with emphasis on colostrum management and fluid therapy. *Acta Vet. Scand.* **2014**, *56*, 75. [CrossRef]
3. Sow, S.O.; Muhsen, K.; Nasrin, D.; Blackwelder, W.C.; Wu, Y.; Farag, T.H.; Panchalingam, S.; Sur, D.; Zaidi, A.K.; Faruque, A.S.; et al. The Burden of *Cryptosporidium* Diarrheal Disease among Children <24 Months of Age in Moderate/High Mortality Regions of Sub-Saharan Africa and South Asia, Utilizing Data from the Global Enteric Multicenter Study (GEMS). *PLoS Negl. Trop. Dis.* **2016**, *10*, e0004729. [CrossRef]
4. Izzo, M.M.; Kirkland, P.D.; Mohler, V.L.; Perkins, N.R.; Gunn, A.A.; House, J.K. Prevalence of major enteric pathogens in Australian dairy calves with diarrhoea. *Aust. Vet. J.* **2011**, *89*, 167–173. [CrossRef]
5. Randhawa, S.S.; Randhawa, S.S.; Zahid, U.N.; Singla, L.D.; Juyal, P.D. Drug combination therapy in control of cryptosporidiosis in Ludhiana district of Punjab. *J. Parasit. Dis.* **2012**, *36*, 269–272. [CrossRef]
6. Niine, T.; Dorbek-Kolin, E.; Lassen, B.; Orro, T. *Cryptosporidium* outbreak in calves on a large dairy farm: Effect of treatment and the association with the inflammatory response and short-term weight gain. *Res. Vet. Sci.* **2018**, *117*, 200–208. [CrossRef]
7. Brar, A.P.S.; Sood, N.K.; Kaur, P.; Singla, L.D.; Sandhu, B.S.; Gupta, K.; Narang, D.; Singh, C.K.; Chandra, M. Periurban outbreaks of bovine calf scours in Northern India caused by *Cryptosporidium* in association with other enteropathogens. *Epidemiol. Infect.* **2017**, *145*, 2717–2726. [CrossRef]
8. Li, N.; Wang, R.; Cai, M.; Jiang, W.; Feng, Y.; Xiao, L. Outbreak of cryptosporidiosis due to *Cryptosporidium parvum* subtype IIdA19G1 in neonatal calves on a dairy farm in China. *Int. J. Parasitol.* **2019**, *49*, 569–577. [CrossRef]
9. Jacobson, C.; Al-Habsi, K.; Ryan, U.; Williams, A.; Anderson, F.; Yang, R.; Abraham, S.; Miller, D. *Cryptosporidium* infection is associated with reduced growth and diarrhoea in goats beyond weaning. *Vet. Parasitol.* **2018**, *260*, 30–37. [CrossRef]

10. Shaw, H.J.; Innes, E.A.; Morrison, L.J.; Katzer, F.; Wells, B. Long-term production effects of clinical cryptosporidiosis in neonatal calves. *Int. J. Parasitol.* **2020**, *50*, 371–376. [CrossRef]

11. Khalil, I.A.; Troeger, C.; Rao, P.C.; Blacker, B.F.; Brown, A.; Brewer, T.G.; Colombara, D.V.; De Hostos, E.L.; Engmann, C.; Guerrant, R.L.; et al. Morbidity, mortality, and long-term consequences associated with diarrhoea from *Cryptosporidium* infection in children younger than 5 years: A meta-analyses study. *Lancet Glob. Health* **2018**, *6*, e758–e768. [CrossRef]

12. Steiner, K.L.; Ahmed, S.; Gilchrist, C.A.; Burkey, C.; Cook, H.; Ma, J.Z.; Korpe, P.S.; Ahmed, E.; Alam, M.; Kabir, M.; et al. Species of Cryptosporidia Causing Subclinical Infection Associated with Growth Faltering in Rural and Urban Bangladesh: A Birth Cohort Study. *Clin. Infect. Dis.* **2018**, *67*, 1347–1355. [CrossRef]

13. Feng, Y. *Cryptosporidium* in wild placental mammals. *Exp. Parasitol.* **2010**, *124*, 128–137. [CrossRef]

14. Feng, Y.; Ryan, U.M.; Xiao, L. Genetic Diversity and Population Structure of *Cryptosporidium*. *Trends Parasitol.* **2018**, *34*, 997–1011. [CrossRef]

15. Feng, Y.; Xiao, L. Molecular Epidemiology of Cryptosporidiosis in China. *Front. Microbiol.* **2017**, *8*, 1701. [CrossRef]

16. Feng, Y.; Alderisio, K.A.; Yang, W.; Blancero, L.A.; Kuhne, W.G.; Nadareski, C.A.; Reid, M.; Xiao, L. *Cryptosporidium* genotypes in wildlife from a new york watershed. *Appl. Environ. Microbiol.* **2007**, *73*, 6475–6483. [CrossRef]

17. Ziegler, P.E.; Wade, S.E.; Schaaf, S.L.; Stern, D.A.; Nadareski, C.A.; Mohammed, H.O. Prevalence of *Cryptosporidium* species in wildlife populations within a watershed landscape in southeastern New York State. *Vet. Parasitol.* **2007**, *147*, 176–184. [CrossRef]

18. Xiao, L.; Sulaiman, I.M.; Ryan, U.M.; Zhou, L.; Atwill, E.R.; Tischler, M.L.; Zhang, X.; Fayer, R.; Lal, A.A. Host adaptation and host-parasite co-evolution in *Cryptosporidium*: Implications for taxonomy and public health. *Int. J. Parasitol.* **2002**, *32*, 1773–1785. [CrossRef]

19. Zhang, W.; Shen, Y.; Wang, R.; Liu, A.; Ling, H.; Li, Y.; Cao, J.; Zhang, X.; Shu, J.; Zhang, L. *Cryptosporidium cuniculus* and *Giardia duodenalis* in rabbits: Genetic diversity and possible zoonotic transmission. *PLoS ONE* **2012**, *7*, e31262. [CrossRef]

20. Deng, L.; Chai, Y.; Luo, R.; Yang, L.; Yao, J.; Zhong, Z.; Wang, W.; Xiang, L.; Fu, H.; Liu, H.; et al. Occurrence and genetic characteristics of *Cryptosporidium* spp. and *Enterocytozoon bieneusi* in pet red squirrels (*Sciurus vulgaris*) in China. *Sci. Rep.* **2020**, *10*, 1026. [CrossRef]

21. Li, X.; Pereira, M.; Larsen, R.; Xiao, C.; Phillips, R.; Striby, K.; McCowan, B.; Atwill, E.R. *Cryptosporidium rubeyi* n. sp. (Apicomplexa: Cryptosporidiidae) in multiple Spermophilus ground squirrel species. *Int. J. Parasitol. Parasites Wildl.* **2015**, *4*, 343–350. [CrossRef]

22. Prediger, J.; Horčičková, M.; Hofmannová, L.; Sak, B.; Ferrari, N.; Mazzamuto, M.V.; Romeo, C.; Wauters, L.A.; McEvoy, J.; Kváč, M. Native and introduced squirrels in Italy host different *Cryptosporidium* spp. *Eur. J. Protistol.* **2017**, *61*, 64–75. [CrossRef]

23. Stenger, B.L.S.; Clark, M.E.; Kváč, M.; Khan, E.; Giddings, C.W.; Prediger, J.; McEvoy, J.M. North American tree squirrels and ground squirrels with overlapping ranges host different *Cryptosporidium* species and genotypes. *Infect. Genet. Evol.* **2015**, *36*, 287–293. [CrossRef]

24. Li, F.; Zhang, Z.; Hu, S.; Zhao, W.; Zhao, J.; Kváč, M.; Guo, Y.; Li, N.; Feng, Y.; Xiao, L. Common occurrence of divergent *Cryptosporidium* species and *Cryptosporidium parvum* subtypes in farmed bamboo rats (*Rhizomys sinensis*). *Parasites Vectors* **2020**, *13*, 149. [CrossRef]

25. Wei, Z.; Liu, Q.; Zhao, W.; Jiang, X.; Zhang, Y.; Zhao, A.; Jing, B.; Lu, G.; Qi, M. Prevalence and diversity of *Cryptosporidium* spp. in bamboo rats (*Rhizomys sinensis*) in South Central China. *Int. J. Parasitol. Parasites Wildl.* **2019**, *9*, 312–316. [CrossRef]

26. Lv, C.; Zhang, L.; Wang, R.; Jian, F.; Zhang, S.; Ning, C.; Wang, H.; Feng, C.; Wang, X.; Ren, X.; et al. *Cryptosporidium* spp. in wild, laboratory, and pet rodents in China: Prevalence and molecular characterization. *Appl. Environ. Microbiol.* **2009**, *75*, 7692–7699. [CrossRef]

27. Liu, X.; Zhou, X.; Zhong, Z.; Zuo, Z.; Shi, J.; Wang, Y.; Qing, B.; Peng, G. Occurrence of novel and rare subtype families of *Cryptosporidium* in bamboo rats (*Rhizomys sinensis*) in China. *Vet. Parasitol.* **2015**, *207*, 144–148. [CrossRef]

28. Li, N.; Xiao, L.; Alderisio, K.; Elwin, K.; Cebelinski, E.; Chalmers, R.; Santin, M.; Fayer, R.; Kvac, M.; Ryan, U.; et al. Subtyping *Cryptosporidium ubiquitum*, a zoonotic pathogen emerging in humans. *Emerg. Infect. Dis.* **2014**, *20*, 217–224. [CrossRef]

29. Kváč, M.; Vlnatá, G.; Ježková, J.; Horčičková, M.; Konečný, R.; Hlásková, L.; McEvoy, J.; Sak, B. *Cryptosporidium occultus* sp. n. (Apicomplexa: Cryptosporidiidae) in rats. *Eur. J. Protistol.* **2018**, *63*, 96–104. [CrossRef]

30. Ng-Hublin, J.S.; Singleton, G.R.; Ryan, U. Molecular characterization of *Cryptosporidium* spp. from wild rats and mice from rural communities in the Philippines. *Infect. Genet. Evol.* **2013**, *16*, 5–12. [CrossRef]

31. Zhao, W.; Zhou, H.; Huang, Y.; Xu, L.; Rao, L.; Wang, S.; Wang, W.; Yi, Y.; Zhou, X.; Wu, Y.; et al. *Cryptosporidium* spp. in wild rats (*Rattus* spp.) from the Hainan Province, China: Molecular detection, species/genotype identification and implications for public health. *Int. J. Parasitol. Parasites Wildl.* **2019**, *9*, 317–321. [CrossRef]

32. Xiao, L.; Feng, Y. Molecular epidemiologic tools for waterborne pathogens *Cryptosporidium* spp. and *Giardia duodenalis*. *Food Waterborne Parasitol.* **2017**, *8*, 14–32. [CrossRef]

33. Chen, L.; Hu, S.; Jiang, W.; Zhao, J.; Li, N.; Guo, Y.; Liao, C.; Han, Q.; Feng, Y.; Xiao, L. *Cryptosporidium parvum* and *Cryptosporidium hominis* subtypes in crab-eating macaques. *Parasites Vectors* **2019**, *12*, 350. [CrossRef]

34. Insulander, M.; Silverlås, C.; Lebbad, M.; Karlsson, L.; Mattsson, J.G.; Svenungsson, B. Molecular epidemiology and clinical manifestations of human cryptosporidiosis in Sweden. *Epidemiol. Infect.* **2013**, *141*, 1009–1020. [CrossRef]

35. Garcia, R.J.; Pita, A.B.; Velathanthiri, N.; French, N.P.; Hayman, D.T.S. Species and genotypes causing human cryptosporidiosis in New Zealand. *Parasitol. Res.* **2020**, *119*, 2317–2326. [CrossRef]

36. Sannella, A.R.; Suputtamongkol, Y.; Wongsawat, E.; Cacciò, S.M. A retrospective molecular study of *Cryptosporidium* species and genotypes in HIV-infected patients from Thailand. *Parasites Vectors* **2019**, *12*, 91. [CrossRef]

37. Fayer, R.; Santin, M.; Trout, J.M. Prevalence of *Cryptosporidium* species and genotypes in mature dairy cattle on farms in eastern United States compared with younger cattle from the same locations. *Vet. Parasitol.* **2007**, *145*, 260–266. [CrossRef]

38. Santín, M.; Trout, J.M.; Fayer, R. A longitudinal study of cryptosporidiosis in dairy cattle from birth to 2 years of age. *Vet. Parasitol.* **2008**, *155*, 15–23. [CrossRef]

39. Mueller-Doblies, D.; Giles, M.; Elwin, K.; Smith, R.P.; Clifton-Hadley, F.A.; Chalmers, R.M. Distribution of *Cryptosporidium* species in sheep in the UK. *Vet. Parasitol.* **2008**, *154*, 214–219. [CrossRef]

40. Geurden, T.; Thomas, P.; Casaert, S.; Vercruysse, J.; Claerebout, E. Prevalence and molecular characterisation of *Cryptosporidium* and *Giardia* in lambs and goat kids in Belgium. *Vet. Parasitol.* **2008**, *155*, 142–145. [CrossRef]

41. Jiang, J.; Alderisio, K.A.; Singh, A.; Xiao, L. Development of procedures for direct extraction of *Cryptosporidium* DNA from water concentrates and for relief of PCR inhibitors. *Appl. Environ. Microbiol.* **2005**, *71*, 1135–1141. [CrossRef]

42. Xiao, L.; Bern, C.; Limor, J.; Sulaiman, I.; Roberts, J.; Checkley, W.; Cabrera, L.; Gilman, R.H.; Lal, A.A. Identification of 5 types of *Cryptosporidium* parasites in children in Lima, Peru. *J. Infect. Dis.* **2001**, *183*, 492–497. [CrossRef]

43. Alves, M.; Xiao, L.; Sulaiman, I.; Lal, A.A.; Matos, O.; Antunes, F. Subgenotype analysis of *Cryptosporidium* isolates from humans, cattle, and zoo ruminants in Portugal. *J. Clin. Microbiol.* **2003**, *41*, 2744–2747. [CrossRef]

44. Sulaiman, I.M.; Lal, A.A.; Xiao, L. Molecular phylogeny and evolutionary relationships of *Cryptosporidium* parasites at the actin locus. *J. Parasitol.* **2002**, *88*, 388–394. [CrossRef]

45. Sulaiman, I.M.; Morgan, U.M.; Thompson, R.C.; Lal, A.A.; Xiao, L. Phylogenetic Relationships of *Cryptosporidium* Parasites Based on the 70-Kilodalton Heat Shock Protein (HSP70) Gene. *Appl. Environ. Microbiol.* **2000**, *66*, 2385–2391. [CrossRef]

46. Li, N.; Neumann, N.F.; Ruecker, N.; Alderisio, K.A.; Sturbaum, G.D.; Villegas, E.N.; Chalmers, R.; Monis, P.; Feng, Y.; Xiao, L. Development and Evaluation of Three Real-Time PCR Assays for Genotyping and Source Tracking *Cryptosporidium* spp. in Water. *Appl. Environ. Microbiol.* **2015**, *81*, 5845–5854. [CrossRef]

Publisher's Note: MDPI stays neutral with regard to jurisdictional claims in published maps and institutional affiliations.

Article

Genetic Characterization of *Cryptosporidium cuniculus* from Rabbits in Egypt

Doaa Naguib [1], Dawn M. Roellig [2], Nagah Arafat [3] and Lihua Xiao [4,*]

1   Department of Hygiene and Zoonoses, Faculty of Veterinary Medicine, Mansoura University, Mansoura 35516, Egypt; doaanaguib246@yahoo.com

2   Division of Foodborne, Waterborne, and Environmental Diseases, National Center for Emerging and Zoonotic Infectious Diseases, Centers for Disease Control and Prevention, Atlanta, GA 30329, USA; iyd4@cdc.gov

3   Department of Poultry Diseases, Faculty of Veterinary Medicine, Mansoura University, Mansoura 35516, Egypt; nagaharafat@yahoo.com

4   Center for Emerging and Zoonotic Diseases, College of Veterinary Medicine, South China Agricultural University, Guangzhou 510642, China

*   Correspondence: lxiao1961@gmail.com; Tel.: +86-183-0173-2862

Abstract: Rabbits are increasingly farmed in Egypt for meat. They are, however, known reservoirs of infectious pathogens. Currently, no information is available on the genetic characteristics of *Cryptosporidium* spp. in rabbits in Egypt. To understand the prevalence and genetic identity of *Cryptosporidium* spp. in these animals, 235 fecal samples were collected from rabbits of different ages on nine farms in El-Dakahlia, El-Gharbia, and Damietta Provinces, Egypt during the period from July 2015 to April 2016. PCR-RFLP analysis of the small subunit rRNA gene was used to detect and genotype *Cryptosporidium* spp. The overall detection rate was 11.9% (28/235). All 28 samples were identified as *Cryptosporidium cuniculus*. The 16 samples successfully subtyped by the sequence analysis of the partial 60 kDa glycoprotein gene belonged to two subtypes, VbA19 (*n* = 1) and VbA33 (*n* = 15). As *C. cuniculus* is increasingly recognized as a cause of human cryptosporidiosis, *Cryptosporidium* spp. in rabbits from Egypt have zoonotic potential.

Keywords: *Cryptosporidium cuniculus*; rabbits; Egypt; *gp60* gene; PCR-RFLP; zoonoses

Citation: Naguib, D.; Roellig, D.M.; Arafat, N.; Xiao, L. Genetic Characterization of *Cryptosporidium cuniculus* from Rabbits in Egypt. *Pathogens* **2021**, *10*, 775. https://doi.org/10.3390/pathogens10060775

Academic Editor: Luiz Shozo Ozaki

Received: 28 May 2021
Accepted: 17 June 2021
Published: 20 June 2021

1. Introduction

Cryptosporidiosis is a common cause of diarrhea in humans and animals [1,2]. It is one of the most important diseases in both developing countries and industrialized nations due to its importance in diarrhea-associated death in young children and foodborne and waterborne outbreaks of illness [1,3–6]. The etiologic agents of cryptosporidiosis, *Cryptosporidium* spp., have over 40 established species and many genotypes of unknown species status [7]. Among them, approximately 20 species and genotypes have been found in humans [8]. Most human cryptosporidiosis cases are caused by *C. parvum* and *C. hominis*. Other human-pathogenic *Cryptosporidium* spp. include *C. meleagridis, C. ubiquitum, C. cuniculus, C. felis, C. canis, C. viatorum*, and *C. muris* [7].

Rabbits are a supply of high-quality protein to humans. Reports of the Food and Agriculture Organization of the United Nations (FAO) showed that Egypt was the fourth largest producer of rabbit meat in the world, with approximately 7.6 million rabbits [9,10]. Results of recent studies indicate that rabbits can serve as reservoirs of many zoonotic pathogens [11–13]. They are commonly infected with several *Cryptosporidium* species, especially *C. cuniculus* [13–20]. In recent years, there have been increasing reports of *C. cuniculus* in humans [21–24]. In the United Kingdom and New Zealand, *C. cuniculus* is the third most common *Cryptosporidium* species in patients with diarrhea [24,25].

In recent years, molecular epidemiological studies have been conducted to understand the transmission of *Cryptosporidium* spp. in humans, livestock, and companion animals in

197

Egypt [26–34]. Although rabbits are commonly farmed in Egypt, to the authors' knowledge there have been no thorough studies on the distribution and genetic identity of *Cryptosporidium* spp. in rabbits in the country. Therefore, this study was conducted to examine the occurrence, genetic characteristics, and zoonotic potential of *Cryptosporidium* spp. in rabbits from three provinces (El-Dakahlia, El-Gharbia, and Damietta) in Egypt.

2. Results

2.1. Cryptosporidium Infection on Rabbit Farms

Cryptosporidium spp. were detected by PCR analysis of the SSU rRNA gene in 28 (11.9%) of the 235 fecal samples analyzed in the study. Eight of the nine farms examined were positive for *Cryptosporidium* spp. Among the eight positive farms, the infection rates ranged from 4% to 24% (Table 1). The farms El-Dakahlia 3 and El-Gharbia 2 had high infection rates of 21% and 24%, respectively, while no infection was detected on farm El-Gharbia 1. By age, *Cryptosporidium* spp. were detected in rabbits of all ages, with a higher infection rate found in rabbits of <3 months (20%; Fisher's exact test = 11.237, p = 0.003 in the overall comparison) (Table 2). By animal breed, *Cryptosporidium* spp. were identified in all breeds, with slightly higher detection rates in Hi-Plus rabbits (15%) than in New Zealand (11%) and Rex (7%), although these rates were not statistically different (Fisher's exact test = 2.283, p = 0.333 in the overall comparison). Farms in El-Dakahlia province recorded higher *Cryptosporidium* occurrence (17%) than El-Gharbia (11%) and Damietta (7%) provinces (Fisher's exact test = 3.685, p = 0.157 in the overall comparison).

Table 1. Distribution of *Cryptosporidium* spp. in rabbits by farm and age group in El Dakahlia, El-Gharbia, and Damietta provinces, Egypt.

Farm	Age (Month)	No. of Samples	*Cryptosporidium* spp. *		
			No. Positive (%)	95% Confidence Interval	
				Lower Limit	Upper Limit
El-Dakahlia 1	<3	10	2	-	-
	3–6	11	1	-	-
	>6	10	1	-	-
	Subtotal	31	4 (13%)	1.10	24.69
El-Dakahlia 2	<3	9	2	-	-
	3–6	7	2	-	-
	>6	7	0	-	-
	Subtotal	23	4 (17%)	1.90	32.89
El-Dakahlia 3	<3	11	3	-	-
	3–6	11	2	-	-
	>6	6	1	-	-
	Subtotal	28	6 (21%)	6.20	36.59
El-Gharbia 1	<3	8	0	-	-
	3–6	8	0	-	-
	>6	10	0	-	-
	Subtotal	26	0 (0%)	0.00	0.00
El-Gharbia 2	<3	7	4	-	-
	3–6	12	2	-	-
	>6	6	0	-	-
	Subtotal	25	6 (24%)	7.25	40.74
El-Gharbia 3	<3	12	2	-	-
	3–6	11	1	-	-
	>6	7	0	-	-
	Subtotal	30	3 (10%)	−0.73	20.73

Table 1. Cont.

Farm	Age (Month)	No. of Samples	No. Positive (%)	95% Confidence Interval Lower Limit	Upper Limit
			Cryptosporidium spp. *		
Damietta 1	<3	9	1	-	-
	3–6	10	0	-	-
	>6	8	0	-	-
	Subtotal	27	1 (4%)	−3.42	10.82
Damietta 2	<3	4	1	-	-
	3–6	8	1	-	-
	>6	8	0	-	-
	Subtotal	20	2 (10%)	−3.14	23.14
Damietta 3	<3	9	1	-	-
	3–6	8	1	-	-
	>6	8	0	-	-
	Subtotal	25	2 (8%)	−2.63	18.63
Total	-	235	28 (11.9%)	-	-

* Fisher's exact test: $t = 12.258$, $p = 0.106$.

Table 2. Factors associated with Cryptosporidium infection in rabbits in Egypt.

Factors		Sample Size	No. Positive (%)	95% Confidence Interval Lower Limit	Upper Limit	Fisher's Exact Test	*p*
			Cryptosporidium spp.				
Age (month)	<3	79	16 (20)	11.38	29.11	11.237	0.003
	3–6	86	10 (12)	4.85	18.40		
	>6	70	2 (3)	−1.03	6.79		
Breed	Rex	57	4 (7)	0.30	13.60	2.283	0.333
	Hi-Plus	98	15 (15)	8.10	22.40		
	New Zealand	80	9 (11)	4.30	18.20		
Locality	El-Dakahlia	82	14 (17)	8.90	25.20	3.685	0.157
	El-Gharbia	81	9 (11)	4.20	17.90		
	Damietta	72	5 (7)	1.00	12.70		

2.2. Cryptosporidium Genotypes and Subtypes

All 28 samples amplified by PCR analysis of the *SSU* rRNA gene had *C. cuniculus* by RFLP analysis (Figure 1). They produced two types of nucleotide sequences. Among them, ten sequences were identical to those in GenBank (AY120901, FJ262724, etc.), while two sequences had an A to T substitution near the $5'$ end of the partial gene. In the phylogenetic analysis of the *SSU* rRNA sequences, *C. cuniculus* sequences obtained from the 12 samples clustered with reference sequences from GenBank (Figure 2). Of the 28 *C. cuniculus* samples, 16 were successfully subtyped by sequence analysis of the *gp60* gene, with two subtypes being identified: VbA19 ($n = 1$) and VbA33 ($n = 15$). These sequences were identical to each other in the non-repeat regions but had one A to T substitution compared to sequences (KU852732, GU097641, GU097647, GU971639, etc.) in GenBank (Figure 3). The VbA19 subtype was found only in a 6-month-old rabbit from farm El-Gharbia 3, while the VbA33 subtype was found on other *Cryptosporidium*-positive farms. Among the samples from four cages of animals with diarrhea, one sample from a cage of 4-month-old rabbits on El-Dakahlia 1 was positive for *C. cuniculus* VbA33 (Table 3).

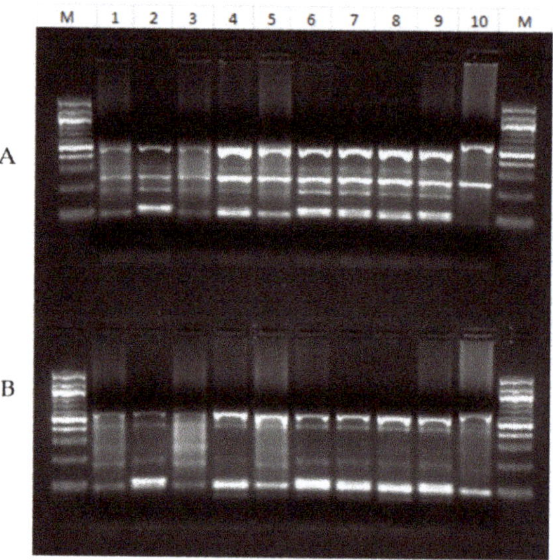

Figure 1. RFLP analysis of PCR products of *SSU rRNA* gene from *Cryptosporidium cuniculus* from rabbits using *Ssp*I (**A**) and *Vsp*I (**B**) restriction enzymes. Lane M: 100 bp molecular markers; Lane 1–9: *C. cuniculus*; Lane 10: positive control (*C. baileyi*).

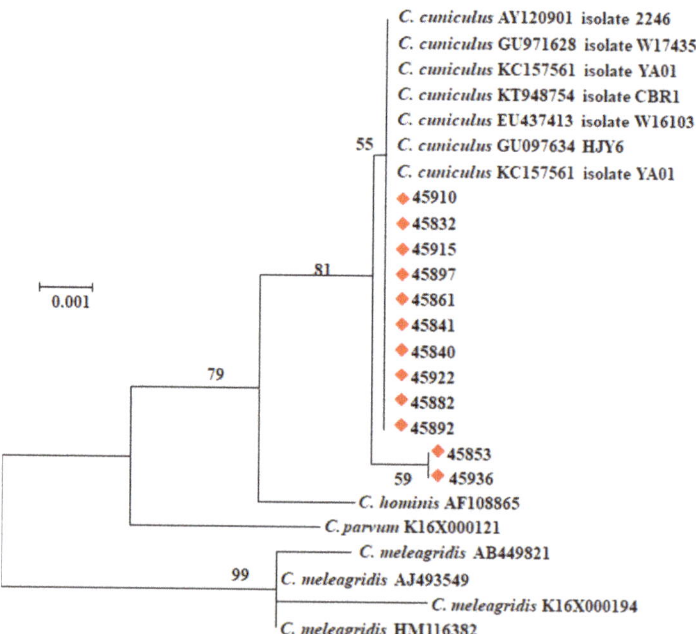

Figure 2. Phylogenetic relationships among *Cryptosporidium* spp. based on the nucleotide sequences of the *SSU* rRNA gene through a maximum likelihood analysis based on substitution rates calculated with the general time reversible model. Numbers at the internodes represent bootstrap values (>50%) from 1000 replicates. The *Cryptosporidium cuniculus* samples identified in this study are labeled with red rhombus.

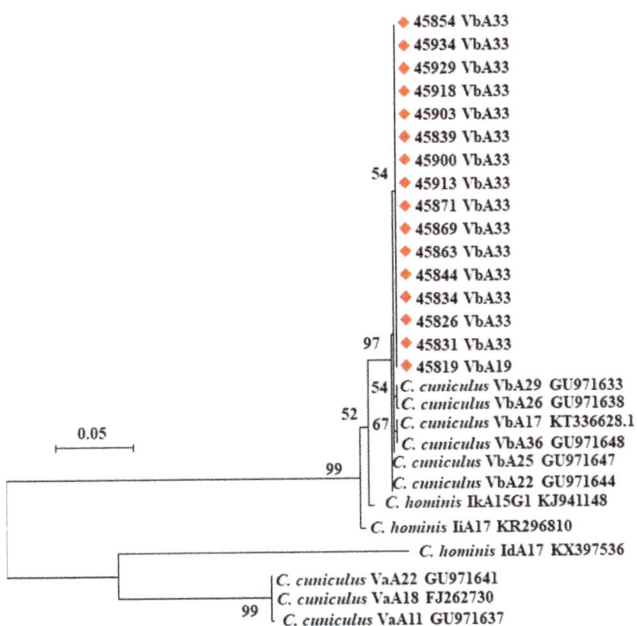

Figure 3. Phylogenetic tree of *Cryptosporidium* spp. based on *gp60* sequences through a maximum likelihood analysis based on substitution rates calculated with the general time reversible model. Numbers at the internodes represent bootstrap values (>70%) from 1000 replicates. The *Cryptosporidium cuniculus* samples identified in this study are labeled with red rhombus.

Table 3. Distribution of 16 *Cryptosporidium cuniculus* subtypes at the *gp60* locus in rabbits in Egypt.

Sample ID	*C. cuniculus* Subtype	Age (Month)	Location	Breed
45819	VbA19	6	El-Gharbia 3	Hi-Plus
45826	VbA33	1	El-Dakahlia 1	Rex
45831 *	VbA33	4	El-Dakahlia 1	Rex
45834	VbA33	4	Damietta 3	New Zealand
45839	VbA33	3	Damietta 2	Hi-Plus
45844	VbA33	4	El-Gharbia 2	Hi-Plus
45854	VbA33	3	El-Gharbia 2	Hi-Plus
45863	VbA33	2	El-Gharbia 2	Hi-Plus
45869	VbA33	2	El-Gharbia 2	Hi-Plus
45871	VbA33	1	El-Gharbia 2	Hi-Plus
45900	VbA33	1	Damietta 3	New Zealand
45903	VbA33	2	El-Dakahlia 3	New Zealand
45913	VbA33	1	El-Dakahlia 3	New Zealand
45918	VbA33	1	El-Dakahlia 3	New Zealand
45929	VbA33	1	El-Dakahlia 2	Hi-Plus
45934	VbA33	2	El-Dakahlia 2	Hi-Plus

* From rabbit with diarrhea.

3. Discussion

The results of the present study suggest a common occurrence of *Cryptosporidium* spp. in rabbits in the study areas. In this study, the overall occurrence of *Cryptosporidium* spp. in rabbits was 11.9% (28/235). This is similar to the infection rates of 11.2% (24/215) in a study of two rabbit farms in Heilongjiang Province, China [35], and 13.2% (14/106) in rabbits residing in Sydney drinking water catchments [36]. It is, however, higher than infection rates found in rabbits from Australia (6.8% or 12/176 and 8.4% or 22/263) [37,38], and Nigeria (3.7% or 4/107) [19], but lower than the infection rate recorded in pet rabbits in Japan (21.9% or 21/96) [16]. Several reports from China showed low infection rates of 1.03% (3/290), 2.4% (9/378), 3.4% (37/1081), and 3.4% (11/321) [13,39–41]. The differences in the infection rates of *Cryptosporidium* spp. among studies may be attributed to differences in sample size, rabbit breeds, management systems, geographic regions, and sample collection seasons. In one study, the infection rate of *Cryptosporidium* spp. in dead juvenile rabbits suffering from diarrhea was significantly higher than healthy ones (30.3% vs 3.3%) [16]. Among the nine farms examined in the present study, the occurrence of *Cryptosporidium* spp. on El-Gharbia 2 (24%) was higher than other farms (0–21%), possibly because of the poor hygiene and management practices on the farm.

Like in other animals, the infection rate of *Cryptosporidium* spp. is significantly higher in rabbits of youngest age. In this study, rabbits of <3 months had a significantly higher *Cryptosporidium* infection rate than older rabbits. Our findings are in agreement with observations in earlier studies, where the highest prevalence of *Cryptosporidium* spp. was recorded in young rabbits [13,39]. Similar age-associated occurrence of *Cryptosporidium* spp. has been reported in humans, cattle, and bamboo rats [33,34,42]. In the present study, a higher detection rate of *Cryptosporidium* spp. was recorded in Hi-Plus rabbits than Rex and New Zealand ones, possibly because of the high number of samples and sampling of many young animals. In contrast to our results, Rex and New Zealand rabbits were more susceptible than other breeds to *Cryptosporidium* infection in some earlier studies in China [13,35].

Generally, few clinical signs have been associated with cryptosporidiosis in rabbits, especially adult ones, and the infection is often not recognized due to the asymptomatic oocyst shedding [39,41]. This is in line with our results, where most rabbits were apparently healthy. Although two reports observed fatality in outbreaks of diarrhea in rabbits due to cryptosporidiosis [16,43], *Cryptosporidium* was detected in only one of the four samples from animals with diarrhea.

All isolates of *Cryptosporidium* spp. detected in our study were genotyped as *C. cuniculus*, which is one of the causes of human cryptosporidiosis and has zoonotic significance [44]. In some countries, such as the UK, Australia and New Zealand, many sporadic cases of human cryptosporidiosis have been attributed to infections with *C. cuniculus* [22,24,25,45]. It was recognized as the third most important *Cryptosporidium* species causing cryptosporidiosis in humans in the UK during 2007 to 2008 and New Zealand during 2009 to 2019 [24,25]. It was also associated with a waterborne outbreak of cryptosporidiosis due to contamination of treated drinking water by wild rabbits [46,47]. Humans may be infected with *C. cuniculus* via contaminated water or direct contact with rabbits [22].

In this study, based on *gp60* sequence analysis, the *C. cuniculus* isolates belong to two subtypes (VbA19 and VbA33) in the Vb subtype family. Previously, the VbA19 subtype was isolated from rabbits in the Czech Republic [46], while the VbA33 subtype was detected in humans in the UK [48]. The two subtypes identified in the present study, however, differed from them by one nucleotide in the non-repeat region. Currently, Va and Vb are the only two subtype families within *C. cuniculus*. Between them, subtypes in the Va subtype family are more commonly seen in humans while those in the Vb subtype family are more commonly seen in rabbits [35]. The occurrence of similar subtypes of *C. cuniculus* in humans and rabbits supports the zoonotic potential of *C. cuniculus* [41].

In conclusion, to the best of our knowledge, this is the first study on the genetic identity of *Cryptosporidium* spp. in rabbits in Egypt. The results of this study suggest a common occurrence of *C. cuniculus* in farm rabbits in several areas of the country. The detection of *C. cuniculus* in this study supports the potential role of rabbits as a source of human infections. Further studies from other localities in Egypt are needed to improve our understanding of the clinical and public health significance of *Cryptosporidium* spp., in rabbits in Egypt.

4. Materials and Methods

4.1. Ethics Statement

Permission was obtained from the owners of the farms before collection of fecal specimens. All fieldwork associated with this study was conducted in compliance with the Guide for the Care and Use of Laboratory Animals in Egypt. The study protocol was approved by the Ethics Committee of the Faculty of Veterinary Medicine, Mansoura University, Egypt.

4.2. Specimen Collections

A total of 235 fresh fecal specimens were collected between July 2015 and April 2016 from nine rabbit farms randomly selected in El-Dakahlia, El-Gharbia, and Damietta provinces in Egypt. All farms sampled in this study were medium-sized farms housing 700–900 rabbits. Fecal specimens were randomly collected from at least 20% of rabbit cages on each farm. Each specimen consisted of 3–5 fresh fecal pellets gathered from each cage. Each collection from each cage (containing 4–7 rabbits) was regarded as one specimen. The fecal pellets were placed into a sterile disposable plastic bag labeled with the age and breed of the animals and sampling date and location. Animals in four cages showed clinical signs of enteric diseases (emaciation, dehydration, and diarrhea) at the time of specimen collection. The rabbits were divided into three age groups: <3-month-old, 3–6-month-old, and >6-month-old. Specimens were stored in 70% ethanol at 4 °C until being transported to the Centers for Disease Control and Prevention, Atlanta, Georgia, USA for DNA extraction and molecular analysis.

4.3. DNA Extraction and PCR Amplification

The fecal specimens were washed twice with distilled water by centrifugation to remove ethanol before DNA extraction. Extraction of genomic DNA from specimens was performed using the FastDNA SPIN Kit for Soil (BIO 101, Carlsbad, CA, USA). The genomic DNA was eluted with 100 μL reagent-grade water and stored at −20 °C until PCR analysis.

4.4. Cryptosporidium Detection, Genotyping and Subtyping

Cryptosporidium spp. in the specimens were detected by nested PCR analysis of a ~830-bp fragment of the small subunit rRNA (*SSU rRNA*) gene as previously described [49]. *Cryptosporidium* species were identified by restriction fragment length polymorphism (RFLP) analysis of the secondary PCR products of *SSU rRNA* gene using *Ssp*I (New England BioLabs, Ipswich, MA, USA) and *Vsp*I (Promega, Madison, WI, USA) restriction enzymes [50]. All *Cryptosporidium*-positive specimens were selected for further subtyping by PCR and sequence analysis of the 60-kDa glycoprotein (*gp60*) gene [51]. Each specimen was analyzed twice for each genetic target, using *C. baileyi* DNA as the positive control for the *SSU rRNA*-based PCR, *C. parvum* DNA as the positive control for *gp60*-based PCR, and reagent-grade water as the negative control for both PCR assays.

4.5. DNA Sequence and Phylogenetic Analysis

Montage PCR filters (Millipore, Bedford, MA, USA) were used to purify all secondary PCR products of both genes. The purified products were sequenced in both directions using the secondary PCR primers and Big Dye® Terminator v3.1 Cycle Sequencing Kit (Applied Biosystems, Foster City, CA, USA) on an ABI 3130 Genetic Analyzer (Applied Biosystems). ChromasPro (version 1.5) (www.technelysium.com.au/ChromasPro.html/, accessed on 22 June 2009) was used to edit and assemble the DNA sequences, while ClustalX 2.0.11 (http://www.clustal.org/, accessed on 1 June 2018) was used to align the obtained nucleotide sequences against each other and reference sequences from GenBank to determine the genetic relatedness of various *C. cuniculus* subtype families. A phylogenetic tree was constructed using the maximum likelihood algorithm implemented in MEGA version 7.0.26 (www.megasoftware.net/, accessed on 1 May 2017) based on substitution rates calculated with the general time reversible model. Bootstrap analysis was applied to evaluate the reliability of cluster formation in the phylogenetic tree with 1000 replicates.

4.6. Statistical Analysis

Differences in infection rates of *Cryptosporidium* spp. among rabbits of different age groups, localities, and breeds were estimated using the Fisher's exact test. The SPSS software version 20.0 (IBM, Armonk, NY, USA) was used in the statistical analysis of the data. Differences were considered significant at $p \leq 0.05$.

Author Contributions: Conceptualization, D.N., L.X.; methodology, D.N., N.A., D.M.R. and L.X.; software, D.N and L.X.; validation, D.N. and L.X.; formal analysis, D.N.; investigation, D.N.; resources, D.N. and L.X.; data curation, D.N.; writing—original draft preparation, D.N.; writing—review and editing, D.N., N.A., D.M.R. and L.X.; visualization, L.X.; supervision, D.M.R. and L.X.; project administration, D.M.R. and L.X.; funding acquisition, D.N. and L.X. All authors have read and agreed to the published version of the manuscript.

Funding: This study was supported partially by the National Natural Science Foundation of China (31820103014), 111 Project (D20008), Innovation Team Project of Guangdong University (2019KCXTD001), and the Centers for Disease Control and Prevention.

Institutional Review Board Statement: Permission was obtained from the owners of the farms before collections of fecal specimens. All fieldwork associated with this study was conducted in compliance with the Guide for the Care and Use of Laboratory Animals in Egypt. The study protocol was approved by the Ethics Committee of the Faculty of Veterinary Medicine, Mansoura University, Egypt.

Informed Consent Statement: Not applicable.

Data Availability Statement: All relevant data are within the paper.

Acknowledgments: The findings and conclusions in this report are those of the authors and do not necessarily represent the represent of the official position of the U.S. Centers for Disease Control and Prevention.

Conflicts of Interest: The authors declare that they have no conflict of interest.

References

1. Kotloff, K.L.; Nataro, J.P.; Blackwelder, W.C.; Nasrin, D.; Farag, T.H.; Panchalingam, S.; Wu, Y.; Sow, S.O.; Sur, D.; Breiman, R.F.; et al. Burden and aetiology of diarrhoeal disease in infants and young children in developing countries (the Global Enteric Multicenter Study, GEMS): A prospective, case-control study. *Lancet* **2013**, *382*, 209–222. [CrossRef]
2. Meganck, V.; Hoflack, G.; Opsomer, G. Advances in prevention and therapy of neonatal dairy calf diarrhoea: A systematical review with emphasis on colostrum management and fluid therapy. *Acta Vet. Scand.* **2014**, *56*, 75. [CrossRef]
3. Platts-Mills, J.A.; Babji, S.; Bodhidatta, L.; Gratz, J.; Haque, R.; Havt, A.; McCormick, B.J.; McGrath, M.; Olortegui, M.P.; Samie, A.; et al. Pathogen-specific burdens of community diarrhoea in developing countries: A multisite birth cohort study (MAL-ED). *Lancet Glob. Health* **2015**, *3*, e564–e575. [CrossRef]
4. Efstratiou, A.; Ongerth, J.E.; Karanis, P. Waterborne transmission of protozoan parasites: Review of worldwide outbreaks—An update 2011–2016. *Water Res.* **2017**, *114*, 14–22. [CrossRef]
5. Ryan, U.; Hijjawi, N.; Xiao, L. Foodborne cryptosporidiosis. *Int. J. Parasitol.* **2018**, *48*, 1–12. [CrossRef]
6. GBD. Estimates of global, regional, and national morbidity, mortality, and aetiologies of diarrhoeal diseases: A systematic analysis for the Global Burden of Disease Study 2015. *Lancet Infect. Dis.* **2017**, *17*, 909–948. [CrossRef]
7. Feng, Y.; Ryan, U.M.; Xiao, L. Genetic diversity and population structure of *Cryptosporidium*. *Trends Parasitol.* **2018**, *34*, 997–1011. [CrossRef]
8. Xiao, L. Molecular epidemiology of cryptosporidiosis: An update. *Exp. Parasitol.* **2010**, *124*, 80–89. [CrossRef] [PubMed]
9. FAO. *Statistical Yearbook*; World Food and Agriculture Organization: Rome, Italy, 2013. Available online: https://reliefweb.int/report/world/fao-statistical-yearbook-2013-world-food-and-agriculture (accessed on 19 June 2013).
10. FAO. FAO Database. 2019. Available online: http://www.faostat.fao.org (accessed on 5 March 2019).
11. Ni, X.; Qin, S.; Lou, Z.; Ning, H.; Sun, X. Seroprevalence and risk factors of *Chlamydia* infection in domestic rabbits (*Oryctolagus cuniculus*) in China. *BioMed Res. Int.* **2015**, *2015*, 460473. [CrossRef] [PubMed]
12. Xie, X.; Bil, J.; Shantz, E.; Hammermueller, J.; Nagy, E.; Turner, P.V. Prevalence of lapine rotavirus, astrovirus, and hepatitis E virus in Canadian domestic rabbit populations. *Vet. Microbiol.* **2017**, *208*, 146–149. [CrossRef] [PubMed]

13. Zhang, X.; Qi, M.; Jing, B.; Yu, F.; Wu, Y.; Chang, Y.; Zhao, A.; Wei, Z.; Dong, H.; Zhang, L. Molecular characterization of *Cryptosporidium* spp., *Giardia duodenalis*, and *Enterocytozoon bieneusi* in rabbits in Xinjiang, China. *J. Eukaryot. Microbiol.* **2018**, *65*, 854–859. [CrossRef]

14. Sturdee, A.P.; Chalmers, R.M.; Bull, S.A. Detection of *Cryptosporidium* oocysts in wild mammals of mainland Britain. *Vet. Parasitol.* **1999**, *80*, 273–280. [CrossRef]

15. Ryan, U.; Xiao, L.; Read, C.; Zhou, L.; Lal, A.A.; Pavlasek, I. Identification of novel *Cryptosporidium* genotypes from the Czech Republic. *Appl. Environ. Microbiol.* **2003**, *69*, 4302–4307. [CrossRef]

16. Shiibashi, T.; Imai, T.; Sato, Y.; Abe, N.; Yukawa, M.; Nogami, S. *Cryptosporidium* infection in juvenile pet rabbits. *J. Vet. Med. Sci.* **2006**, *68*, 281–282. [CrossRef]

17. Robinson, G.; Chalmers, R.M. The European rabbit (*Oryctolagus cuniculus*), a source of zoonotic cryptosporidiosis. *Zoonoses Public Health* **2010**, *57*, e1–e13. [CrossRef] [PubMed]

18. Mosier, D.A.; Cimon, K.Y.; Kuhls, T.L.; Oberst, R.D.; Simons, K.R. Experimental cryptosporidiosis in adult and neonatal rabbits. *Vet. Parasitol.* **1997**, *69*, 163–169. [CrossRef]

19. Ayinmode, A.B.; Agbajelola, V.I. Molecular identification of *Cryptosporidium parvum* in rabbits (*Oryctolagus cuniculus*) in Nigeria. *Ann. Parasitol.* **2019**, *65*, 237–243. [CrossRef] [PubMed]

20. Yang, Z.; Yang, F.; Wang, J.; Cao, J.; Zhao, W.; Gong, B.; Yan, J.; Zhang, W.; Liu, A.; Shen, Y. Multilocus sequence typing and population genetic structure of *Cryptosporidium cuniculus* in rabbits in Heilongjiang Province, China. *Infect. Genet. Evol.* **2018**, *64*, 249–253. [CrossRef]

21. Molloy, S.F.; Kirwan, P.; Asaolu, S.O.; Holland, C.V.; Nichols, R.A.B.; Connelly, L.; Smith, H.V. Identification of a high diversity of *Cryptosporidium* species genotypes and subtypes in a pediatric population in Nigeria. *Am. J. Trop. Med. Hyg.* **2010**, *82*, 608–613. [CrossRef] [PubMed]

22. Koehler, A.V.; Whipp, M.J.; Haydon, S.R.; Gasser, R.B. *Cryptosporidium cuniculus*–new records in human and kangaroo in Australia. *Parasites Vectors* **2014**, *7*, 492. [CrossRef]

23. Martinez-Ruiz, R.; de Lucio, A.; Fuentes, I.; Carmena, D. Autochthonous *Cryptosporidium cuniculus* infection in Spain: First report in a symptomatic paediatric patient from Madrid. *Enferm. Infecc. Microbiol. Clin.* **2016**, *34*, 532–534. [CrossRef]

24. Garcia, R.J.; Pita, A.B.; Velathanthiri, N.; French, N.P.; Hayman, D.T.S. Species and genotypes causing human cryptosporidiosis in New Zealand. *Parasitol. Res.* **2020**, *119*, 2317–2326. [CrossRef]

25. Chalmers, R.M.; Elwin, K.; Hadfield, S.J.; Robinson, G. Sporadic human cryptosporidiosis caused by *Cryptosporidium cuniculus*, United Kingdom, 2007–2008. *Emerg. Infect. Dis.* **2011**, *17*, 536–538. [CrossRef]

26. Amer, S.; Honma, H.; Ikarashi, M.; Tada, C.; Fukuda, Y.; Suyama, Y.; Nakai, Y. *Cryptosporidium* genotypes and subtypes in dairy calves in Egypt. *Vet. Parasitol.* **2010**, *169*, 382–386. [CrossRef]

27. Amer, S.; Zidan, S.; Adamu, H.; Ye, J.; Roellig, D.; Xiao, L.; Feng, Y. Prevalence and characterization of *Cryptosporidium* spp. in dairy cattle in Nile River delta provinces, Egypt. *Exp. Parasitol.* **2013**, *135*, 518–523. [CrossRef]

28. El-Badry, A.A.; Al-Antably, A.S.; Hassan, M.A.; Hanafy, N.A.; Abu-Sarea, E.Y. Molecular seasonal, age and gender distributions of *Cryptosporidium* in diarrhoeic Egyptians: Distinct endemicity. *Eur. J. Clin. Microbiol. Infect. Dis.* **2015**, *34*, 2447–2453. [CrossRef]

29. Ibrahim, M.A.; Abdel-Ghany, A.E.; Abdel-Latef, G.K.; Abdel-Aziz, S.A.; Aboelhadid, S.M. Epidemiology and public health significance of *Cryptosporidium* isolated from cattle, buffaloes, and humans in Egypt. *Parasitol. Res.* **2016**, *115*, 2439–2448. [CrossRef] [PubMed]

30. Helmy, Y.A.; Krucken, J.; Nockler, K.; von Samson-Himmelstjerna, G.; Zessin, K.H. Molecular epidemiology of *Cryptosporidium* in livestock animals and humans in the Ismailia province of Egypt. *Vet. Parasitol.* **2013**, *193*, 15–24. [CrossRef]

31. Mahfouz, M.E.; Mira, N.; Amer, S. Prevalence and genotyping of *Cryptosporidium* spp. in farm animals in Egypt. *J. Vet. Med. Sci.* **2014**, *76*, 1569–1575. [CrossRef]

32. Gharieb, R.M.A.; Merwad, A.M.A.; Saleh, A.A.; Abd El-Ghany, A.M. Molecular screening and genotyping of *Cryptosporidium* species in household dogs and in-contact children in Egypt: Risk factor analysis and zoonotic importance. *Vector Borne Zoonotic Dis.* **2018**, *18*, 424–432. [CrossRef]

33. Naguib, D.; El-Gohary, A.H.; Roellig, D.; Mohamed, A.A.; Arafat, N.; Wang, Y.; Feng, Y.; Xiao, L. Molecular characterization of *Cryptosporidium* spp. and *Giardia duodenalis* in children in Egypt. *Parasites Vectors* **2018**, *11*, 403. [CrossRef] [PubMed]

34. Naguib, D.; El-Gohary, A.H.; Mohamed, A.A.; Roellig, D.M.; Arafat, N.; Xiao, L. Age patterns of *Cryptosporidium* species and *Giardia duodenalis* in dairy calves in Egypt. *Parasitol. Int.* **2018**, *67*, 736–741. [CrossRef] [PubMed]

35. Yang, Z.; Zhao, W.; Shen, Y.; Zhang, W.; Shi, Y.; Ren, G.; Yang, D.; Ling, H.; Yang, F.; Liu, A.; et al. Subtyping of *Cryptosporidium cuniculus* and genotyping of *Enterocytozoon bieneusi* in rabbits in two farms in Heilongjiang Province, China. *Parasite* **2016**, *23*, 52. [CrossRef]

36. Zahedi, A.; Monis, P.; Aucote, S.; King, B.; Paparini, A.; Jian, F.; Yang, R.; Oskam, C.; Ball, A.; Robertson, I.; et al. Zoonotic *Cryptosporidium* species in animals inhabiting Sydney water catchments. *PLoS ONE* **2016**, *11*, e0168169. [CrossRef]

37. Nolan, M.J.; Jex, A.R.; Koehler, A.V.; Haydon, S.R.; Stevens, M.A.; Gasser, R.B. Molecular-based investigation of *Cryptosporidium* and *Giardia* from animals in water catchments in southeastern Australia. *Water Res.* **2013**, *47*, 1726–1740. [CrossRef]

38. Nolan, M.J.; Jex, A.R.; Haydon, S.R.; Stevens, M.A.; Gasser, R.B. Molecular detection of *Cryptosporidium cuniculus* in rabbits in Australia. *Infect. Genet. Evol.* **2010**, *10*, 1179–1187. [CrossRef]

39. Shi, K.; Jian, F.; Lv, C.; Ning, C.; Zhang, L.; Ren, X.; Dearen, T.K.; Li, N.; Qi, M.; Xiao, L. Prevalence, genetic characteristics, and zoonotic potential of *Cryptosporidium* species causing infections in farm rabbits in China. *J. Clin. Microbiol.* **2010**, *48*, 3263–3266. [CrossRef]

40. Liu, X.; Zhou, X.; Zhong, Z.; Chen, W.; Deng, J.; Niu, L.; Wang, Q.; Peng, G. New subtype of *Cryptosporidium cuniculus* isolated from rabbits by sequencing the gp60 gene. *J. Parasitol.* **2014**, *100*, 532–536. [CrossRef] [PubMed]

41. Zhang, W.; Shen, Y.; Wang, R.; Liu, A.; Ling, H.; Li, Y.; Cao, J.; Zhang, X.; Shu, J.; Zhang, L. *Cryptosporidium cuniculus* and *Giardia duodenalis* in rabbits: Genetic diversity and possible zoonotic transmission. *PLoS ONE* **2012**, *7*, e31262. [CrossRef]

42. Li, F.; Zhang, Z.; Hu, S.; Zhao, W.; Zhao, J.; Kvac, M.; Guo, Y.; Li, N.; Feng, Y.; Xiao, L. Common occurrence of divergent Cryptosporidium species and *Cryptosporidium parvum* subtypes in farmed bamboo rats (*Rhizomys sinensis*). *Parasites Vectors* **2020**, *13*, 149. [CrossRef]

43. Kaupke, A.; Kwit, E.; Chalmers, R.M.; Michalski, M.M.; Rzezutka, A. An outbreak of massive mortality among farm rabbits associated with *Cryptosporidium* infection. *Res. Vet. Sci.* **2014**, *97*, 85–87. [CrossRef]

44. Ryan, U.; Fayer, R.; Xiao, L. *Cryptosporidium* species in humans and animals: Current understanding and research needs. *Parasitology* **2014**, *141*, 1667–1685. [CrossRef]

45. Elwin, K.; Hadfield, S.J.; Robinson, G.; Chalmers, R.M. The epidemiology of sporadic human infections with unusual cryptosporidia detected during routine typing in England and Wales, 2000–2008. *Epidemiol. Infect.* **2012**, *140*, 673–683. [CrossRef]

46. Chalmers, R.M.; Robinson, G.; Elwin, K.; Hadfield, S.J.; Xiao, L.; Ryan, U.; Modha, D.; Mallaghan, C. *Cryptosporidium* sp. rabbit genotype, a newly identified human pathogen. *Emerg. Infect. Dis.* **2009**, *15*, 829–830. [CrossRef]

47. Puleston, R.L.; Mallaghan, C.M.; Modha, D.E.; Hunter, P.R.; Nguyen-Van-Tam, J.S.; Regan, C.M.; Nichols, G.L.; Chalmers, R.M. The first recorded outbreak of cryptosporidiosis due to *Cryptosporidium cuniculus* (formerly rabbit genotype), following a water quality incident. *J. Water Health* **2014**, *12*, 41–50. [CrossRef]

48. Robinson, G.; Wright, S.; Elwin, K.; Hadfield, S.J.; Katzer, F.; Bartley, P.M.; Hunter, P.R.; Nath, M.; Innes, E.A.; Chalmers, R.M. Re-description of *Cryptosporidium cuniculus* Inman and Takeuchi, 1979 (Apicomplexa: Cryptosporidiidae): Morphology, biology and phylogeny. *Int. J. Parasitol.* **2010**, *40*, 1539–1548. [CrossRef] [PubMed]

49. Xiao, L.; Singh, A.; Limor, J.; Graczyk, T.K.; Gradus, S.; Lal, A. Molecular characterization of *Cryptosporidium* oocysts in samples of raw surface water and wastewater. *Appl. Env. Microbiol.* **2001**, *67*, 1097–1101. [CrossRef]

50. Xiao, L.; Lal, A.A.; Jiang, J. Detection and differentiation of *Cryptosporidium* oocysts in water by PCR-RFLP. *Methods Mol. Biol.* **2004**, *268*, 163–176. [CrossRef] [PubMed]

51. Alves, M.; Xiao, L.; Sulaiman, I.; Lal, A.A.; Matos, O.; Antunes, F. Subgenotype analysis of *Cryptosporidium* isolates from humans, cattle, and zoo ruminants in Portugal. *J. Clin. Microbiol.* **2003**, *41*, 2744–2747. [CrossRef] [PubMed]

MDPI

St. Alban-Anlage 66

4052 Basel

Switzerland

Tel. +41 61 683 77 34

Fax +41 61 302 89 18

www.mdpi.com

Pathogens Editorial Office

E-mail: pathogens@mdpi.com

www.mdpi.com/journal/pathogens